Selected Titles in This Series

(Continued in the back of this publication)

CONTEMPORARY
MATHEMATICS

204

Recent Developments in Optimization Theory and Nonlinear Analysis

AMS/IMU Special Session on
Optimization and Nonlinear Analysis
May 24–26, 1995
Jerusalem, Israel

Yair Censor
Simeon Reich
Editors

American Mathematical Society
Providence, Rhode Island

This volume contains the refereed proceedings of the Special Session on Optimization and Nonlinear Analysis held as part of the Joint Meeting of the American Mathematical Society and the Israel Mathematical Union in Jerusalem, Israel, on May 24–26, 1995. The volume also contains papers by several mathematicians who were invited to speak at the meeting but were unable to attend.

1991 *Mathematics Subject Classification.* Primary 35Bxx, 47Hxx, 49Jxx, 49Mxx, 58Exx, 65Jxx, 65Kxx, 90Cxx.

Library of Congress Cataloging-in-Publication Data

AMS/IMU Special Session on Optimization and Nonlinear Analysis (1995 : Jerusalem)
Recent developments in optimization theory and nonlinear analysis : AMS/IMU Special Session on Optimization and Nonlinear Analysis, May 24–26, 1995, Jerusalem, Israel / Yair Censor, Simeon Reich, editors.
p. cm.—(Contemporary mathematics : 204)
Includes bibliographical references.
ISBN 0-8218-0515-0 (alk. paper)
1. Mathematical optimization—Congresses. 2. Nonlinear functional analysis—Congresses. I. Censor, Yair. II. Reich, Simeon. III. Title. IV. Series: Contemporary mathematics (American Mathematical Society) ; v. 204.
QA402.5.A456 1995
515′.7242—dc21 — 96-29934
CIP

∞ The paper used in this book is acid-free and falls within the guidelines
established to ensure permanence and durability.

10 9 8 7 6 5 4 3 2 1 02 01 00 99 98 97

CONTENTS

Preface

This volume contains the refereed proceedings of the Special Session on Optimization and Nonlinear Analysis held at the Joint Meeting of the American Mathematical Society and the Israel Mathematical Union which took place at the Hebrew University of Jerusalem on May 24–26, 1995. In addition to the undersigned, the organizing committee of the special session also included Zvi Artstein (Weizmann Institute of Science, Rehovot), Alexander Ioffe (The Technion–Israel Institute of Technology, Haifa), Victor J. Mizel (Carnegie Mellon University, Pittsburgh, PA) and David Shoikhet (International College of Technology, Karmiel).

Most of the papers that appear here originated from the lectures delivered at this special session. Some of the lectures described research that had already been submitted elsewhere and so could not be included in this volume. In a few cases, however, the corresponding topic does appear in the volume as an exposition of the subject that is not available elsewhere. Some participants who did not give a talk at the special session, as well as several invited speakers who were unable to attend the joint meeting, also contributed their work to these proceedings. All the papers were carefully refereed by experts and revised whenever necessary.

The fields of optimization theory and nonlinear analysis continue to be very active. A glance at the table of contents will reveal not only the wide spectrum and diversity of the results presented in this volume, but also their manifold connections to other areas such as differential equations, functional analysis, operator theory, calculus of variations, numerical analysis, and mathematical programming. Thus one encounters papers that deal, for instance, with convex, quasiconvex and generalized convex functions, fixed and periodic points, fractional-linear transformations, moduli of convexity, monotone operators, Morse lemmas, Navier-Stokes equations, nonexpansive maps, nonsmooth analysis, numerical stability, products of projections, steepest descent, the Leray-Schauder degree, the turnpike property, and variational inequalities.

We are very grateful to the other members of the organizing committee for all their work, to the speakers and participants of the special session who attended it despite the fact that no support was available, to the authors who graciously agreed to take all the comments made by the referees and the editors into account, and to the referees who generously provided expert advice that enhanced the value of this volume. Thanks are also due to the organizing committe of the joint meeting itself. We are especially pleased to thank its Chair, Moshe Jarden (Tel Aviv University), and Yoav Benyamini (The Technion–Israel Institute of Technology) for their kind help. Finally, we are indebted to Donna Harmon, Ralph Sizer, and Christine Thivierge of the American Mathematical Society for their patient cooperation in the publication of this volume.

Yair Censor (University of Haifa)
Simeon Reich (The Technion–Israel Institute of Technology)
Haifa, October 1996

Contemporary Mathematics
Volume **204**, 1997

The method of cyclic projections for closed convex sets in Hilbert space

Heinz H. Bauschke, Jonathan M. Borwein, and Adrian S. Lewis

ABSTRACT. The method of cyclic projections is a powerful tool for solving convex feasibility problems in Hilbert space. Although in many applications, in particular in the field of image reconstruction (electron microscopy, computed tomography), the convex constraint sets do not necessarily intersect, the method of cyclic projections is still employed. Results on the behaviour of the algorithm for this general case are improved, unified, and reviewed. The analysis relies on key concepts from convex analysis and the theory of nonexpansive mappings. The notion of the angle of a tuple of subspaces is introduced. New linear convergence results follow for the case when the constraint sets are closed subspaces whose orthogonal complements have a closed sum; this holds, in particular, for hyperplanes or in Euclidean space.

1. The problem

1.1. Introduction and formulation. A frequent problem in various areas of mathematics and physical sciences consists of determining a "solution" in form of a point satisfying some convex "constraints", i.e. lying in the intersection of certain convex sets. This problem is usually referred to as the *Convex Feasibility Problem.*

Often these constraint sets are closed subspaces in some Hilbert space; an immensely successful algorithm for solving this problem is the method of *cyclic projections*, where a sequence is generated by projecting cyclically onto the constraint subspaces. The fundamental result is due to von Neumann and Halperin:

FACT 1.1.1. Suppose $C_1, \ldots, C_N$ are intersecting closed affine subspaces with corresponding projections $P_1, \ldots, P_N$ in some Hilbert space H. If $x_0 \in H$, then the sequence

$$(x_0, x_1 = P_1x_0, x_2 = P_2x_1, \ldots, x_N = Px_{N-1}, x_{N+1} = P_1x_N, \ldots)$$

converges in norm to the projection of x_0 onto $\bigcap_{i=1}^N C_i$.

1991 *Mathematics Subject Classification.* Primary 47H09, 49M45, 90C25; Secondary 46N10, 47N10, 52A05, 65F10, 92C55.

Research partially supported by NSERC.

In 1933, von Neumann [**81**, Theorem 13.7] proved this result for two closed subspaces; it was later extended to finitely many closed subspaces by Halperin [**52**]. (The generalization to intersecting closed affine subspaces is easy.)

An impressive survey on applications of the method of cyclic projections for intersecting closed affine subspaces can be found in Deutsch's [**28**]. In particular, if each C_i is a hyperplane, then one obtains the well-known method of Kaczmarz [**57**] for solving systems of linear equations.

In some important applications, however, Fact 1.1.1 cannot be used since either the constraints are not affine (i.e. the C_i are not all closed affine subspaces) or the constraints are inconsistent (i.e. the intersection $\bigcap_{i=1}^N C_i$ is empty) or even both.

Non-affine constraints occur, for instance, when the desired solution is supposed to be nonnegative, or bounded by some prescribed value, or a solution of a linear inequality; we refer the interested reader to the surveys [**8, 24, 25**].

In theory the mathematical models employed almost always possess consistent constraints; in practice, however, due (unavoidably) to inaccurate measurements, one is often confronted with *inconsistent constraints.*

Inconsistent (non-affine) constraints constitute a typical example of an *ill-posed problem.* They occur frequently in the field of *image reconstruction*: for instance in *signal reconstruction* (Goldburg and Marks [**48**]), in *electron microscopy* (Sezan [**75**]), in *limited-angle tomography* (Boyd and Little [**14**]), and in *medical imaging* (Viergever [**80**]); in particular in *computed tomography* (Censor [**21**]).

Our objective in this paper is to improve, unify, and review results on the method of cyclic projections for possibly non-intersecting closed convex nonempty sets in Hilbert space. We will also speculate a little on some open questions and hope very much to transmit some of our interest and enthusiasm to the reader.

The mathematical model we analyze is described as follows:

SETTING. Throughout the paper, we assume that H is a real Hilbert space and $C_1, \dots, C_N$ are closed convex nonempty subsets of H with corresponding projections $P_1, \dots, P_N$. (We discuss projections in Section 2.)
For an arbitrary *starting point* $x^0 \in H$, the *method of cyclic projections* generates N sequences $(x_i^n)_n$ by

$$\boxed{\begin{array}{llll} x_1^1 := P_1 x^0, & x_2^1 := P_2 x_1^1, & \dots, & x_N^1 := P_N x_{N-1}^1, \\ x_1^2 := P_1 x_N^1, & x_2^2 := P_2 x_1^2, & \dots, & x_N^2 := P_N x_{N-1}^2, \\ x_1^3 := P_1 x_N^2, & \dots . & & \end{array}}$$

We collect these sequences cyclically in one sequence $(x^0, x_1^1, x_2^1, \dots, x_N^1, x_1^2, x_2^2, \dots)$ to which we refer as the *orbit generated by* x^0 or the *orbit with starting point* x^0. We further define

$$\boxed{Q_1 := P_1 P_N P_{N-1} \cdots P_2,\ Q_2 := P_2 P_1 P_N \cdots P_3,\ \dots,\ Q_N := P_N P_{N-1} \cdots P_1,}$$

which allows us to write more concisely

$$x_i^n := Q_i^{n-1} x_i^1, \text{ for all } n \geq 1 \text{ and every } i;$$

for convenience, we set $P_0 := P_N$, $P_{N+1} := P_1$, $x_0^n := x_N^{n-1}$, and $x_{N+1}^n := x_1^{n+1}$, so that

$$x_{i+1}^n = P_{i+1}x_i^n, \text{ for all } n \geq 1 \text{ and every } i.$$

When appropriate, we will similarly identify $i = 0$ with $i = N$ and $i = N + 1$ with $i = 1$.

OVERVIEW.

Basic facts on *projections* and some examples are contained in Section 2. We touch on *quasi-projections*, a useful concept related to projections and recently introduced by Baillon and Bruck [**5**].

In Section 3, we provide a "toolbox" which makes the subsequent analysis clearer: some of our results rely indirectly on the notion of a *polar cone*. *Recession cones* are intimately connected with convex sets and their projections. One important example of a convex set is a *convex polyhedron*; here the study of the mappings Q_i naturally involves *polyhedral multifunctions*. We give some basic facts on convex *Kadec/Klee sets*, a notion which generalize the better-known concept of (locally) uniformly convex sets. A discussion on *(bounded) (linear) regularity* of N-tuples of closed convex sets follows; this key condition is sufficient for norm convergence of any orbit in the consistent case. We then introduce the notion of the *angle of an N-tuple $(L_1, \ldots, L_N)$ of closed subspaces* and show that the angle is positive if and only if $L_1^\perp + \cdots + L_N^\perp$ is closed. This strengthens results on the well-known two set case (Deutsch [**27**], Bauschke and Borwein [**9**]) and is used to yield new linear convergence results. The section concludes with a short review on *Fejér monotone sequences*.

Each projection P_i is *nonexpansive* and so are the compositions Q_i. Moreover, they actually satisfy various stronger conditions introduced by Bruck and Reich [**19**], by De Pierro and Iusem [**71**], and by Bauschke and Borwein [**8**]. These conditions, which yield significant information on the behaviour of orbits, are discussed in Section 4.

Section 5 contains our main results. After considering the *geometry of the problem*, we give a *dichotomy result on orbits* which roughly says that if each Q_i is fixed point free, then the orbit has no bounded subsequence; otherwise, each subsequence (x_i^n) converges weakly to some fixed point of Q_i. (Throughout this section, we specialize our general results to the *two set case* ($N = 2$) and to the *consistent case* ($\bigcap_{i=1}^N C_i \neq \emptyset$) to allow comparison with known results for these important special cases.) Two central questions arise:

1. When does each Q_i have a fixed point?
2. If each Q_i has a fixed point, when do the subsequences (x_i^n) converge in norm (or even linearly)?

Concerning Question 1, we provide sufficient conditions for the existence of fixed points or of *approximate fixed points* (that is, $\inf_{x \in H} \|x - Q_i x\| = 0$, for each i). It follows that while fixed points of Q_i need not exist for non-intersecting closed affine subspaces, approximate fixed points must.

In respect to Question 2, a variety of conditions guaranteeing norm convergence (in the presence of fixed points for each Q_i) is offered: one of the sets C_i has to be *(boundedly) compact* or all sets are *Kadec/Klee*, or *convex polyhedra*, or *closed*

affine subspaces. In the affine subspace case, each sequence (x_i^n) converges to the fixed point of Q_i nearest to x^0. Moreover, the convergence will be linear, whenever the angle of the N-tuple of the associated closed subspaces is positive. In contrast, if $N = 2$ and the angle is zero, then the convergence will be "arbitrarily slow". We draw connections to the theory of generalized inverses by specializing to hyperplanes with linearly independent normal vectors.
The analysis in Section 5 extends results by Bauschke and Borwein [**9**], by Cheney and Goldstein [**23**], by Gubin et al. [**51**], by De Pierro and Iusem [**71**], by Kosmol [**60**], and by Kosmol and Zhou [**61, 62**]. We end the paper by referring the reader to related algorithms.

1.2. Interpretations and notation. Because the orbit (x_i^n) is generated by the rule $x_i^n := P_i x_{i-1}^n$, the problem described in the previous subsection allows interpretations in the following (related) mathematical areas:

- *Difference Equations:* the orbit (x_i^n) is given by a *nonautonomous difference equation* ([**63**]).
- *Discrete Dynamical Systems:* the orbit (x_i^n) is the *orbit of a discrete dynamical system* ([**66**]).
- *Discrete Processes:* the orbit (x_i^n) is the *positive orbit of a difference equation* ([**64**]).

Less seriously, one can also think about our problem as a *deterministic magnetic billiard*: the starting point x^0 is a *metal billiard-ball* and the sets C_i are *electric magnets.* Now switch the magnets on and off cyclically: the trajectory of the billard ball corresponds exactly to the orbit (x_i^n).

NOTATION. The *norm* of a point $x \in H$ is given by $\|x\| := \sqrt{\langle x, x\rangle}$, where $\langle \cdot, \cdot \rangle$ is the *inner product* of H. We write I for the *identity mapping* on H, B_H for the *unit ball* $\{x \in H : \|x\| \leq 1\}$, and $B(x_0, r)$ for $x_0 + rB_H$, where $x_0 \in H$ and $r \geq 0$. Suppose C is a nonempty subset of H. The *interior* (resp. *intrinsic core, boundary, closure, span*) *of* C is denoted $\mathrm{int}(C)$ (resp. $\mathrm{icr}(C)$, $\mathrm{bd}(C)$, $\overline{C}$, $\mathrm{span}(C)$). The *indicator function of a* C, ι_C, is defined by $\iota_C(x) = 0$, if $x \in C$; $+\infty$, otherwise; its Fenchel conjugate, the *support function of* C, is $\iota_C^* = \sup\langle C, \cdot \rangle$. The *normal cone* $N_C(x)$ *at* $x \in C$ is given by $N_C(x) = \{x^* \in H : \langle x^*, C - x\rangle \leq 0\}$. If C is a subspace, then we write $C^\perp$ for its *orthogonal complement.* The subdifferential of a convex function f from H to $\mathbb{R}$ at $x \in H$ is abbreviated $\partial f(x) = \{x^* \in H : \langle x^*, h\rangle \leq f(x + h) - f(x)\ \forall h \in H\}$. If D is another nonempty subset of H, then the *gap between* C *and* D is defined by $d(C, D) := \inf_{c \in C,\ d \in D} \|c - d\|$. For sequences, we use the symbol $\rightarrow$ (resp. $\rightharpoonup$) to indicate *norm* (resp. *weak*) *convergence.* Finally, we write $[\cdot]$ for the *mod* function with values in $\{1, \ldots, N\}$.

2. Projections

2.1. Basic properties.

DEFINITION 2.1.1. Suppose C is a closed convex nonempty subset of H. Then for every $x \in H$, there exists a unique nearest point, say $P_C x$, to x in C:

$$\|x - P_C x\| \leq \|x - c\|, \text{ for all } c \in C.$$

The associated mapping P_C from H to C which sends every point to its nearest point in C is called the *projection onto* C. The function

$$d(\cdot, C) : H \to \mathbb{R} : x \mapsto \|x - P_C x\|$$

is called the *distance function (to* C*)*.

We list basic facts on projections. (i) and (ii) are folklore (see for instance [**84**, Section 1]) whereas (iii) is a simple but useful expansion.

PROPOSITION 2.1.2. Suppose C is a closed convex nonempty subset of H. Then:

(i) The projection onto C is well-defined and the distance function to C is continuous and convex, hence weakly lower semicontinuous.

(ii) For any two points $x, c^* \in H$, we have $c^* = P_C x$ if and only if

$$c^* \in C \text{ and } \langle C - c^*, x - c^* \rangle \leq 0.$$

(iii) For any $x, y \in H$, the formula below contains exclusively nonnegative terms:

$$\begin{aligned} \|x - y\|^2 &= \|(I - P_C)x - (I - P_C)y\|^2 + \|P_C x - P_C y\|^2 \\ &\quad + 2\langle (I - P_C)x - (I - P_C)y, P_C x - P_C y\rangle \\ &= \|(x - y) - (P_C x - P_C y)\|^2 + \|P_C x - P_C y\|^2 \\ &\quad + 2\langle x - P_C x, P_C x - P_C y\rangle + 2\langle y - P_C y, P_C y - P_C x\rangle; \end{aligned}$$

in particular, $\langle x - y, P_C x - P_C y\rangle \geq \|P_C x - P_C y\|^2$.

COROLLARY 2.1.3. Suppose (x_i^n) (resp. (y_i^n)) is the orbit generated by some x^0 (resp. y^0) in H. Then the sequences

$$(\|(x_i^n - y_i^n) - (x_{i+1}^n - y_{i+1}^n)\|^2),\ (\langle x_i^n - x_{i+1}^n, x_{i+1}^n - y_{i+1}^n\rangle),\ (\langle y_i^n - y_{i+1}^n, y_{i+1}^n - x_{i+1}^n\rangle)$$

are nonnegative and tend to zero for each i. In fact, the series generated by the terms of these sequences is convergent.

2.2. Examples.

EXAMPLES 2.2.1. Suppose C is a closed convex nonempty subset of H and x is an arbitrary point in H.

(i) If $C = B_H$, then $P_C x = \begin{cases} x, & \text{if } \|x\| \leq 1; \\ x/\|x\|, & \text{otherwise.} \end{cases}$

(ii) If C is a hyperplane, say $C = \{x \in H : \langle a, x\rangle = b\}$ with $a \in H \setminus \{0\}$, then $P_C x = x - ((\langle a, x\rangle - b)/\|a\|^2)a$.

(iii) If C is a halfspace, say $C = \{x \in H : \langle a, x\rangle \leq b\}$ with $a \in H \setminus \{0\}$, then $P_C x = x - ((\langle a, x\rangle - b)^+/\|a\|^2)a$.

(iv) If H is a Hilbert lattice with lattice cone H^+, then $P_C x = x^+$.

(v) Suppose C is a closed convex cone. If the negative polar cone is denoted $C^\ominus$ (see Subsection 3.1), then $P_{C^\ominus}(x) = x - P_C x$. In particular, if C is a closed subspace, then $P_{C^\perp}(x) = x - P_C x$.

(vi) Suppose Y is another Hilbert space and $A : H \to Y$ is a continuous linear operator with closed range. Let $C = \text{range}(A^*)$. If the generalized inverse of A is denoted $A^\dagger$ (see Definiton 5.7.10), then $P_C x = A^\dagger A x$.

(vii) Suppose $Y = \mathbb{R}^N$ and a continuous linear operator A is given by $Ah = (\langle a_1, h\rangle, \dots, \langle a_N, h\rangle)$ for some vectors $a_1, \dots, a_N \in H$. Suppose C is equal to $\text{span}\{a_1, \dots, a_N\}$. Then $P_C x = A^\dagger A x$. Furthermore, if the vectors $a_1, \dots, a_N$ are linearly independent, then $P_C x = A^*(AA^*)^{-1} A x$.

PROOF. (i), (ii), (iii), and (v) are easy to check. (iv) is proved by Borwein and Yost's [**13**, Theorem 8]. (vi) is due to Desoer and Whalen; see [**26**]. (vii) is a special case of (vi) and was used in [**9**, Section 5]. □

The following proposition is useful for building more examples of projections. We omit the simple proof.

PROPOSITION 2.2.2. Suppose C is a closed convex nonempty subset of H and x is an arbitrary point in H. Then:

(i) (translation formula) If $z \in H$, then $P_{z+C}(x) = z + P_C(x - z)$.
(ii) (dilation formula) If $\rho \neq 0$, then $P_{\rho C}(x) = \rho P_C(x/\rho)$.

2.3. Baillon and Bruck's quasi-projection.

DEFINITION 2.3.1. (Baillon and Bruck's quasi-projection [**5**]) Suppose C is a closed convex nonempty subset of H. For every $x \in H$, define the *quasi-projection of x onto C*, $\mathcal{Q}_C x$, by

$$\mathcal{Q}_C x = \{c^* \in C : \|c^* - c\| \leq \|x - c\|, \text{ for all } c \in C\}.$$

It is easy to verify the following proposition (see also [**7**]):

PROPOSITION 2.3.2. Suppose C is a closed convex nonempty subset of H and x is a point in H. Then:

(i) $\mathcal{Q}_C x$ is a bounded closed convex nonempty subset of C; moreover:

$$P_C x \in \mathcal{Q}_C x \subseteq C \cap B(P_C x, d(x, C)).$$

Consequently, if $x \in C$, then $\mathcal{Q}_C x = \{P_C x\}$.
(ii) (translation formula) $\mathcal{Q}_{z+C} x = z + \mathcal{Q}_C(x - z)$, for all $z \in H$.
(iii) If C is a closed affine subspace, then $\mathcal{Q}_C x = \{P_C x\}$.

3. A toolbox

3.1. Polar cones.

DEFINITION 3.1.1. The *positive* (resp. *negative*) *polar cone* $S^{\oplus}$ (resp. $S^{\ominus}$) of a nonempty subset S of H is given by

$$S^{\oplus} := \{x \in H : \langle x, S\rangle \geq 0\} \text{ (resp. } S^{\ominus} := \{x \in H : \langle x, S\rangle \leq 0\} = -S^{\oplus}).$$

The next proposition is readily verified:

PROPOSITION 3.1.2.

(i) Every positive (resp. negative) polar cone is a closed convex nonempty cone.
(ii) If $S_1, S_2, \dots, S_M$ are nonempty subsets of H, then $\overline{S_1^{\oplus} + S_2^{\oplus} + \cdots + S_M^{\oplus}} \subseteq (S_1 \cap S_2 \cap \cdots \cap S_M)^{\oplus}$.

The class of closed subspaces can be characterized in the class of closed convex nonempty cones:

PROPOSITION 3.1.3. Suppose S is a closed convex nonempty cone in H. Then S is a subspace if and only if $S \cap S^{\oplus} = \{0\}$.

PROOF. "$\Longrightarrow$": If S is a subspace, then $S^{\oplus} = S^{\perp}$ and hence $S \cap S^{\oplus} = \{0\}$.
"$\Longleftarrow$": By Examples 2.2.1.(v) and Proposition 2.2.2, $x = P_S x - P_{S^{\oplus}}(-x)$, for all $x \in H$. Let us assume that S is not a subspace. Then there exists some vector $x \in S$ such that $-x \notin S$, that is $P_S(-x) \neq -x$. Thus $-x = P_S(-x) - P_{S^{\oplus}} x$, where $P_{S^{\oplus}} x \neq 0$. Since $P_{S^{\oplus}} x = P_S(-x) + x \in S$, we have found $P_{S^{\oplus}} x \in (S \cap S^{\oplus}) \setminus \{0\}$. The proof is complete. □

3.2. Recession cones.

DEFINITION 3.2.1. The *recession cone* $\mathrm{rec}(C)$ of a closed convex nonempty subset C of H is given by

$$\mathrm{rec}(C) := \{x \in H : x + C \subseteq C\}.$$

PROPOSITION 3.2.2. Suppose C is a closed convex nonempty subset of H. Then:

(i) (Zarantonello's [**84**, Section 3]) $\mathrm{rec}(C)$ is a closed convex nonempty cone.
(ii) A point $x \in H$ belongs to $\mathrm{rec}(C)$ if and only if there exists a sequence (c_n) in C and a sequence of reals (r_n) tending to $+\infty$ such that (c_n/r_n) converges weakly to x.
(iii) If D is a closed convex superset of C, then $\mathrm{rec}(C) \subseteq \mathrm{rec}(D)$.

PROOF. (ii): "$\Longrightarrow$": Since $\mathrm{rec}(C)$ is cone, points of the form $c + nx$ belong to C, for all $c \in C$ and $n \geq 1$. Let us choose $c_n := c + nx$ and $r_n := n$. Then the sequence $(c_n/r_n) = (c/n + x)$ converges to x, as desired.
"$\Longleftarrow$": Fix any $c \in C$; then $(1 - 1/r_n)c + c_n/r_n \in C$, for all large n. Since C is weakly closed, the weak limit $c + x$ lies in C. Finally, (iii) is obvious from (ii). □

There is an intimate connection between the projection, the recession cone, and the indicator function of a given closed convex nonempty set:

FACT 3.2.3. Suppose C is a closed convex nonempty subset of H. Then:

(i) (Castaing and Valadier's [**20**, Proposition I-7]) $\mathrm{rec}(C) = (\mathrm{domain}(\iota_C^*))^{\ominus}$.
(ii) (Zarantonello's [**84**, Theorem 3.1]) $\overline{\mathrm{range}}(P_C - I) = (\mathrm{rec}(C))^{\oplus}$.

REMARK 3.2.4. The set $\mathrm{domain}(\iota_C^*) = \{x \in H : \sup\langle x, C\rangle < \infty\}$ is also known as the *barrier cone of* C; see for instance [**4**].

3.3. Convex polyhedra.

DEFINITION 3.3.1. The intersection of finitely many halfspaces is called a *convex polyhedron.*

PROPOSITION 3.3.2. Suppose each C_i is a convex polyhedron given as $\{x \in H : \langle a_{i,j}, x\rangle \leq b_{i,j}\rangle\}$ for finitely many vectors $a_{i,j}$ and real numbers $b_{i,j}$. Suppose further K is a closed subspace of $\bigcap_{i,j} \mathrm{kernel}(a_{i,j})$. Define $D_i := C_i \cap K^{\perp}$. Then:

(i) Each D_i is a convex polyhedron in $K^{\perp}$.
(ii) $P_i = P_K + P_{D_i} P_{K^{\perp}}$, for each i.
(iii) $I - Q_N = P_{K^{\perp}} - P_{D_N} \cdots P_{D_1} P_{K^{\perp}}$.
(iv) $\inf_{x \in H} \|x - Q_N x\|$ is attained if and only if $\inf_{y \in K^{\perp}} \|y - P_{D_N} \cdots P_{D_1} y\|$ is.

PROOF. (i) is obvious. (ii): Fix $x \in H$ and an index i. We check Proposition 2.1.2.(ii): since $C_i = K \oplus D_i$, the point $P_K x + P_{D_i} P_{K^{\perp}} x$ belongs to C_i. Pick any $k \in K$ and $d_i \in D_i$. Then

$$\begin{aligned} & \langle k + d_i - (P_K x + P_{D_i} P_{K^\perp} x), x - (P_K x + P_{D_i} P_{K^\perp} x) \rangle \\ = & \langle k - P_K x, (x - P_K x) - P_{D_i} P_{K^\perp} x \rangle + \langle d_i - P_{D_i} P_{K^\perp} x, P_{K^\perp} x - P_{D_i} P_{K^\perp} x \rangle \\ \leq & 0, \end{aligned}$$

as desired. Hence (ii) holds. (iii) follows from (ii) inductively whereas (iv) is obvious from (iii). □

REMARK 3.3.3. It follows that range$(I - P_i)$ is a subset of $K^\perp$. Ky Fan ([**40**, Theorem 14]) showed that range$(I - P_i)$ is actually contained in the cone generated by the $a_{i,j}$; this will, however, not be needed in the sequel.

3.4. Polyhedral multifunctions.

DEFINITION 3.4.1. Suppose X, Y are Euclidean spaces and $\Omega : X \rightrightarrows Y$ is a multifunction. If the graph of Ω can be written as the union of finitely many convex polyhedra, then Ω is called a *polyhedral multifunction*.

FACT 3.4.2. (Robinson [**73**, Section 1]) The class of polyhedral multifunctions is closed under addition, scalar multiplication, composition, and inversion.

FACT 3.4.3. (Robinson [**73**, Section 1]) If A is a linear mapping between the Euclidean spaces X and Y, then A is a polyhedral multifunction. If D is a convex polyhedron, then $\partial \iota_D(\cdot) = N_D(\cdot)$ is a polyhedral multifunction.

COROLLARY 3.4.4. If $D_1, \dots, D_N$ are convex polyhedra in the Euclidean space X, then $I - P_{D_N} \cdots P_{D_1}$ is a single-valued polyhedral multifunction.

PROOF. $y = P_{D_i} x \Leftrightarrow x \in (I + \partial \iota_{D_i})(y) \Leftrightarrow y \in (I + \partial \iota_{D_i})^{-1}(x)$. Now I and $\partial \iota_{D_i}$ are polyhedral multifunctions by Fact 3.4.3; hence so is $I - P_{D_N} \cdots P_{D_1}$ by repeated application of Fact 3.4.2 . □

FACT 3.4.5. (Robinson's [**73**, Proposition 2]) If Ω is a polyhedral multifunction between the Euclidean spaces X and Y and if W is a bounded subset of range(Ω), then there exists some bounded subset V of X such that $\Omega(V) \supseteq W$.

COROLLARY 3.4.6. If $D_1, \dots, D_N$ are convex polyhedra in the Euclidean space X, then the infimum $\inf_{x \in X} \|x - P_{D_N} \cdots P_{D_1} x\|$ is attained.

PROOF. $\Omega := I - P_{D_N} \cdots P_{D_1}$ is polyhedral and single-valued (Corollary 3.4.4). Let $\beta > \inf_{x \in X} \|\Omega x\|$ and set $W :=$ range$(\Omega) \cap \beta B_X$. By the previous fact, we get $\alpha > 0$ such that $\Omega(\alpha B_X) \supseteq W$. Thus $\inf_{x \in X} \|\Omega x\| = \inf_{x \in \alpha B_X} \|\Omega x\|$; however, this latter infimum is attained, since Ω is continuous and αB_X is compact. □

REMARK 3.4.7. In particular, if each D_i is a halfspace in the Euclidean space X, then the infimum $\inf_{x \in X} \|x - P_{D_N} \cdots P_{D_1} x\|$ is attained — this also follows from De Pierro and Iusem's [**71**, Proposition 13].

Combining Proposition 3.3.2.(iv) and Corollary 3.4.6 thus yields:

COROLLARY 3.4.8. If each C_i is a convex polyhedron, then the infimum $\inf_{x \in H} \|x - Q_N x\|$ is attained.

3.5. Kadec/Klee sets. The following definition generalizes well-known notions of rotundity of unit balls in Banach spaces; see Deville et al.'s recent monograph [**33**].

DEFINITION 3.5.1. Suppose C is a closed convex nonempty subset of H. We say that C is:

(i) *Kadec/Klee*, if $\left.\begin{array}{c}(x_n) \text{ in } C,\\ x_n \rightharpoonup x \in \operatorname{bd}(C)\end{array}\right\}$ implies $x_n \to x$.

(ii) *uniformly convex*, if $\left.\begin{array}{c}(x_n),(y_n) \text{ in } C,\\ d((x_n+y_n)/2, \operatorname{bd}(C)) \to 0\end{array}\right\}$ implies $x_n - y_n \to 0$.

(iii) *locally uniformly convex*, if $\left.\begin{array}{c}(x_n) \text{ in } C,\ x \in C,\\ d((x_n+x)/2, \operatorname{bd}(C)) \to 0\end{array}\right\}$ implies $x_n \to x$.

(iv) *strictly convex*, if $\left.\begin{array}{c}x, y \in C,\\ (x+y)/2 \in \operatorname{bd}(C)\end{array}\right\}$ implies $x = y$.

PROPOSITION 3.5.2. Suppose C is a closed convex nonempty subset of H. Then:

(i) C uniformly convex $\Rightarrow$ C locally uniformly convex $\Rightarrow$ C strictly convex.
(ii) C strictly convex and compact $\Rightarrow$ C uniformly convex.
(iii) If $\operatorname{int}(C) = \emptyset$, then: C is strictly convex $\Leftrightarrow$ C is singleton.
(iv) If $\operatorname{int}(C) = \emptyset$, then: C is Kadec/Klee $\Leftrightarrow$ C is boundedly compact.
(v) C boundedly compact $\Rightarrow$ C Kadec/Klee $\Rightarrow$ $\operatorname{int}(C) \neq \emptyset$ or C boundedly compact.
(vi) C locally uniformly convex $\Rightarrow$ C Kadec/Klee.
(vii) The function

$$\delta_C(t) := \sup\{r \geq 0 : B((x+y)/2, r) \subseteq C, \text{ for all } x, y \in C \text{ with } \|x-y\| = t\}$$

is increasing. Moreover: C is uniformly convex $\Leftrightarrow$ $\delta_C(t) > 0$, for all $t > 0$.

PROOF. We prove here only (vi); the rest is straightforward. Suppose C is locally uniformly convex. In view of (iii) and (iv), we can assume that $\operatorname{int}(C) \neq \emptyset$. Pick a sequence (x_n) in C that converges weakly to some $x \in \operatorname{bd}(C)$. Then x is a support point of C and there is a support functional f with $f(x) \geq f(C)$ and $\|f\| = 1$. Let $t_n := 1/n + (f(x) - f(x_n))/2$, for all n; then $t_n \to 0^+$. Also, $f(t_n f + (x_n + x)/2) = 1/n + f(x) > f(x)$ so that $t_n f + (x_n + x)/2 \notin C$, for all n. Hence $d((x_n + x)/2, \operatorname{bd}(C)) \leq \|(x_n + x)/2 - ((x_n + x)/2 + t_n f)\| = t_n \to 0$. Therefore, since C is locally uniformly convex, $x_n \to x$ and (vi) holds. □

REMARKS 3.5.3.

- Let $H = \mathbb{R}^2$ and suppose $C = \{(x, y) \in H : y \geq 1/x > 0\}$. Then C is locally uniformly convex but not uniformly convex. Hence the assumption of compactness in (ii) is important.
- Note that every non-singleton boundedly compact closed convex nonempty set is Kadec/Klee; thus there are many Kadec/Klee sets which are not strictly convex.
- (Borwein and Fitzpatrick's [**11**, Example following Corollary 3.5]) Let $H = \ell_2 \times \mathbb{R}$ and define a new norm $|||\cdot|||$ on H by $|||(x,r)|||^2 := \sum_n (x_n/n)^2 + r^2 + \max\{\sum_n x_n^2, r^2\}$. Suppose $C := \{(x, r) \in H : |||(x,r)||| \leq 1\}$. Then C is strictly convex. Denote the nth unit vector of ℓ_2 by e_n. Then the sequence $((e_n, 1)/\sqrt{2 + 1/n^2})$ lies in the boundary of C and converges weakly to $(0, 1/\sqrt{2}) \in \operatorname{bd}(C)$, but not in norm. Therefore, C is an strictly convex set

with nonempty interior but C is neither Kadec/Klee nor locally uniformly convex.

- Borwein and Lewis [**12**] showed that the notions of locally uniform convexity and strict convexity coincide in Euclidean spaces.

DEFINITION 3.5.4. Suppose C is a uniformly convex subset of H. If there is some $\gamma > 0$ such that $\delta_C(t) \geq \gamma t^2$, then C is called *strongly convex.*

REMARKS 3.5.5.

- If $C = \rho B_H$, where $\rho > 0$, then $\delta_C(t) \geq t^2/(8\rho)$ (use the Parallelogram Law) and hence C is strongly convex.
- Let $H = \mathbb{R}^2$ and suppose $C = \{(x,y) \in H : x^4 + y^4 \leq 1\}$. Fix an arbitrary $0 < s < 1$ and set $\mathbf{x}_s = (s, (1-s^4)^{(1/4)})$, $\mathbf{y}_s = (s, -(1-s^4)^{(1/4)})$. Then $\mathbf{x}_s, \mathbf{y}_s \in C$, $(\mathbf{x}_s + \mathbf{y}_s)/2 = (s, 0)$, and $d((\mathbf{x}_s+\mathbf{y}_s)/2, \mathrm{bd}(C)) \leq \|(\mathbf{x}_s+\mathbf{y}_s)/2 - (1,0)\| = 1-s$. Also, $\|\mathbf{x}_s - \mathbf{y}_s\| = 2(1-s^4)^{(1/4)}$, and thus $\delta_C(2(1-s^4)^{(1/4)}) \leq 1-s$. Changing variables, we obtain $\delta_C(t) \leq 1-(1-(t/2)^4)^{(1/4)} = t^4/64 + \mathcal{O}(t^8)$. Therefore, $\delta_C(t)/t^2 \leq t^2/64 + \mathcal{O}(t^6)$ and C is not strongly convex.

3.6. Regularities.

The two set case.

DEFINITION 3.6.1. Suppose $N = 2$ and $d_1 := P_{\overline{C_2 - C_1}}(0) \in C_2 - C_1$. We say that (C_1, C_2) is:

(i) *boundedly regular,* if

$$\left.\begin{array}{r}(x_n) \text{ is a bounded sequence in } H,\\ \max\{d(x_n, C_1), d(x_n, C_2 - d_1)\} \to 0\end{array}\right\} \text{ implies } d(x_n, C_1 \cap (C_2 - d_1)) \to 0.$$

(ii) *regular,* if

$$\left.\begin{array}{r}(x_n) \text{ is a sequence in } H,\\ \max\{d(x_n, C_1), d(x_n, C_2 - d_1)\} \to 0\end{array}\right\} \text{ implies } d(x_n, C_1 \cap (C_2 - d_1)) \to 0.$$

Obviously, regularity implies bounded regularity. These conditions have turned out to be sufficient for norm convergence of orbits; see Subsection 5.5.

FACT 3.6.2. Suppose $N = 2$.

(i) ([**9**, Theorem 3.9]) If C_1 or C_2 is boundedly compact and $d_1 \in C_2 - C_1$, then (C_1, C_2) is boundedly regular.

(ii) ([**9**, Corollary 3.14]) If C_1 or C_2 is bounded and (C_1, C_2) is boundedly regular, then (C_1, C_2) is regular.

REMARK 3.6.3. The first two authors wish to point out the following lacunae in their paper [**9**]: the definition of bounded regularity [**9**, Definiton 3.6] needs in addition the assumption $P_{\overline{C_2 - C_1}}(0) \in C_2 - C_1$ and is then equivalent to the present, more handy Definition 3.6.1; similarly, [**9**, Theorems 3.7, 3.8, and Corollary 3.10] need in addition the assumption $P_{\overline{C_2 - C_1}}(0) \in C_2 - C_1$.

The consistent case.

DEFINITION 3.6.4. Suppose $\bigcap_{i=1}^N C_i \neq \emptyset$. We say that $(C_1, \ldots, C_N)$ is:

(i) *boundedly regular,* if $\max_i d(x_n, C_i) \to 0$ implies $d(x_n, \bigcap_i C_i) \to 0$, for all bounded sequences (x_n) in H.

(ii) *regular,* if $\max_i d(x_n, C_i) \to 0$ implies $d(x_n, \bigcap_i C_i) \to 0$, for all sequences (x_n) in H.

(iii) *boundedly linearly regular*, if for every bounded subset S of H, there exists $\kappa_S > 0$ such that $d(x, \bigcap_i C_i) \leq \kappa_S \max_i d(x, C_i)$, for all $x \in S$.
(iv) *linearly regular*, if there exists $\kappa > 0$ such that $d(x, \bigcap_i C_i) \leq \kappa \max_i d(x, C_i)$, for all $x \in H$.

Clearly,

$$\begin{array}{ccc} \text{linearly regular} & \Longrightarrow & \text{boundedly linearly regular} \\ \Downarrow & & \Downarrow \\ \text{regular} & \Longrightarrow & \text{boundedly regular,} \end{array}$$

and again we prefer the more intuitive definition of (bounded) regularity to the equivalent [**8**, Definition 5.1].

THEOREM 3.6.5.

(i) ([**8**, Proposition 5.4.(i)]) If some C_i is boundedly compact, then $(C_1, \ldots, C_N)$ is boundedly regular.
(ii) ([**8**, Proposition 5.4.(iii)]) If H is finite-dimensional, then $(C_1, \ldots, C_N)$ is always boundedly regular.
(iii) If every C_i, except possibly one, is Kadec/Klee, then $(C_1, \ldots, C_N)$ is boundedly regular.
(iv) ([**8**, Proposition 5.4.(ii)]) If $(C_1, \ldots, C_N)$ is boundedly regular and some C_i is bounded, then $(C_1, \ldots, C_N)$ is regular.
(v) (Gubin et al.'s [**51**, Proof of Lemma 5]) If every C_i, except possibly one, is uniformly convex, then $(C_1, \ldots, C_N)$ is regular.

PROOF. (iii): After relabeling if necessary, we assume that $C_1, \ldots, C_{N-1}$ are Kadec/Klee. Now pick a bounded sequence with $\max_{i=1}^N d(x_n, C_i) \to 0$. After passing to a subsequence if necessary, we assume that $(d(x_n, \bigcap_{i=1}^N C_i))$ converges to some limit, say L. It suffices to show that L equals 0. After passing once more to a subsequence, we also assume that (x_n) converges weakly to some x, where x lies necessarily in $\bigcap_{i=1}^N C_i$ (Proposition 2.1.2.(i)). If $x \in \text{bd}(C_i)$, for some $i \in \{1, \ldots, N-1\}$, then $(P_i x_n)$ converges in norm to x (because $x_n - P_i x_n \to 0$ and C_i is Kadec/Klee) and so does (x_n) implying $L = 0$, as desired. Otherwise, $x \in C_N \cap \text{int}(\bigcap_{i=1}^{N-1} C_i)$. Then, by Fact 3.6.9, $(C_1, \ldots, C_N)$ is even boundedly linearly regular. The proof is complete. □

REMARK 3.6.6.

- Gubin et al. ([**51**, Lemma 5]) claimed only bounded regularity in (iv); their proof, however, works for regularity.
- For an example of two intersecting sets C_1 and C_2 where (C_1, C_2) is not boundedly regular, see Bauschke and Borwein's [**9**, Example 5.5]. Hence the assumption of bounded compactness in (i) is important.

We now give some useful results on (bounded) linear regularity; see also [**8**].

FACT 3.6.7. ([**8**, Theorem 5.19]) Suppose $L_1, \ldots, L_N$ are closed subspaces of H. Then $L_1^\perp + \cdots + L_N^\perp$ is closed if and only if $(L_1, \ldots, L_N)$ is regular in any of of the four senses. Consequently, if each C_i is a closed affine subspace, where $C_i = c + L_i$ and L_i is linear, then $(C_1, \ldots, C_N)$ is regular in any of the four senses if and only if $L_1^\perp + \cdots + L_N^\perp$ is closed.

FACT 3.6.8. ([**8**, Corollary 5.26]) If each C_i is a convex polyhedron, then the N-tuple $(C_1, \ldots, C_N)$ is linearly regular.

FACT 3.6.9. ([**8**, Corollaries 5.13 and 5.14]) If $0 \in \bigcap_{i=1}^{N-1} \operatorname{icr}((\bigcap_{j=1}^{i} C_j) - C_{i+1})$, then $(C_1, \dots, C_N)$ is boundedly linearly regular. In particular, this happens whenever $C_N \cap \operatorname{int}(\bigcap_{i=1}^{N-1} C_i) \neq \emptyset$.

REMARK 3.6.10. Since the definition of (bounded) (linear) regularity does not depend on the order of the sets, the last fact has corresponding formulations for all other orders of the sets.

The general case. In general, without understanding the geometry of the problem (see Subsection 5.1), it seems to be hard even to suggest useful notions of regularity. We feel that a positive answer to Conjecture 5.1.6 would yield the appropriate general notion of regularity.

3.7. The angle of a tuple of subspaces. In this subsection, we generalize the notion of the angle between two subspaces to finitely many subspaces. The results are also interesting from an operator-theoretic point of view and will lead to some new linear convergence results in Subsection 5.7.

Throughout this subsection, we assume that

$$L_1, L_2, \dots, L_N \text{ are closed subspaces and } L := \bigcap_{i=1}^{N} L_i.$$

FACT 3.7.1. (Kayalar and Weinert's [**58**, Theorem 1])

$$P_{L_N} \cdots P_{L_1} P_{L^\perp} = P_{L_N \cap L^\perp} \cdots P_{L_1 \cap L^\perp}.$$

The next proposition is easily verified.

PROPOSITION 3.7.2. Suppose B is a closed subspace and A is a subset of $B^\perp$. Then A is closed if and only if $A \oplus B$ is.

PROPOSITION 3.7.3. The following are equivalent:

(i) $(L_1 \cap L^\perp, \dots, L_N \cap L^\perp)$ is regular in any of the four senses.
(ii) $L_1^\perp + \cdots + L_N^\perp + L$ is closed.
(iii) $(L_1, \dots, L_N)$ is regular in any of the four senses.
(iv) $L_1^\perp + \cdots + L_N^\perp$ is closed.

PROOF. In view of Fact 3.6.7, it suffices to establish the equivalence of (ii) and (iv). By Proposition 3.7.2, $(L_i \cap L^\perp)^\perp = \overline{L_i^\perp + L} = L_i^\perp + L$, for each i. Using Proposition 3.7.2 once more, we get: $L_1^\perp + \cdots + L_N^\perp$ is closed $\Leftrightarrow (L_1^\perp + \cdots + L_N^\perp) + L$ is closed $\Leftrightarrow (L_1^\perp + L) + \cdots + (L_N^\perp + L)$ is closed $\Leftrightarrow (L_1 \cap L^\perp)^\perp + \cdots + (L_N \cap L^\perp)^\perp$ is closed. $\square$

THEOREM 3.7.4. The Hilbert space operator norm $\|P_{L_N} \cdots P_{L_1} P_{L^\perp}\|$ is strictly less than 1 if and only if the sum $L_1^\perp + \cdots + L_N^\perp$ is closed.

PROOF. "$\Longrightarrow$": We prove the contrapositive and assume that $L_1^\perp + \cdots + L_N^\perp$ is not closed. By Proposition 3.7.3, $(L_1 \cap L^\perp, \dots, L_N \cap L^\perp)$ is not boundedly regular. Hence there is some bounded sequence (x_n) with $\max_i d(x_n, L_i \cap L^\perp) \to 0$, but $d(x_n, \bigcap_i L_i \cap L^\perp) = \|x_n\| \not\to 0$. After passing to a subsequence and normalizing if necessary, we may assume $\|x_n\| = 1$, for all n. Now $x_n - P_{L_1 \cap L^\perp} x_n \to 0$, which implies, by nonexpansivity of $P_{L_2 \cap L^\perp}$, that $P_{L_2 \cap L^\perp} x_n - P_{L_2 \cap L^\perp} P_{L_1 \cap L^\perp} x_n \to 0$.

Since $x_n - P_{L_2 \cap L^\perp} x_n \to 0$, we obtain $x_n - P_{L_2 \cap L^\perp} P_{L_1 \cap L^\perp} x_n \to 0$. Repeating this line of thought yields eventually

$$x_n - P_{L_N \cap L^\perp} \cdots P_{L_1 \cap L^\perp} x_n \to 0.$$

The triangle inequality implies $\|x_n\| - \|P_{L_N \cap L^\perp} \cdots P_{L_1 \cap L^\perp} x_n\| \to 0$; thus $\|P_{L_N \cap L^\perp} \cdots P_{L_1 \cap L^\perp} x_n\| \to 1$. Therefore, by Fact 3.7.1, $\|P_{L_N \cap L^\perp} \cdots P_{L_1 \cap L^\perp}\| = \|P_{L_N} \cdots P_{L_1} P_{L^\perp}\| = 1$.
"$\Longleftarrow$": By Proposition 3.7.3, $(L_1 \cap L^\perp, \ldots, L_N \cap L^\perp)$ is linearly regular. Hence there is some $\kappa > 0$ such that

$$\|x\| = d(x, \textstyle\bigcap_i L_i \cap L^\perp) \leq \kappa \max_i d(x, L_i \cap L^\perp), \text{ for all } x \in H.$$

On the other hand, by Fact 4.5.2 below,

$$\begin{aligned} d^2(x, L_i \cap L^\perp) &\leq \|x - P_{L_i \cap L^\perp} \cdots P_{L_1 \cap L^\perp} x\|^2 \\ &\leq 2^{i-1}(\|x\|^2 - \|P_{L_i \cap L^\perp} \cdots P_{L_1 \cap L^\perp} x\|^2) \\ &\leq 2^{N-1}(\|x\|^2 - \|P_{L_N \cap L^\perp} \cdots P_{L_1 \cap L^\perp} x\|^2), \end{aligned}$$

for all $x \in H$ and each i. Altogether,

$$\|x\|^2 \leq \kappa^2 2^{N-1}(\|x\|^2 - \|P_{L_N \cap L^\perp} \cdots P_{L_1 \cap L^\perp} x\|^2), \text{ for all } x \in H,$$

which implies $\|P_{L_N \cap L^\perp} \cdots P_{L_1 \cap L^\perp}\| \leq (1 - \kappa^{-2} 2^{-(N-1)})^{1/2} < 1$. The theorem now follows from Fact 3.7.1. ☐

DEFINITION 3.7.5. We define the *angle* $\gamma := \gamma(L_1, \ldots, L_N) \in [0, \pi/2]$ *of the* N*-tuple* $(L_1, \ldots, L_N)$ by

$$\cos \gamma = \|P_{L_N} \cdots P_{L_1} P_{L^\perp}\|.$$

REMARKS 3.7.6.

(i) The classical definition of the angle $\bar{\gamma} := \bar{\gamma}(L_1, L_2)$ between *two* subspaces was given by Friedrichs [**43**] in 1937:

$$\cos \bar{\gamma} = \sup\{\langle l_1, l_2 \rangle : l_1 \in L_1 \cap L^\perp;\ l_2 \in L_2 \cap L^\perp;\ \|l_1\|, \|l_2\| \leq 1\}.$$

Deutsch ([**27**, Proof of Lemma 2.5.(iii)]) showed that $\|P_{L_2} P_{L_1} P_{(L_2 \cap L_1)^\perp}\|$ is equal to $\cos \bar{\gamma}$; therefore, $\gamma = \bar{\gamma}$ and our Definition 3.7.5 of the angle is consistent with Friedrichs' original one.

(ii) The importance of the angle between two subspaces stems from the fact that a positive angle guarantees linear convergence of the sequence of alternating projections; see, for instance, [**3**, pages 378ff.], [**9**, Theorem 4.11], [**27**, Theorem 2.3], and [**58**, Theorem 2]. In Subsection 5.7, we will show among other things that a positive angle of an N-tuple of subspaces yields linear convergence of the sequence of cyclic projections as well.
Very recently, Deutsch and Hundal [**29, 31**] studied the angle (in the sense of Friedrichs) and another notion of an angle (due to Dixmier) in great detail; they explored the relationships between these angles and error bounds for the method of cyclic projections for closed subspaces. Deutsch and Hundal [**32**] also carefully analyzed Dykstra's algorithm (see subsection 5.8) for intersecting halfspaces and obtained finite and linear convergence results. It is interesting to note that they provide an upper bound for the rate of convergence which depends on the angle of tuples of subspaces as in Definition 3.7.5.

(iii) By (i), the angle between two subspaces is independent of the order, that is $\gamma(L_1, L_2) = \gamma(L_2, L_1)$. For $N \geq 3$, the angle of the N-tuple $(L_1, \ldots, L_N)$ may well depend on the order of the subspaces; to see this, we pick three unit vectors a, b, c with $\langle b, c\rangle \neq 0$ and $|\langle a, b\rangle| \neq |\langle a, c\rangle|$. (This requires H to be of dimension at least 2.) Let $L_1 := L_2 := \cdots := L_{N-2} := \mathrm{span}(a)$, $L_{N-1} := \mathrm{span}(b)$, $L_N := \mathrm{span}(c)$; then $L_{N-1} \cap L_N = \{0\}$. Hence $L = \{0\}$ and one easily calculates:
$\|P_{L_N} P_{L_{N-1}} P_{L_{N-2}} \cdots P_{L_1} P_{L^\perp}\| = \|P_{L_N} P_{L_{N-1}} P_{L_{N-2}} \cdots P_{L_1}\| = |\langle a, b\rangle \langle b, c\rangle|$
$\|P_{L_{N-1}} P_{L_N} P_{L_{N-2}} \cdots P_{L_1} P_{L^\perp}\| = \|P_{L_{N-1}} P_{L_N} P_{L_{N-2}} \cdots P_{L_1}\| = |\langle a, c\rangle \langle c, b\rangle|$.
Thus $\gamma(L_1, \ldots, L_{N-2}, L_{N-1}, L_N) \neq \gamma(L_1, \ldots, L_{N-2}, L_N, L_{N-1})$.
In general, it follows from Theorem 3.7.4 that $\gamma(L_1, \ldots, L_N)$ is positive if and only if $\gamma(L_{\tau(1)}, \ldots, L_{\tau(N)})$ is, for every permutation τ of $\{1, \ldots, N\}$.

(iv) Although the angle of the N-tuple $(L_1, \ldots, L_N)$ depends on the order, it is true that $\gamma(L_1, \ldots, L_N) = \gamma(L_N, \ldots, L_1)$, because (see Fact 3.7.1) $\|P_{L_N} \cdots P_{L_1} P_{L^\perp}\| = \|P_{L_N \cap L^\perp} \cdots P_{L_1 \cap L^\perp}\| = \|(P_{L_N \cap L^\perp} \cdots P_{L_1 \cap L^\perp})^*\| = \|P_{L_1 \cap L^\perp} \cdots P_{L_N \cap L^\perp}\| = \|P_{L_1} \cdots P_{L_N} P_{L^\perp}\|$. This was observed by Kayalar and Weinert ([**58**, Corollary 1]) and it explains once more why $\gamma(L_1, L_2)$ equals $\gamma(L_2, L_1)$.

PROPOSITION 3.7.7. The angle of the N-tuple $(L_1, \ldots, L_N)$ is positive if and only if the sum $L_1^\perp + \cdots + L_N^\perp$ is closed. In particular, this holds whenever one of the following conditions is satisfied:

(i) Some $L_i \cap L^\perp$ is finite-dimensional.
(ii) Some L_i is finite-dimensional.
(iii) H is finite-dimensional.
(iv) All L_i, except possibly one, are finite-codimensional.
(v) Each L_i is a hyperplane.
(vi) Each angle $\gamma(L_i, L_i \cap \cdots \cap L_N)$ is positive.

PROOF. The first statement is just a reformulation of Theorem 3.7.4. (i): By assumption, $L_i \cap L^\perp$ is boundedly compact. Hence Theorem 3.6.5.(i) implies bounded regularity of the N-tuple $(L_1 \cap L^\perp, \ldots, L_N \cap L^\perp)$ which is equivalent to the closedness of $L_1^\perp + \cdots + L_N^\perp$ (Proposition 3.7.3). (ii) and (iii) clearly imply (i). Condition (iv) yields linear regularity of $(L_1, \ldots, L_N)$ and hence closedness of $L_1^\perp + \cdots + L_N^\perp$ by [**8**, Corollary 5.21.(ii)] and Proposition 3.7.3. Obviously, (v) implies (iv). Finally, (vi) follows from [**8**, Example 6.18] . □

REMARKS 3.7.8.

- Gaffke and Mathar [**44**] suggested condition (i) in their study of Dykstra's algorithm. Condition (vi) is popular in computed tomography; see Smith et al.'s [**76**, Theorem 2.2].
- For $N = 2$, the angle between L_1 and L_2 is positive if and only if $L_1^\perp + L_2^\perp$ is closed which in turn is equivalent to the closedness of $L_1 + L_2$ (see, for instance, [**56**, Corollary 35.6]). For $N \geq 3$, however, the closedness of the sum $L_1 + \cdots + L_N$ is *independent* of the closedness of the sum $L_1^\perp + \cdots + L_N^\perp$ ([**8**, Remarks 5.20]).

3.8. Fejér monotone sequences.

DEFINITION 3.8.1. Suppose C is a closed convex nonempty subset of H. A sequence $(x_n)_{n\geq 0}$ is called *Fejér monotone w.r.t.* C, if

$$\|x_{n+1} - c\| \leq \|x_n - c\|, \text{ for every } c \in C \text{ and all } n \geq 0.$$

Fejér monotone sequences are a useful concept when dealing with projection algorithms or sequences of iterates of nonexpansive mappings. In our context, the most important property of Fejér monotone sequences is the following (others can be found in [**8**]):

PROPOSITION 3.8.2. Suppose $(x_n)_{n\geq 0}$ is Fejér monotone w.r.t. C. Then:

(i) (x_n) converges weakly to some point in C if and only if all weak cluster points of (x_n) lie in C. In this case, the weak limit of (x_n) belongs to $\mathcal{Q}_C x_0$.
(ii) (x_n) converges in norm to some point in C if and only if (x_n) has at least one norm cluster point in C.

PROOF. (i): Combine [**8**, Theorem 2.16.(ii)] and [**7**, Proposition 2.6.(vi)]. (ii) is [**8**, Theorem 2.16.(v)]. □

4. Nonexpansive mappings

4.1. Some bits of general theory.

DEFINITION 4.1.1. A selfmapping T of H is called *nonexpansive*, if

$$\|Tx - Ty\| \leq \|x - y\|, \text{ for all } x, y \in H.$$

A point $x \in H$ is a *fixed point of* T, if $x = Tx$; the set of all fixed points of T is denoted $\operatorname{Fix} T$.

Obviously, the class of nonexpansive mappings is closed under composition. In view of Proposition 2.1.2, we thus have:

COROLLARY 4.1.2. Each P_i and each Q_i is nonexpansive.

PROPOSITION 4.1.3. (minimal displacement vector) Suppose T is a nonexpansive selfmapping of H. Then:

(i) (Pazy's [**68**, Lemma 4]) The set $\overline{\text{range}}(T - I)$ is convex and consequently has a unique element of minimal norm (namely the projection of 0 onto $\overline{\text{range}}(T-I)$) which we call the *minimal displacement vector of* T and denote $v(T)$.
(ii) The set $\{x \in H : Tx - x = v(T)\}$ is closed and convex. In particular, $\operatorname{Fix} T$ is a closed convex (possibly empty) subset of H.
(iii) (Pazy's [**68**, Corollary 2]) For every $x \in H$, the sequence $((1/n)T^n x)$ converges in norm to $v(T)$.
(iv) The minimal displacement vector $v(T)$ is contained in $\operatorname{rec}(\overline{\operatorname{conv}}(\operatorname{range}(T)))$.

PROOF. (ii): Let $F := (I + T)/2$. Then F is *firmly nonexpansive* (see the following subsection). Since $\|Fx - x\| = \|Tx - x\|/2$ for all x, we have $\{x \in H : Tx - x = v(T)\} = \{x \in H : Fx - x = v(F)\}$. Now the latter set is convex by Corollary 4.2.4.
(iv): $T^n x \in \overline{\operatorname{conv}}(\operatorname{range}(T))$ for all n, thus $v(T) \in \operatorname{rec}(\overline{\operatorname{conv}}(\operatorname{range}(T)))$ by (iii) and Proposition 3.2.2.(ii). □

PROPOSITION 4.1.4. Suppose T is a nonexpansive selfmapping of H with $\operatorname{Fix} T \neq \emptyset$. Then the sequence of iterates $(T^n x)_{n\geq 0}$ is Fejér monotone w.r.t. $\operatorname{Fix} T$.

PROOF. By Proposition 4.1.3.(ii), Fix T is closed and convex. If $f \in \text{Fix } T$ and $n \geq 0$, then $\|T^{n+1}x - f\| = \|T^{n+1}x - Tf\| \leq \|T^n x - f\|$; the result follows. □

COROLLARY 4.1.5. If Fix Q_i is nonempty, then the sequence (x_i^n) is Fejér monotone w.r.t. Fix Q_i.

FACT 4.1.6. (Browder [**17**]/Göhde [**47**]/Kirk [**59**]) Every nonexpansive selfmapping of a bounded closed convex nonempty subset of the Hilbert space H has at least one fixed point.

REMARK 4.1.7. Numerous generalizations of this result have been established (see, for instance, Goebel and Kirk's [**45**] or Goebel and Reich's [**46**]); however, here we need only the Hilbert space case for which Pazy [**68**, Corollary 5] gave a very instructive proof.

4.2. Firmly nonexpansive mappings.

DEFINITION 4.2.1. A nonexpansive selfmapping T of H is called *firmly nonexpansive*, if

$$\langle Tx - Ty, x - y \rangle \geq \|Tx - Ty\|^2, \text{ for all } x, y \in H.$$

FACT 4.2.2. (Goebel and Kirk's [**45**, Theorem 12.1], Goebel and Reich's [**46**, Chapter 1, Section 11], Rockafellar's [**74**, Proposition 1], Zarantonello's [**84**, Lemma 1.3]) Suppose T is a selfmapping of H. Then the following are equivalent:

(i) T is firmly nonexpansive.
(ii) $I - T$ is firmly nonexpansive.
(iii) $\langle (I-T)x - (I-T)y, Tx - Ty \rangle \geq 0$, for all $x, y \in H$.
(iv) $2T - I$ is nonexpansive.

COROLLARY 4.2.3. Projections are strongly nonexpansive. Consequently, each P_i is firmly nonexpansive.

PROOF. Immediate from Proposition 2.1.2.(iii) and Fact 4.2.2. □

COROLLARY 4.2.4. Suppose T is a firmly nonexpansive selfmapping of H and x is a point in H. Then $Tx - x = v(T)$ if and only if $\langle y - Ty + v(T), Ty - v(T) - x \rangle \geq 0$, for all $y \in H$. Consequently, the set $\{x \in H : Tx - x = v(T)\}$ is closed and convex.

PROOF. Follows easily from Fact 4.2.2.(iii). □

Among all nonexpansive mappings, the class of firmly nonexpansive mappings is probably the closest to the class of projections. It possesses, however, a serious drawback:

EXAMPLE 4.2.5. The class of firmly nonexpansive mapping is not closed under composition: consider in $H := \mathbb{R}^2$ the mappings $T_1(x,y) := P_{\text{span}\{(1,0)\}}(x,y) = (x,0)$ and $T_2(x,y) := P_{\text{span}\{(1,1)\}}(x,y) = (x+y, x+y)/2$. Then $T_2T_1(x,y) = (x,x)/2$ and $\langle T_2T_1(1,-2) - T_2T_1(0,0), (1,-2) - (0,0) \rangle = -1/2 \not\geq 0$; thus T_2T_1 is not firmly nonexpansive.
Conclusion: The composition of two projections can fail even to be firmly nonexpansive.

Useful substitutes to "repair this defect" were provided by Bruck and Reich, by De Pierro and Iusem, and by Bauschke and Borwein: they introduced classes of nonexpansive mappings which are closed under composition and preserve characteristic features of projections.

4.3. Bruck and Reich's strongly nonexpansive mappings.

DEFINITION 4.3.1. (Bruck and Reich [**19**]) A nonexpansive selfmapping T of H is called *strongly nonexpansive*, if for every pair of sequences $(x_n), (y_n)$,

$$\left.\begin{array}{r}(x_n - y_n) \text{ is bounded,}\\ \\ \|x_n - y_n\| - \|Tx_n - Ty_n\| \to 0\end{array}\right\} \text{ implies } (x_n - y_n) - (Tx_n - Ty_n) \to 0.$$

PROPOSITION 4.3.2. Projections are strongly nonexpansive. Consequently, each P_i is strongly nonexpansive.

PROOF. Let P be a projection. If $(x_n - y_n)$ is bounded, then so is $(\|Px_n - Py_n\|)$ and hence $(\|x_n - y_n\| + \|Px_n - Py_n\|)$. Thus if $\|x_n - y_n\| - \|Px_n - Py_n\| \to 0$, then $\|x_n - y_n\|^2 - \|Px_n - Py_n\|^2 \to 0$ and Proposition 2.1.2 yields $(x_n - y_n) - (Px_n - Py_n) \to 0$. □

REMARK 4.3.3. In fact, Bruck and Reich [**19**, Proposition 2.1] proved that every firmly nonexpansive mapping is strongly nonexpansive whenever the underlying space is *uniformly convex* (which is the case in our Hilbert space setting).

FACT 4.3.4. (Bruck and Reich's [**19**, Lemma 2.1]) Suppose $T_1, \ldots, T_N$ are strongly nonexpansive selfmappings of H. Then $T_N T_{N-1} \cdots T_1$ is strongly nonexpansive. Moreover, $\operatorname{Fix} T_N T_{N-1} \cdots T_1 = \bigcap_{j=1}^N \operatorname{Fix} T_j$ whenever the latter intersection is nonempty.

COROLLARY 4.3.5. Each Q_i is strongly nonexpansive. Also, $\operatorname{Fix} Q_i = \bigcap_{j=1}^N C_j$ provided the latter intersection is nonempty.

REMARK 4.3.6. In view of the last remark and Example 4.2.5, we see that the class of firmly nonexpansive mappings is a proper subclass of the class of strongly nonexpansive mappings.

FACT 4.3.7. (Bruck and Reich's [**19**, Corollaries 1.3, 1.4, and 1.5]) Suppose T is a strongly nonexpansive selfmapping of H. Then:

(i) $\operatorname{Fix} T \neq \emptyset$ if and only if $(T^n x)$ converges weakly to some fixed point of T, for all $x \in H$.
(ii) $\operatorname{Fix} T = \emptyset$ if and only if $\lim_n \|T^n x\| = +\infty$, for all $x \in H$.
(iii) $\lim_n T^{n+1}x - T^n x = v(T)$.

THEOREM 4.3.8. The minimal displacement vectors $v(Q_1) = \lim_n x_1^{n+1} - x_1^n, \ldots,$ $v(Q_N) = \lim_n x_N^{n+1} - x_N^n$ all coincide and are contained in

$$\bigcap_{i=1}^N \operatorname{rec}(C_i) \cap \left(\bigcap_{i=1}^N \operatorname{rec}(C_i)\right)^{\oplus}.$$

PROOF. By Fact 4.3.7.(iii), $x_i^{n+1} - x_i^n = Q_i^n x_i^1 - Q_i^{n-1} x_i^1 \longrightarrow v_i$, for each i. Corollary 2.1.3 implies (after choosing $y^0 := x_N^1$)

$$0 \longleftarrow (x_i^{n+1} - x_i^n) - (x_{i+1}^{n+1} - x_{i+1}^n), \text{ for each } i;$$

consequently, $v(Q_1) = \cdots = v(Q_N)$. On the one hand, by Proposition 4.1.3.(iv) and Proposition 3.2.2.(iii), $v(Q_i) \in \operatorname{rec}(\overline{\operatorname{conv}}(\operatorname{range}(Q_i))) \subseteq \operatorname{rec}(\overline{\operatorname{conv}}(\operatorname{range}(P_i))) = \operatorname{rec}(C_i)$, for each i. On the other hand,

$$\begin{aligned} x_N^{n+1} - x_N^n &= (x_N^{n+1} - x_{N-1}^{n+1}) + (x_{N-1}^{n+1} - x_{N-2}^{n+1}) + \cdots + (x_1^{n+1} - x_N^n) \\ &\in \operatorname{range}(P_N - I) + \operatorname{range}(P_{N-1} - I) + \cdots + \operatorname{range}(P_1 - I), \end{aligned}$$

hence $v(Q_N) \in \overline{\text{range}(P_N - I) + \text{range}(P_{N-1} - I) + \cdots + \text{range}(P_1 - I)}$. Therefore, by Fact 3.2.3.(ii) and Proposition 3.1.2.(ii),

$$v(Q_N) \in \overline{(\text{rec}(C_N))^{\oplus} + \cdots + (\text{rec}(C_1))^{\oplus}} \subseteq \left(\bigcap_{i=1}^N \text{rec}(C_i)\right)^{\oplus}.$$

The proof is complete. □

DEFINITION 4.3.9. We abbreviate $v(Q_1)$ $(= v(Q_2) = \cdots = v(Q_N))$ by v and refer to v as the *minimal displacement vector.* The corresponding nonnegative real number $\mu := \|v\| = \inf_x \|x - Q_1 x\|$ $(= \inf_x \|x - Q_2 x\| = \cdots = \inf_x \|x - Q_N x\|)$ is called the *minimal displacement.* We say that *the minimal displacement* μ *is attained* (resp. *unattained*), if the infimum $\inf_x \|x - Q_1 x\|$ is (resp. is not) attained. Finally, we write shorter F_i instead of Fix Q_i, for each i.

PROPOSITION 4.3.10.

(i) Each F_i is a closed convex subset of C_i. If $\bigcap_{j=1}^N C_j \neq \emptyset$, then each F_i is equal to this intersection.
(ii) μ is attained if and only if each infimum $\inf_x \|x - Q_i x\|$ is.
(iii) Either each F_i is nonempty or each F_i is empty.

PROOF. (i) follows from Proposition 4.1.3.(ii) and Corollary 4.3.5.
(ii), (iii): Suppose μ is attained; that is $\mu = v(Q_1) = \|\bar{x} - Q_1 \bar{x}\|$, for some $\bar{x} \in H$. Then $\mu = v(Q_2) \leq \|Q_2 P_2 \bar{x} - P_2 \bar{x}\| = \|P_2 Q_1 \bar{x} - P_2 \bar{x}\| \leq \|Q_1 \bar{x} - \bar{x}\| = \mu$; hence $\inf_x \|x - Q_2 x\|$ is attained (at $P_2 \bar{x}$) and so are the remaining infima. □

REMARK 4.3.11. As we will see in the next subsection, the following is actually true: if μ is attained, then $\mu = 0$.

4.4. Nonexpansive mappings à la De Pierro and Iusem.

DEFINITION 4.4.1. (De Pierro and Iusem [**71**]) We say a nonexpansive selfmapping T of H is *DPI-nonexpansive,* if for all $x, y \in H$:

$$\|Tx - Ty\| = \|x - y\| \text{ implies } \begin{cases} Tx - Ty = x - y, \\ \langle x - y, Ty - y \rangle = 0. \end{cases}$$

FACT 4.4.2. (De Pierro and Iusem's [**71**, Proposition 1]) The class mappings that are DPI-nonexpansive is closed under composition.

COROLLARY 4.4.3. (De Pierro and Iusem's [**71**, Propositions 8 and 9]) Projections are DPI-nonexpansive. Consequently, each P_i and each Q_i is DPI-nonexpansive.

PROOF. Combine Proposition 2.1.2.(iii) and the preceding fact. □

FACT 4.4.4. (De Pierro and Iusem's [**71**, Proof of Lemma 1]) Suppose T is a DPI-nonexpansive selfmapping of H. If $\inf_x \|x - Tx\|$ is attained, then it is 0.

COROLLARY 4.4.5. If the minimal displacement μ is positive, it is unattained.

REMARK 4.4.6. De Pierro and Iusem formulated their results in Euclidean space — their proofs, nevertheless, work in general Hilbert space.

4.5. Strongly attracting mappings. Strongly attracting mappings are a useful tool for studying projection algorithms; see [**8**, Section 2].

DEFINITION 4.5.1. ([**8**]) A nonexpansive selfmapping T of H with $\operatorname{Fix} T \neq \emptyset$ is called *strongly attracting* or κ-attracting, if there exists $\kappa > 0$ such that

$$\kappa\|x - Tx\|^2 \leq \|x - y\|^2 - \|Tx - y\|^2, \text{ for all } x \in H,\ y \in \operatorname{Fix} T.$$

FACT 4.5.2.

(i) ([**8**, Corollary 2.5]) Projections are 1-attracting. In particular, each P_i is 1-attracting.

(ii) ([**8**, Proposition 2.10.(ii)]) If $T_1, \dots, T_N$ are strongly attracting with constants $\kappa_1, \dots, \kappa_N$ and if $\bigcap_{i=1}^N \operatorname{Fix} T_i \neq \emptyset$, then the composition $T_N \cdots T_1$ is $\min\{\kappa_1, \dots, \kappa_N\}/2^{N-1}$-attracting. Consequently, if $\bigcap_{i=1}^N C_i \neq \emptyset$, then each Q_i is $2^{-(N-1)}$-attracting.

5. Main results

5.1. Geometry of the problem.

DEFINITION 5.1.1. Suppose each F_i is nonempty. Fix any $f_1 \in F_1$ and recursively define $f_{i+1} := P_{i+1} f_i$ so that $f_i \in F_i$, for each i. We refer to the N-tuple $(f_1, \dots, f_N)$ as a *cycle*. The ith *difference vector* d_i is defined by $d_i := f_{i+1} - f_i$, for each i.

THEOREM 5.1.2. Suppose each F_i is nonempty. Then the difference vectors are well-defined, i.e. they do not depend on the choice of f_1. They satisfy $\langle C_{i+1} - F_{i+1}, -d_i \rangle \leq 0$, $\langle F_{i+1} - F_{i+1}, d_i \rangle = 0$, for each i and $d_1 + d_2 + \cdots + d_N = 0$. The restriction $P_{i+1}|_{F_i}$ is a bijection between F_i and F_{i+1}; it is given by $d_i + I$ and the sets F_i are all translates of each other. Finally, $P_{F_{i+1}} = d_i + P_{F_i}$.

PROOF. Suppose the difference vectors d_i arise from a cycle (f_i). Let (g_i) be another cycle. Then Corollary 2.1.3 implies $d_i = g_{i+1} - g_i$ (the difference vectors are thus well-defined). Proposition 2.1.2.(ii) yields $\langle C_{i+1} - F_{i+1}, -d_i \rangle \leq 0$, hence $\langle F_{i+1} - F_{i+1}, d_i \rangle = 0$. The rest follows readily. □

REMARK 5.1.3. Youla and Velasco already showed that the difference vectors are well-defined; see [**83**, Theorem 4].

The two set case. In this situation, the geometry is well-understood:

FACT 5.1.4. Suppose $N = 2$. Then

(i) (Cheney and Goldstein's [**23**, Theorem 2]) $F_1 = \{c_1 \in C_1 : d(c_1, C_2) = d(C_1, C_2)\}$, $F_2 = \{c_2 \in C_2 : d(c_2, C_1) = d(C_2, C_1)\}$.

(ii) ([**10**, Lemmata 2.1 and 2.3]) $d_1 = P_{\overline{C_2 - C_1}}(0)$, $d_2 = P_{\overline{C_1 - C_2}}(0) = -d_1$. $F_1 = C_1 \cap (C_2 + d_2)$, $F_2 = C_2 \cap (C_1 + d_1)$.

REMARKS 5.1.5.

- Goldburg and Marks rediscovered (i); see [**48**, Theorem 3]. Gubin et al. stated (i) without proof; see [**51**, Note 1 to Theorem 2].
- It is important to note that (ii) gives a description of the difference vectors *which is independent of* F_1 *and* F_2. This alternative representation is the key to the understanding of the cyclic projection algorithm (and also of Dykstra's algorithm) for two sets.

The consistent case. Here Theorem 5.1.2 reduces to Proposition 4.3.10: if $\bigcap_{i=1}^N C_i \neq \emptyset$, then $F_1 = \cdots = F_N = \bigcap_{i=1}^N C_i$ and $d_1 = \cdots = d_N = 0$.

The general case again. The general case is likely to be much more complicated: consider, for instance, in $\mathbb{R}^2$ the case when $N = 3$ and the sets C_i are the sides of an equilateral triangle. Then each F_i is a singleton. If (f_1, f_2, f_3) denotes the unique cycle, then $\|f_2 - f_1\| + \|f_3 - f_2\| + \|f_1 - f_3\|$ is *not* the infimal length of a closed polygonal path with a single vertex in each set (and other similar measures) — in stark distinction from the two set case (Fact 5.1.4.(i))! A similar example was given by Kosmol [**60**].

We conclude this subsection with a conjecture which we saw holds true for the two set and for the consistent case:

CONJECTURE 5.1.6. (geometry conjecture) The difference vectors can be described without referring to the fixed point sets F_i and thus they exist even when each F_i is empty. Moreover: $d_1 + \cdots + d_N = 0$ and

$$F_N = C_N \cap (C_{N-1} + d_{N-1}) \cap \cdots \cap (C_1 + d_1 + \cdots d_{N-1});$$

corresponding formulae hold for $F_1, \ldots, F_{N-1}$.

REMARK 5.1.7. By Theorem 5.1.2, we always have $F_N \subseteq C_N \cap (C_{N-1} + d_{N-1}) \cap \cdots \cap (C_1 + d_1 + \cdots d_{N-1})$.

5.2. Dichotomy.

THEOREM 5.2.1. Exactly one of the following alternatives holds:

(i) Each F_i is empty. Then $\lim_n \|x_i^n\| = +\infty$, for all i.
(ii) Each F_i is nonempty. Then the N-tuple of sequence $(x_1^n, x_2^n, \ldots, x_N^n)$ converges weakly to some cycle $(f_1, f_2, \ldots, f_N)$ and each f_i is an element of $\mathcal{Q}_{F_i}(x_i^m)$, for all m. The sequence of differences $(x_{i+1}^n - x_i^n)$ converges in norm to the ith difference vector d_i, for each i.

PROOF. (i): Use Corollary 4.3.5, Fact 4.3.7.(ii), and Proposition 4.3.10.(iii). (ii): If (i) doesn't hold, then each F_i is nonempty, by Proposition 4.3.10.(iii). Comparison of the orbit (x_i^n) with the orbit of any cycle yields $x_{i+1}^n - x_i^n - d_i \to 0$, for all i (Corollary 2.1.3). Also, by Fact 4.3.7.(i), $x_i^n \rightharpoonup f_i$ for some $f_i \in F_i$, for all i. Hence $f_{i+1} = f_i + d_i$, for each i and thus $(f_i)_{i=1}^N$ is identified as a cycle. Finally, $(x_i^n)_{n\geq m}$ is Fejér monotone w.r.t. F_i (Corollary 4.1.5) and converges weakly to $f_i \in F_i$; thus $f_i \in \mathcal{Q}_{F_i}(x_i^m)$ by Proposition 3.8.2.(i). □

The two set case.

FACT 5.2.2. ([**10**, Theorem 4.8]) Suppose $N = 2$. Then

$$x_2^n - x_1^n \to d_1 = P_{\overline{C_2 - C_1}}(0), \; x_1^{n+1} - x_2^n \to d_2 = P_{\overline{C_1 - C_2}}(0) = -d_1.$$

Moreover, exactly one of the following alternatives holds:

(i) F_1 and F_2 are empty. Then $\|x_1^n\|, \|x_2^n\| \to +\infty$.
(ii) F_1 and F_2 are nonempty. Then $x_1^n \rightharpoonup f_1$ and $x_2^n \rightharpoonup f_2$, where $f_1 \in F_1$ and $f_2 = f_1 + d_1 \in F_2$.

The consistent case. Here Theorem 5.2.1.(ii) results in the following:

THEOREM 5.2.3. Suppose $\bigcap_{i=1}^{N} C_i \neq \emptyset$. Then the entire orbit (x_i^n) converges weakly to some point in $\bigcap_{i=1}^{N} C_i$. Also, $x_{i+1}^n - x_i^n \to 0$ which implies $d(x_i^n, C_j) \to 0$, for all i, j.

REMARK 5.2.4. The weak convergence of (x_i^n) was already known to Bregman ([**16**, Theorem 1]).

The general case again. Nurtured by the results for the two set case and the consistent case, we formulate two conjectures:

CONJECTURE 5.2.5. (convergent differences conjecture) The sequence of differences $(x_{i+1}^n - x_i^n)$ is norm convergent for all starting points and every i.

REMARK 5.2.6. If the convergent differences conjecture is true, then it follows that $\lim_n x_{i+1}^n - x_i^n$ is independent of the starting point (Corollary 2.1.3).

The second conjecture is formally weaker that the first one:

CONJECTURE 5.2.7. (bounded differences conjecture) The sequences of differences $(x_{i+1}^n - x_i^n)$ are bounded for all starting points and every i.

We would like to mention the following long-standing open problem:

OPEN PROBLEM 5.2.8. Even when $N = 2$ and $C_1 \cap C_2 \neq \emptyset$, it is still not known whether the convergence of the entire orbit (x_i^n) to some point in $C_1 \cap C_2$ can actually be only weak.

5.3. Minimal displacement.

THEOREM 5.3.1.

(i) $\mu = 0$ whenever $\bigcap_{i=1}^{N} \operatorname{rec}(C_i)$ is a subspace.

(ii) $\mu = 0$ and is attained if and only if there exists a bounded closed convex nonempty subset of H that is left invariant by some Q_j. In particular, this holds whenever some C_j is bounded.

(iii) $\mu = 0$ and is attained whenever $\bigcap_{i=1}^{N} \operatorname{rec}(C_i)$ is a subspace and there exists some j such that the "normalization" $\hat{C}_j := \{c_j/\|c_j\| : c_j \in C_j \setminus \{0\}\}$ of C_j is relatively compact. The latter "normalization" condition holds whenever H is finite-dimensional.

PROOF. (i): By Proposition 3.1.3, $\bigcap_{i=1}^{N} \operatorname{rec}(C_i)$ is a subspace if and only if $\bigcap_{i=1}^{N} \operatorname{rec}(C_i) \cap (\bigcap_{i=1}^{N} \operatorname{rec}(C_i))^{\oplus} = \{0\}$. (i) is now obvious from Theorem 4.3.8.
(ii): For the "only if" part, simply choose $f_j \in F_j$ so that $\{f_j\}$ is left invariant by Q_j. The "if" and "In particular" parts follow from Corollary 4.1.2 and Fact 4.1.6.
(iii): Suppose to the contrary μ is not simultaneously 0 and attained. In view of Corollary 4.4.5 and Theorem 5.2.1, each Q_i is fixed point free and all orbits (x_i^n) tend in norm to infinity. We may also assume (after relabeling, if necessary) that $j = N$. Fix an orbit (x_i^n) and extract a subsequence (n_k) of (n) such that

$$\|x_N^{n_k+1}\| = \|Q_N x_N^{n_k}\| > \|x_N^{n_k}\| =: r_k, \text{ for all } k.$$

Then (r_k) is strictly increasing and tends to infinity. Let us abbreviate

$$y_0^k := x_N^{n_k} \text{ and } y_i^k := P_i y_{i-1}^k, \text{ for all } k \text{ and every } i.$$

Now choose a positive integer m so large that $y_i^n \neq y_i^m$, for all $n \geq m$ and every i. (If this weren't possible, then we would conclude $y_N^n = y_N^m$ infinitely often, which would be absurd because $\|y_N^n\| > r_n \to +\infty$.)

Claim: $\lim_n \|y_i^n - y_i^m\|/\|y_{i-1}^n - y_{i-1}^m\| = 1$, for all i.

Since all of these quotients are at most 1, it suffices to show that $\lim_n \|y_N^n - y_N^m\|/\|y_0^n - y_0^m\| = 1$. This, however, follows from

$$1 \geq \frac{\|y_N^n - y_N^m\|}{\|y_0^n - y_0^m\|} \geq \frac{\|y_N^n\| - \|y_N^m\|}{\|y_0^n\| + \|y_0^m\|} > \frac{r_n - \|y_N^m\|}{r_n + \|y_0^m\|} \to 1;$$

the claim is thus verified.

Since $\|y_0^n\| = r_n \to +\infty$, we reach the following:

Conclusion: $\lim_n \|y_i^n\| = +\infty$, for all i.

After passing to another subsequence if necessary, we can also assume that the sequence $((y_i^n - y_i^m)/\|y_i^n - y_i^m\|)$ converges weakly to some point q_i, for each i. The Claim and Proposition 2.1.2.(iii) imply

$$\frac{y_{i-1}^n - y_{i-1}^m}{\|y_{i-1}^n - y_{i-1}^m\|} - \frac{y_i^n - y_i^m}{\|y_i^n - y_i^m\|}\frac{\|y_i^n - y_i^m\|}{\|y_{i-1}^n - y_{i-1}^m\|} \to 0, \text{ for all } i.$$

Hence the weak limits q_i all coincide, say $q := q_1 = \cdots = q_N$. Next, by the Conclusion and Proposition 3.2.2,

$$q_i = \operatorname{weak\,lim}_n \frac{y_i^n - y_i^m}{\|y_i^n - y_i^m\|} = \operatorname{weak\,lim}_n \frac{y_i^n}{\|y_i^n\|} \in \operatorname{rec}(C_i), \text{ for all } i.$$

Thus

$$q \in \textstyle\bigcap_{i=1}^N \operatorname{rec}(C_i).$$

Now $\hat{C}_N$ is relatively compact (by hypothesis), hence we can assume that $(y_N^n/\|y_N^n\|)$ converges in norm to q (after passing to yet another subsequence, if necessary) which implies

$$\|q\| = 1.$$

On the other hand, for the orbit (x_i^n), we certainly have (Fact 3.2.3.(ii) and Proposition 3.1.2.(ii)) $x_N^{n+1} - x_N^n \in \sum_{i=1}^N \operatorname{range}(P_i - I) \subseteq (\bigcap_{i=1}^N \operatorname{rec}(C_i))^{\oplus}$, for all n. Hence $x_N^{n+l} - x_N^n \in (\bigcap_{i=1}^N \operatorname{rec}(C_i))^{\oplus}$, for all positive integers n, l. Remembering the special structure of (y_i^n), we conclude $(y_N^n - y_N^m)/\|y_N^n - y_N^m\| \in (\bigcap_{i=1}^N \operatorname{rec}(C_i))^{\oplus}$, for all n, which yields

$$q \in (\textstyle\bigcap_{i=1}^N \operatorname{rec}(C_i))^{\oplus}.$$

Altogether, we obtain

$$0 \neq q \in (\textstyle\bigcap_{i=1}^N \operatorname{rec}(C_i)) \cap (\bigcap_{i=1}^N \operatorname{rec}(C_i))^{\oplus}.$$

Finally, $\bigcap_{i=1}^N \operatorname{rec}(C_i)$ is a subspace so that $(\bigcap_{i=1}^N \operatorname{rec}(C_i)) \cap (\bigcap_{i=1}^N \operatorname{rec}(C_i))^{\oplus} = \{0\}$ (Proposition 3.1.3). Therefore, q has to be 0 which is the desired contradiction. □

COROLLARY 5.3.2. If each C_i is a closed affine subspace, then $\mu = 0$.

REMARK 5.3.3. Theorem 5.3.1.(iii) is sharp in the following sense:

- Kosmol and Zhou [**62**, Theorem] and Bauschke and Borwein [**10**, Example 4.3] constructed examples where $N = 2$ and each C_i is a closed affine subspace but $\mu = 0$ and is *unattained*. Hence the preceding corollary cannot be improved and the assumption of relative compactness in Theorem 5.3.1.(iii) is important.

- The following simple example shows that the assumption on the recession cones in Theorem 5.3.1.(iii) is important: suppose $H = \mathbb{R}^2$, $N = 2$, and let $C_1 = \{(r,0) : r \geq 0\}$, $C_2 = \{(x,y) : y \geq 1/x > 0\}$. Then $\text{rec}(C_1) \cap \text{rec}(C_2) = C_1$ is not linear, $\mu = 0$ (see the subsubsection on the two set case below) and unattained.

REMARKS 5.3.4. Suppose C is a closed convex nonempty subset of H and let $\hat{C} := \{c/\|c\| : c \in C \setminus \{0\}\}$.

- $\hat{C}$ need not be closed: consider in $\mathbb{R}^2$ the epigraph of the function $1/x$ restricted to the positive reals, i.e. $C = \{(x,y) \in \mathbb{R}^2 : y \geq 1/x > 0\}$. Then $\hat{C} = \{(x,y) \in \mathbb{R}^2 : x^2 + y^2 = 1;\ x > 0;\ y > 0\}$ is not closed.
- It is easy to check that if $\hat{C}$ is relatively compact, then C is boundedly compact.
- In general, if C is boundedly compact, then $\hat{C}$ need not be relatively compact: consider in $H = \ell_2$ the compact convex nonempty set $C := \overline{\text{conv}}\{e^n/n : n \geq 1\}$, where e^n denotes the nth unit vector. Then $\hat{C}$ contains $\{e^1, e^2, \dots\}$ and thus cannot be relatively compact.

The two set case.

THEOREM 5.3.5. If $N = 2$, then $\mu = 0$.

PROOF. By Fact 5.2.2, $x_2^n - x_1^n \to d_1$ and $x_1^{n+1} - x_2^n \to d_2 = -d_1$. Adding both sequences yields $x_1^{n+1} - x_1^n = Q_1 x_1^n - x_1^n \to 0$ and therefore $\mu = \inf_x \|Q_1 x - x\| = 0$. □

The general case again. In view of the results of this subsection, the following conjecture is very plausible (and true for the two set and the consistent cases):

CONJECTURE 5.3.6. (zero displacement conjecture) The minimal displacement μ always equals 0.

THEOREM 5.3.7. The bounded differences conjecture is stronger than the zero displacement conjecture

PROOF. Suppose the bounded differences conjecture is true. Comparison of the orbits generated by x^0 and x_N^1 via Corollary 2.1.3 yields $\lim_n \langle x_i^n - x_{i+1}^n, x_{i+1}^n - x_{i+1}^{n+1} \rangle = 0$, for all i. Hence, by assumption and Theorem 4.3.8, $\lim_n \langle x_i^n - x_{i+1}^n, -v \rangle = 0$, for all i. Addition yields $0 = \lim_n \langle x_1^n - x_1^{n+1}, -v \rangle = \langle -v, -v \rangle = \|v\|^2$; therefore, $\mu = \|v\| = 0$ and the proof is complete. □

5.4. Convergence results I.

THEOREM 5.4.1.
(weak convergence) If some set C_j is bounded, then $(x_1^n, \dots, x_N^n)$ converges weakly to some cycle.
(norm convergence) $(x_1^n, \dots, x_N^n)$ converges in norm to some cycle whenever (at least) one of the following conditions holds:

(i) Each F_i is nonempty and some C_j is boundedly compact.
(ii) Some C_i is bounded and some C_j is boundedly compact.
(iii) Some C_j is compact.
(iv) Some C_j is bounded and H is finite-dimensional.

(v) Some normalization $\hat{C}_j = \{c_j/\|c_j\| : c_j \in C_j \setminus \{0\}\}$ is relatively compact and $\bigcap_{i=1}^N \mathrm{rec}(C_i)$ is a subspace.

(vi) H is finite-dimensional and $\bigcap_{i=1}^N \mathrm{rec}(C_i)$ is a subspace.

In all situations, $\lim_n x_{i+1}^n - x_i^n = d_i$, for each i.

PROOF. (weak convergence): Combine Theorems 5.3.1.(ii) and 5.2.1.(ii).
(norm convergence).(i): On the one hand, by Theorem 5.2.1.(ii), (x_j^n) is weakly convergent to some $f_j \in F_j$. Since (x_j^n) is bounded and C_j is boundedly compact, f_j is actually a norm cluster point of (x_j^n). On the other hand, (x_j^n) is Fejér monotone w.r.t. F_j (Corollary 4.1.5). Thus, by Proposition 3.8.2.(ii), (x_j^n) converges in norm to f_j. Therefore, by continuity of each P_i, each (x_i^n) is norm convergent. The rest follows since: (iv) $\Rightarrow$ (iii) $\Rightarrow$ (ii) $\Rightarrow$ (i) $\Leftarrow$ (v) $\Leftarrow$ (vi). (For "(v) $\Rightarrow$ (i)", use Theorem 5.3.1.(iii) and Remarks 5.3.4.)
"In all situations" is nothing but Theorem 5.2.1.(ii). □

REMARKS 5.4.2.

- A slightly less general form of the weak convergence part of the last theorem is Gubin et al.'s [**51**, weak convergence part of Theorem 2]; they also observed (iv).
- The two set version of Theorem 5.4.1.(i) and (iii) is due to Cheney and Goldstein; see [**23**, Theorem 4]
- Consider the second example in the remark following Theorem 5.3.1. Since F_1 and F_2 are empty, the sequences (x_1^n) and (x_2^n) tend to infinity. Therefore, the assumption on the recession cones in Theorem 5.4.1.(v) is important.
- The following consistent case is interesting: if each set C_i is *symmetric* ($C_i = -C_i$), then Bruck and Reich's [**19**, Theorem 2.2] or Baillon et al.'s [**6**, Corollary 2.4] implies that the entire orbit (x_i^n) converges in norm to some point in $\bigcap_i C_i$.

The two set case. In this case, there is another sufficient condition for weak convergence:

THEOREM 5.4.3. If $C_2 - C_1$ is closed, then (x_1^n, x_2^n) converges weakly to some cycle.

PROOF. By Fact 5.1.4.(ii), F_1 and F_2 are nonempty. The result now follows from Theorem 5.2.1.(ii). □

5.5. Convergence results II.

THEOREM 5.5.1. Suppose each C_i, except possibly one, is Kadec/Klee and each F_i is nonempty. Then $(x_1^n, \dots, x_N^n)$ converges in norm to some cycle.

PROOF. By Theorem 5.2.1.(ii), $(x_1^n, \dots, x_N^n)$ converges weakly to some cycle $(f_1, \dots, f_N)$. *Case 1:* $\bigcap_{i=1}^N C_i = \emptyset$. Then there is some i such that $f_{i+1} = P_{i+1} f_i \neq f_i$ and C_{i+1} is Kadec/Klee. Hence $f_{i+1} \in \mathrm{bd}(C_{i+1})$. Since $x_{i+1}^n \rightharpoonup f_{i+1}$, we conclude $x_{i+1}^n \to f_{i+1}$ and thus $(x_1^n, \dots, x_N^n) \to (f_1, \dots, f_N)$. *Case 2:* $\bigcap_{i=1}^N C_i \neq \emptyset$. Then $(C_1, \dots, C_N)$ is boundedly regular (Theorem 3.6.5.(iii)) and the result follows from Fact 5.5.5.(iii) below. □

COROLLARY 5.5.2. (Gubin et al.'s [**51**, part of Theorem 2]) Suppose each C_i, except possibly one, is uniformly convex and some C_j is bounded. Then $(x_1^n, \dots, x_N^n)$ converges in norm to some cycle.

PROOF. Clear from Proposition 3.5.2, Theorem 5.4.1, and Theorem 5.5.1. □

The two set case.

FACT 5.5.3. ([**9**, Theorem 3.7 and Corollary 3.10]) Suppose $N = 2$. If (C_1, C_2) is boundedly regular, then (x_1^n, x_2^n) converges in norm to some cycle. In particular, this happens whenever C_1 or C_2 is boundedly compact and each F_i is nonempty.

REMARK 5.5.4. The first two authors' [**9**, Theorem 3.7 and Corollary 3.10] lack the assumption of nonemptiness of each F_i; see Remark 3.6.3.

The consistent case. Our observations on (linear) regularity in Subsection 3.6 combined with convergence results in [**8**, Section 5] yield the following two facts:

FACT 5.5.5. ([**8**, Theorem 5.2]) Suppose $\bigcap_i C_i \neq \emptyset$ and $(C_1, \dots, C_N)$ is boundedly regular. Then the entire orbit (x_i^n) converges in norm to some point in $\bigcap_i C_i$. In particular, this happens if (i) some C_i is boundedly compact; (ii) H is finite-dimensional; (iii) every C_i, except possibly one, is Kadec/Klee; or (iv) every C_i, except possibly one, is uniformly convex.

REMARK 5.5.6. Fact 5.5.5.(i),(ii) give a different explanation for the consistent case of Theorem 5.4.1.(i). Fact 5.5.5.(iv) is due to Gubin et al.; see [**51**, Theorem 1.(b)].

FACT 5.5.7. ([**8**, Theorem 5.7]) Suppose $\bigcap_i C_i \neq \emptyset$ and $(C_1, \dots, C_N)$ is boundedly linearly regular. Then: (a) the entire orbit (x_i^n) converges linearly to some point in $\bigcap_i C_i$; moreover, (b) the rate of convergence is independent of the starting point whenever $(C_1, \dots, C_N)$ is linearly regular. In particular, (a) happens whenever (i) $0 \in \bigcap_{i=1}^{N-1} \operatorname{icr}((\bigcap_{j=1}^{i} C_j) - C_{i+1})$ or (ii) $C_N \cap \operatorname{int}(\bigcap_{i=1}^{N-1} C_i) \neq \emptyset$.

REMARKS 5.5.8. Fact 5.5.7.(a).(ii) is also due to Gubin et al.; see [**51**, Theorem 1.(a)]. Corresponding formulations of Fact 5.5.7.(a),(i),(ii) with different orders hold, too; see Remark 3.6.10. We discuss some more special cases of Fact 5.5.7 in Subsections 5.6 and 5.7.

FACT 5.5.9. ([**9**, Theorem 5.3.(iv)]) Suppose $N = 2$ and H is a Hilbert lattice with lattice cone H^+. If C_1 is a hyperplane, $C_2 = H^+$, and $C_1 \cap C_2 \neq \emptyset$, then the entire orbit (x_i^n) converges in norm to some point in $C_1 \cap C_2$.

REMARKS 5.5.10.

- Classical examples are $H = \ell_2$ with $H^+ = \{(s_n) \text{ in } H : s_n \geq 0, \text{ for all } n\}$ and $H = L_2[0,1]$ with $H^+ = \{f \in H : f \geq 0 \text{ almost everywhere}\}$.
- Fact 5.5.9 relies heavily on the order structure of H. It would be interesting to know whether this result holds when C_1 is a closed affine finite-codimensional subspace instead of a hyperplane.

The inconsistent case.

FACT 5.5.11. (Gubin et al.'s [**51**, part of Theorem 2]) Suppose each C_i, except possibly one, is strongly convex, some C_j is bounded, and $\bigcap_{i=1}^N C_i = \emptyset$. Then $(x_1^n, \dots, x_N^n)$ converges linearly to some cycle.

REMARK 5.5.12. The simple example of two discs in $\mathbb{R}^2$ with precisely one common point shows that the assumption of emptiness of the intersection is important.

5.6. The convex polyhedral case.

THEOREM 5.6.1. If each C_i is a convex polyhedron, then each F_i is nonempty and the N-tuple of sequences $(x_1^n, \dots, x_N^n)$ *converges in norm* to some cycle.

PROOF. Corollary 3.4.8, Corollary 4.4.5, and Theorem 5.2.1.(ii) imply the weak convergence of $(x_1^n, \dots, x_N^n)$ to some cycle. Now suppose that each C_i is given as $\{x \in H : \langle a_{i,j}, x\rangle \leq b_{i,j}\}$. Let $K := \bigcap_{i,j} \operatorname{kernel}(a_{i,j})$ and let $D_i := C_i \cap K^\perp$, for each i. Then, by Proposition 3.3.2,

$$x_{i+1}^n - x_i^n = P_K x_i^n + P_{D_{i+1}} P_{K^\perp} x_i^n - x_i^n = -P_{K^\perp} x_i^n + P_{D_{i+1}} P_{K^\perp} x_i^n \in K^\perp,$$

for all n and each i. Telescoping yields

$$x_i^n \in x^0 + K^\perp, \text{ for all } n \text{ and every } i;$$

hence the entire orbit (x_i^n) lies in an affine finite-dimensional subspace. Here, however, weak and norm convergence coincide and the result follows. □

REMARK 5.6.2. Theorem 5.6.1 is related to a result by De Pierro and Iusem [**71**, Proposition 13 and Theorem 1] which states that (x_N^n) converges in norm to some point in F_N when H is finite-dimensional and each C_i is a halfspace; their algorithm, however, is more general.

The consistent case. Combining Fact 3.6.8 with Fact 5.5.7.(b) yields:

THEOREM 5.6.3. Suppose each C_i is a convex polyhedron and $\bigcap_{i=1}^N C_i \neq \emptyset$. Then the entire orbit (x_i^n) converges linearly to some point in $\bigcap_{i=1}^N C_i$ with a rate independent of the starting point.

REMARK 5.6.4. In the setting of the problem, special cases of this theorem were obtained by Gubin et al. ([**51**, Theorem 1.(d)]) and by Herman et al. ([**53**, Theorem 1]); see also Mandel's [**65**, Theorem 3.1] for an upper bound for the rates of convergence and [**8**, Section 6] for more general algorithms.

The general case again. Encouraged by the last fact, we believe the following is true:

CONJECTURE 5.6.5. If each C_i is a convex polyhedron, then $(x_1^n, \dots, x_N^n)$ *converges linearly* to some cycle *with a rate independent* of the starting point.

5.7. The affine subspace case. Throughout this subsection, we assume that

$$\boxed{C_1 = c_1 + L_1, \ \dots, \ C_N = c_N + L_N \text{ are closed affine subspaces,}}$$

where, without loss of generality,

$$\boxed{L_i \text{ is linear, } L := \textstyle\bigcap_{i=1}^N L_i, \text{ and } c_i \in L_i^\perp, \text{ for each } i.}$$

We investigate the behaviour of the method of cyclic projections in this context and restrict ourselves to the subsequence (x_N^n) of an arbitrary but fixed orbit. Proposition 2.2.2.(i) yields $P_i x = c_i + P_{L_i} x$, for each i and all x; thus if we let $\mathbf{c} := (c_1, \dots, c_N)$ and define the operator

$$\boxed{\mathbf{T} : L_1^\perp \times \cdots \times L_N^\perp \to H : \mathbf{y} = (y_1, \dots, y_N) \mapsto \textstyle\sum_{i=1}^N P_{L_N} \cdots P_{L_{i+1}} y_i,}$$

it then follows that $Q_N x = P_{L_N} \cdots P_{L_1} x + \mathbf{Tc}$, and hence inductively

$$x_N^n = Q_N^n x^0 = (P_{L_N} \cdots P_{L_1})^n x^0 + \textstyle\sum_{k=0}^{n-1} (P_{L_N} \cdots P_{L_1})^k (\mathbf{Tc}), \text{ for all } n \geq 1.$$

In passing, we note that the first term on the right hand side of the last equation, $(P_{L_N} \cdots P_{L_1})^n x^0$, tends to $P_L x^0$ by the Fact 1.1.1.

PROPOSITION 5.7.1.

(i) $L^\perp$ is an invariant subspace of P_L and $P_{L_N} \cdots P_{L_i}$, for each i.
(ii) $\mathbf{Tc} \in L^\perp$.
(iii) $(P_{L_N} \cdots P_{L_1})^k = P_L \oplus (P_{L_N} \cdots P_{L_1} P_{L^\perp})^k$, for all $k \geq 1$.
(iv) $P_L = P_L P_{L_i}$, for each i.

PROOF. (i): Obviously, $L^\perp$ is an invariant subspace of P_L. Now fix $l^\perp \in L^\perp$ and $l \in L$. Then $\langle l, P_{L_N} \cdots P_{L_i} l^\perp \rangle = \langle (P_{L_N} \cdots P_{L_i})^* l, l^\perp \rangle = \langle P_{L_i} \cdots P_{L_N} l, l^\perp \rangle = \langle l, l^\perp \rangle = 0$, which implies that $L^\perp$ is an invariant subspaces of $P_{L_N} \cdots P_{L_i}$.
(ii): Since $c_i \in L_i^\perp \subseteq L^\perp$, for all i, (i) and the definition of $\mathbf{T}$ imply $\mathbf{Tc} \in L^\perp$.
(iii) is proved by induction on k: $P_{L_N} \cdots P_{L_1} = P_{L_N} \cdots P_{L_1}(P_L \oplus P_{L^\perp}) = P_L \oplus P_{L_N} \cdots P_{L_1} P_{L^\perp}$; hence (iii) holds for $k = 1$. If it holds for some $k \geq 1$, then $(P_{L_N} \cdots P_{L_1})^{k+1} = (P_{L_N} \cdots P_{L_1})(P_L \oplus (P_{L_N} \cdots P_{L_1} P_{L^\perp})^k) =$
$P_L \oplus (P_{L_N} \cdots P_{L_1} P_{L^\perp})^{k+1}$, because $L^\perp$ is an invariant subspace of $P_{L_N} \cdots P_{L_1}$ by (i).
(iv): Fix $x \in H$ and $l \in L$. Then $\langle l - P_L P_{L_i} x, x - P_L P_{L_i} x \rangle = \langle l - P_L P_{L_i} x, P_{L_i} x - P_L P_{L_i} x \rangle + \langle l - P_L P_{L_i} x, P_{L_i^\perp} x \rangle = 0$; since this is true for all $l \in L$, we conclude (Proposition 2.1.2.(ii)) that $P_L x = P_L P_{L_i} x$ and the result follows. ☐

COROLLARY 5.7.2. The sequence (x_N^n) satisfies

$$x_N^n = P_L x^0 + (P_{L_N} \cdots P_{L_1} P_{L^\perp})^n x^0 + \textstyle\sum_{k=0}^{n-1} (P_{L_N} \cdots P_{L_1} P_{L^\perp})^k (\mathbf{Tc}), \text{ for all } n \geq 1.$$

This observation is useful, since the operator norm of $P_{L_N} \cdots P_{L_1} P_{L^\perp}$ is less than or equal to the operator norm of $P_{L_N} \cdots P_{L_1}$. Moreover, $\|P_{L_N} \cdots P_{L_1} P_{L^\perp}\|$ is nothing but the cosine of the angle of the N-tuple $(L_1, \ldots, L_N)$; see Definition 3.7.5.

The next proposition provides information on the set of fixed points F_N:

PROPOSITION 5.7.3.

(i) $F_N = (I - P_{L_N} \cdots P_{L_1})^{-1}(\mathbf{Tc})$.
(ii) (Kosmol and Zhou's [**62**, Proposition 4]) If F_N is nonempty, then $F_N = f_N^* + L$, for all $f_N^* \in F_N$.

PROOF. (i): $x \in F_N \Leftrightarrow x - Q_N x = 0 \Leftrightarrow x - (P_{L_N} \cdots P_{L_1})x = \mathbf{Tc}$.
(ii): Pick $f_N, f_N^* \in F_N$. Then, using (i), the linearity of $P_{L_N} \cdots P_{L_1}$, and Corollary 4.3.5, we get: $f_N - P_{L_N} \cdots P_{L_1} f_N = \mathbf{Tc} = f_N^* - P_{L_N} \cdots P_{L_1} f_N^* \Leftrightarrow f_N - f_N^* = (P_{L_N} \cdots P_{L_1})(f_N - f_N^*) \Leftrightarrow f_N - f_N^* \in \operatorname{Fix}(P_{L_N} \cdots P_{L_1}) \Leftrightarrow f_N \in f_N^* + L$. ☐

The following useful identity is due to Kosmol and Zhou; it is proved inductively using Proposition 5.7.3.(i).

FACT 5.7.4. ([**61**, Proof of Theorem 3]) If F_N is nonempty and $f_N^* \in F_N$, then

$$x_N^n = (P_{L_N} \cdots P_{L_1})^n (x^0 - f_N^*) + f_N^*, \text{ for all } n \geq 1.$$

THEOREM 5.7.5. (norm dichotomy) Exactly one of the following alternatives holds:

(i) F_N is empty. Then $(\|x_N^n\|)$ tends to $+\infty$.
(ii) F_N is nonempty. Then the sequence (x_N^n) converges in norm:

$$\lim_n x_N^n = P_{F_N} x^0 = P_L x^0 + \textstyle\sum_{k=0}^{\infty} (P_{L_N} \cdots P_{L_1} P_{L^\perp})^k (\mathbf{Tc}).$$

Consequently, $\sum_{k=0}^{\infty} (P_{L_N} \cdots P_{L_1} P_{L^\perp})^k (\mathbf{Tc})$ is equal to $P_{F_N} 0$, the minimum norm fixed point of Q_N.

PROOF. In view of Theorem 5.2.1, we assume that F_N is nonempty and fix $f_N^* \in F_N$. By Fact 5.7.4 and Fact 1.1.1, $x_N^n \to P_L(x^0 - f_N^*) + f_N^* = P_L x^0 + P_{L^\perp} f_N^*$. Also, by Proposition 2.2.2.(i) and Proposition 5.7.3.(ii), $P_L(x^0 - f_N^*) + f_N^* = P_{f_N^* + L}(x^0) = P_{F_N} x^0$ and $P_{L^\perp} f_N^* = f_N^* + P_L(0 - f_N^*) = P_{f_N^* + L}(0) = P_{F_N} 0$. Hence, by Corollary 5.7.2, $P_{L^\perp} f_N^* = P_{F_N} 0 = \sum_{k=0}^\infty (P_{L_N} \cdots P_{L_1} P_{L^\perp})^k (\mathbf{Tc})$. □

COROLLARY 5.7.6. If all F_i are empty, then $\lim_n \|x_i^n\| = +\infty$, for each i. Otherwise, all F_i are nonempty, each sequence (x_i^n) converges in norm to $P_{F_i} x^0$, and $d_i = P_{L^\perp}(F_{i+1} - F_i)$.

PROOF. In view of Theorem 5.2.1, we only have to verify the "otherwise" part. Theorem 5.7.5.(ii) implies $x_N^n \to P_{F_N} x^0$; if we apply the theorem once more (this time to (x_1^n)), we get $x_1^n \to P_{F_1} x_1^1$. Fix $f_1 \in F_1$. Then, by Proposition 5.7.3.(ii), Proposition 2.2.2.(i), and Proposition 5.7.1, $P_{F_1} x_1^1 = P_{f_1 + L}(P_1 x^0) = f_1 + P_L(c_1 + P_{L_1} x^0 - f_1) = f_1 + P_L(x^0 - f_1) = P_{f_1+L}(x^0) = P_{F_1} x^0$. Also, $x_2^n \to P_{F_2} x_2^1$, and, similarly to what we just did, $P_{F_2} x_2^1 = P_{F_2} x_1^1 = P_{F_2} x^0$. Analogously, $P_{F_3} x_3^1 = P_{F_3} x^0$, ..., up to $P_{F_{N-1}} x_{N-1}^1 = P_{F_{N-1}} x^0$. Finally, by Theorem 5.1.2, Proposition 5.7.3.(ii), and Proposition 2.2.2.(i), $d_i = P_{F_{i+1}}(x^0) - P_{F_i}(x^0) = P_{f_{i+1}+L}(x^0) - P_{f_i + L}(x^0) = f_{i+1} + P_L(x^0 - f_{i+1}) - (f_i + P_L(x^0 - f_i)) = P_{L^\perp}(f_{i+1} - f_i)$, for all $f_i \in F_i$ and $f_{i+1} \in F_{i+1}$. □

REMARKS 5.7.7.

- Kosmol and Zhou obtained the "otherwise" part of Corollary 5.7.6; see [**61**, Theorem 3].
- The fact that the limit of the entire orbit (x_i^n) has to be $P_{F_i} x^0$ (provided it exists) can be explained differently by using Fejér monotone sequences and quasi-projections. Indeed, the sequence (x_n^i) is Fejér monotone w.r.t. F_i (Corollary 4.1.5), hence its limit lies in $\mathcal{Q}_{F_i} x_i^1$ (Proposition 3.8.2) which is — in this setting — the singleton $\{P_{F_i} x_i^1\}$ (Proposition 2.3.2.(iii)). However, as we saw in the proof of Corollary 5.7.6, $P_{F_i} x_i^1 = P_{F_i} x^0$.

THEOREM 5.7.8. (angle and linear convergence) If $L_1^\perp + \cdots + L_N^\perp$ is closed, then F_N is nonempty and (x_N^n) converges linearly with a rate no worse than the cosine of the angle of the N-tuple $(L_1, \ldots, L_N)$:

$$\lim_n x_N^n = P_{F_N} x^0 = P_L x^0 + P_{F_N} 0 = P_L x^0 + (I - P_{L_N} \cdots P_{L_1} P_{L^\perp})^{-1} (\mathbf{Tc}).$$

In particular, this holds whenever:

(i) Some $L_i \cap L^\perp$ is finite-dimensional.
(ii) Some L_i is finite-dimensional.
(iii) H is finite-dimensional.
(iv) All L_i, except possibly one, are finite-codimensional.
(v) Each C_i is a hyperplane.
(vi) Each angle $\gamma(L_i, L_{i+1} \cap \cdots \cap L_N)$ is positive.

Consequently, each sequence (x_i^n) converges linearly to $P_{F_i} x^0$ with a rate no worse than the cosine of the angle $\gamma(L_{i+1}, \ldots, L_N, L_1, \ldots, L_i)$.

PROOF. Proposition 3.7.7 implies that the angle $\gamma(L_1, \ldots, L_N)$ of the N-tuple $(L_1, \ldots, L_N)$ is positive, in other words: $\|P_{L_N} \cdots P_{L_1} P_{L^\perp}\| < 1$. Hence, by Corollary 5.7.2, (x_N^n) converges linearly with a rate no worse than $\cos \gamma(L_1, \ldots, L_N)$; its limit is $P_L x^0 + (I - P_{L_N} \cdots P_{L_1} P_{L^\perp})^{-1} (\mathbf{Tc})$. The main statement of the theorem now follows with Theorem 5.7.5. Conditions (i) through (vi) ensure, again by Proposition 3.7.7, $\gamma(L_1, \dots, L_N) > 0$. Finally, the "consequently" part follows from Corollary 5.7.6 and the first statement of this theorem. □

REMARKS 5.7.9.

- If $N = 2$ and $L_1^\perp + L_2^\perp$ is not closed, then convergence can be "arbitrarily slow"; see Theorem 5.7.16.
- Kosmol proved a less general version of (iii); see [**60**, Satz 3].

We now specialize to the case when each C_i is a hyperplane:

$$C_i = \{x \in H : \langle a_i, x\rangle = b_i\}, \text{ for some } a_i \in H \setminus \{0\} \text{ and } b_i \in \mathbb{R}.$$

In the terminology of this subsection,

$$c_i = \frac{b_i}{\|a_i\|^2} a_i,\ L_i = \text{kernel}(a_i) = \{a_i\}^\perp,\ L_i^\perp = \text{span}\{a_i\}, \text{ for each } i.$$

Theorem 5.7.8.(v) implies that the angle $\gamma(L_1, \dots, L_N)$ is positive; hence $\|P_{L_N} \cdots P_{L_1} P_{L^\perp}\| < 1$ and each subsequence (x_i^n) converges linearly.

The mapping

$$\Psi : \mathbb{R}^N \to L_1^\perp \times \cdots \times L_N^\perp : \mathbf{y} = (y_1, \dots, y_N) \mapsto \left(\frac{y_1}{\|a_1\|^2} a_1, \dots, \frac{y_N}{\|a_N\|^2} a_N\right)$$

not only defines a linear isomorphism between $\mathbb{R}^N$ and $L_1^\perp \times \cdots \times L_N^\perp$, but also allows us to think of $\mathbf{T}$ as being defined on $\mathbb{R}^N$; this will yield a nice connection to generalized inverses.

DEFINITION 5.7.10. Suppose Y is another Hilbert space and $A : H \to Y$ is a continuous linear operator with closed range. The *generalized inverse of* A, denoted $A^\dagger$, is the unique continuous linear operator from Y to H satisfying (i) $A^\dagger A = I$ on $(\text{kernel}(A))^\perp$ and (ii) $A^\dagger = 0$ on $(\text{range}(A))^\perp$.

REMARK 5.7.11. Here, we have chosen the definition of the generalized inverse due to Desoer and Whalen [**26**] which is equivalent to better known definitions by Moore [**67**] and by Penrose [**69**]. A proof of these equivalences and more can be found in Groetsch's monograph [**50**].

Now consider

$$A : H \to \mathbb{R}^N : x \mapsto (\langle a_1, x\rangle, \dots, \langle a_N, x\rangle).$$

Then $A^* : \mathbb{R}^N \to H : \mathbf{y} = (y_1, \dots, y_N) \mapsto \sum_{i=1}^N y_i a_i$. Because of $(\text{kernel}(A))^\perp = \text{range}(A^*)$ and $(\text{range}(A))^\perp = \text{kernel}(A^*)$, we have

$$\text{span}\{a_1, \dots, a_N\} = \text{range}(A^*) = (\text{kernel}(A))^\perp = \left(\bigcap_{i=1}^N \text{kernel}(a_i)\right)^\perp = L^\perp,$$

and

$$(\text{range}(A))^\perp = \text{kernel}(A^*) = \{\mathbf{y} = (y_1, \dots, y_N) \in \mathbb{R}^N : \textstyle\sum_{i=1}^N y_i a_i = 0\}.$$

PROPOSITION 5.7.12. If the vectors $a_1, \dots, a_N$ are linearly independent, then $(I - P_{L_N} \cdots P_{L_1} P_{L^\perp})^{-1} \mathbf{T}\Psi$ is the generalized inverse $A^\dagger$ of A.

PROOF. We have to check conditions (i) and (ii) of Definition 5.7.10.
(i): Fix $x \in L^\perp = (\text{kernel}(A))^\perp$. Then

$$\begin{aligned}
\mathbf{T}\Psi Ax &= \mathbf{T}\Psi(\langle a_1, x\rangle, \dots, \langle a_N, x\rangle) \\
&= \mathbf{T}\left(\frac{\langle a_1, x\rangle}{\|a_1\|^2}a_1, \dots, \frac{\langle a_N, x\rangle}{\|a_N\|^2}a_N\right) \\
&= \mathbf{T}(P_{L_1^\perp}x, \dots, P_{L_N^\perp}x) \\
&= \textstyle\sum_{i=1}^N (P_{L_N}\cdots P_{L_{i+1}})P_{L_i^\perp}x \\
&= \textstyle\sum_{i=1}^N (P_{L_N}\cdots P_{L_{i+1}})(x - P_{L_i}x) \\
&= x - P_{L_N}\cdots P_{L_1}x \\
&= (I - P_{L_N}\cdots P_{L_1}P_{L^\perp})x,
\end{aligned}$$

so that $(I - P_{L_N}\cdots P_{L_1}P_{L^\perp})^{-1}\mathbf{T}\Psi Ax = x$, as promised.
(ii): Since the vectors $a_1, \dots, a_N$ are linearly independent, $(\text{range}(A))^\perp = \{0\}$ and hence (ii) clearly holds. □

The following example shows that the assumption of linear independence in the last theorem is important:

EXAMPLE 5.7.13. Let $H = \mathbb{R}$, $N = 2$, $a_1 = a_2 = 1$. A simple calculation yields

$$A = \begin{pmatrix} 1 \\ 1 \end{pmatrix}, \ (I - P_{L_2}P_{L_1}P_{L^\perp})^{-1}\mathbf{T}\Psi = (0,1), \text{ but } A^\dagger = \left(\tfrac{1}{2}, \tfrac{1}{2}\right).$$

THEOREM 5.7.14. *Suppose $a_1, \dots, a_N$ are nonzero vectors in H and let $A : H \to \mathbb{R}^N : x \mapsto (\langle a_i, x\rangle)_{i=1}^N$. Given any vector $\mathbf{b} = (b_i)_{i=1}^N \in \mathbb{R}^N$, let $C_i = \{x \in H : \langle a_i, x\rangle = b_i\}$, for all i. Then:*

(i) *The subsequence (x_N^n) of the orbit with starting point x^0 converges linearly to $(I - P_{L_N}\cdots P_{L_1}P_{L^\perp})^{-1}\mathbf{T}\Psi\mathbf{b} + (I - A^\dagger A)x^0$.*
(ii) *(generalized inverses via cyclic projections) If the vectors $a_1, \dots, a_N$ are furthermore linearly independent, then the subsequence (x_N^n) of the orbit with starting point $x^0 = 0$ converges linearly to $A^\dagger\mathbf{b}$.*

PROOF. (i): On the one hand, (x_N^n) converges linearly to

$$P_Lx^0 + (I - P_{L_N}\cdots P_{L_1}P_{L^\perp})^{-1}\mathbf{T}\mathbf{c} = P_Lx^0 + (I - P_{L_N}\cdots P_{L_1}P_{L^\perp})^{-1}\mathbf{T}\Psi\mathbf{b}.$$

On the other hand, by Examples 2.2.1.(v),(vii), $P_L = I - P_{L^\perp} = I - P_{\text{span}\{a_1,\dots,a_N\}} = I - A^\dagger A$. Hence (i) holds. (ii) follows from the last proposition and (i). □

REMARKS 5.7.15.

- Related finite-dimensional versions of (i) were obtained by Censor et al. [**22**, Lemma 2], De Pierro and Iusem [**71**, Theorem 1 and Proposition 10], Eggermont et al. [**38**, Theorem 1.1], Tanabe [**77**, Corollary 9], and Trummer [**78**, Theorem 14].
- In view of Example 5.7.13, the assumption of linear independence in (ii) is important.

The two set case. We proved the two set version of Theorem 5.7.8 in [**9**, Theorem 4.11]. In addition, it should be noted that

$$d_1 = P_{L_1^\perp \cap L_2^\perp}(c_2 - c_1),\ d_2 = P_{L_1^\perp \cap L_2^\perp}(c_1 - c_2) = -d_1;$$

see [**9**, Example 2.2].

We conclude with a striking trichotomy result:

THEOREM 5.7.16. (trichotomy) Suppose $N = 2$. Exactly one of the following alternatives holds:

(i) F_1, F_2 are empty. Then $\gamma(L_1, L_2) = 0$ is zero and $\lim_n \|x_2^n\| = +\infty$.

(ii) F_1, F_2 are nonempty and $\gamma(L_1, L_2) = 0$. Then for every starting point $x^0 \in H$, the sequence (x_2^n) converges in norm to $P_{F_2}x^0$. Convergence is "arbitrarily slow" in the following sense: for each sequence (λ_k) of real numbers with $1 > \lambda_1 \geq \lambda_2 \geq \cdots \geq \lambda_{k-1} \geq \lambda_k \to 0$, there exists some starting point $x^0 \in H$ such that $\|x_2^k - P_{F_2}x^0\| \geq \lambda_k$, for all k.

(iii) F_1, F_2 are nonempty and $\gamma(L_1, L_2) > 0$. Then for every starting point x^0, the sequence (x_2^n) converges linearly to $P_{F_2}x^0$ with a rate no worse than $\cos\gamma(L_1, L_2)$.

PROOF. By Theorems 5.7.5, 5.7.8, we have only to prove (ii). In view of Fact 5.7.4, we assume without loss of generality that $C_1 = L_1$, $C_2 = L_2$, so that $F_2 = F_1 = C_1 \cap C_2$. Set $A := C_1 \cap (C_1 \cap C_2)^\perp$, $B := C_2 \cap (C_1 \cap C_2)^\perp$.

Step 1: Select a sequence of integers (s_k) such that

$$s_k\lambda_k < 1 \leq (s_k + 1)\lambda_k, \text{ for all } k.$$

Then (s_k) is increasing and every term s_k occurs only finitely often. Let (t_n) be the strictly increasing sequence of integers with $\{t_1, t_2, \dots\} = \{s_1, s_2, \dots\}$. Note that $\sum_n 1/t_n^2$ is convergent.

Step 2: Define $k_0(n) := \min\{k : s_k = t_n\}$, $k_1(n) := \max\{k : s_k = t_n\}$ so that

$$s_{k_0(n)-1} = t_{n-1} < t_n = s_{k_0(n)} = s_{k_0(n)+1} = \cdots = s_{k_1(n)} < t_{n+1} = s_{k_1(n)+1};$$

and further set

$$\alpha_n := \left(\lambda_{k_0(n)} t_n\right)^{\frac{1}{2k_1(n)}}, \text{ for all } n.$$

Since $1 > \lambda_{k_0(n)}s_{k_0(n)} = \lambda_{k_0(n)}t_n \geq 1 - \lambda_{k_0(n)}$ and $\lambda_k \to 0$, we get $\alpha_n \to 1^-$.

Claim: There exist sequences (e'_n) in A, (f'_n) in B such that for all n, m with $n \neq m$:

$$\langle e'_n, e'_m\rangle = \langle e'_n, f'_m\rangle = \langle f'_n, f'_m\rangle = 0,\ \|e'_n\|, \|f'_n\| \leq 1,\ \langle e'_n, f'_n\rangle \geq \alpha_n.$$

By Remarks 3.7.6.(i), $\sup\{\langle a, b\rangle : a \in A,\ b \in B,\ \|a\|, \|b\| \leq 1\} = 1$. We construct the sequences inductively. Clearly, e'_1, f'_1 exist. Now suppose we have already found $e'_1, \dots, e'_m$ and $f'_1, \dots, f'_m$. Set $E := \operatorname{span}\{e'_1, \dots, e'_m\}, F := \operatorname{span}\{f'_1, \dots, f'_m\}$. Pick sequences (a_n) in A, (b_n) in B with $\|a_n\| = \|b_n\| = 1$, $\langle a_n, b_n\rangle \to 1$, and $a_n \rightharpoonup a$, $b_n \rightharpoonup b$. Then $a_n - b_n \to 0$ and hence $a = b = 0$. Because P_E, P_F are compact operators, we conclude $P_E a_n, P_F b_n \to 0$ and further

$$1 = \lim_n\langle a_n, b_n\rangle = \lim_n\langle P_E a_n + P_{E^\perp}a_n, P_F b_n + P_{F^\perp}b_n\rangle = \lim_n\langle P_{E^\perp}a_n, P_{F^\perp}b_n\rangle.$$

Consequently, for n sufficiently large, $e'_{m+1} := P_{E^\perp}a_n, f'_{m+1} := P_{F^\perp}b_n$ do the job and the claim is verified.

Step 3: Define $e_n := e'_n/\|e'n\|$, $f_n := f'_n/\|f'_n\|$, for all n and $E := \overline{\operatorname{span}}\{e_1, e_2, \dots\}$, $F := \overline{\operatorname{span}}\{f_1, f_2, \dots\}$. Then the sequences (e_n), (f_n) have the same properties as the sequences (e'_n), (f'_n); moreover, their terms are norm one. Since $E = A \cap (E+F)$, $F = B \cap (E+F)$, $(E+F)^\perp = E^\perp \cap F^\perp$, we obtain

$$C_1 = (C_1 \cap C_2) \oplus E \oplus (A \cap E^\perp \cap F^\perp), C_2 = (C_1 \cap C_2) \oplus F \oplus (B \cap E^\perp \cap F^\perp).$$

If $z \in F$, then $P_1 z = P_{(C_1 \cap C_2)} z + P_E z + P_{(A \cap E^\perp \cap F^\perp)} z = P_E z$, because F is orthogonal to $C_1 \cap C_2$ and to $A \cap E^\perp \cap F^\perp$. Similarly, if $z \in F$, then $P_2 z = P_E z$. Thus $Q_2|_F \equiv P_2 P_1|_F \equiv P_F P_E|_F$.

Step 4: $P_{C_1 \cap C_2}|_F \equiv P_{E \cap F}|_F \equiv 0$. This follows from Fact 1.1.1 and Step 3: if $z \in F$, then $P_{C_1 \cap C_2} z = \lim_n (P_2 P_1)^n z = \lim_n (P_F P_E)^n z = P_{E \cap F} z = P_{\{0\}} z = 0$.

Step 5: Set $x^0 := \sum_n (1/t_n) f_n$. Then $P_{F_2} x^0 = P_{C_1 \cap C_2} x^0 = 0$ and one readily verifies that $(P_2 P_1)^k x^0 = \sum_n (1/t_n) \langle e_n, f_n \rangle^{2k} f_n$ which yields

$$\|x_2^k - P_{F_2} x^0\| \geq \frac{1}{t_n} \alpha_n^{2k}, \text{ for all } n, k.$$

Last Step: Fix an arbitrary k and get $\bar{n}$ such that $k_0(\bar{n}) \leq k \leq k_1(\bar{n})$. Then

$$\|x_2^k - P_{F_2} x^0\| \geq \frac{1}{t_{\bar{n}}} \alpha_{\bar{n}}^{2k} \geq \frac{\alpha_{\bar{n}}^{2k_1(\bar{n})}}{t_{\bar{n}}} \geq \lambda_{k_0(\bar{n})} \geq \lambda_k;$$

the proof is complete. □

REMARKS 5.7.17. For examples of case (i), see Kosmol and Zhou's [**62**, Theorem] and Bauschke and Borwein's [**10**, Example 4.3]. Franchetti and Light have an example of case (ii) in [**42**, Section 4]; we give another example below. Theorem 5.7.8 provides many examples of case (iii).

EXAMPLE 5.7.18. Let $H := \ell_2$ and denote the unit vectors by (u_n). Suppose (α_n) is a sequence of positive real numbers with $1 > \alpha_n > -1$ and $\lim_n \alpha_n = 1$. Set $e_n := u_{2n-1}$, $f_n := \alpha_n u_{2n-1} + \sqrt{1-\alpha_n^2} u_{2n}$, $C_1 := \overline{\text{span}}\{e_1, e_2, \ldots\}$, and $C_2 := \overline{\text{span}}\{f_1, f_2, \ldots\}$. Then $C_1 \cap C_2 = \{0\}$ and the sum $C_1^\perp + C_2^\perp$ is dense in H but not closed. By choosing (α_n) and (x^0) appropriately, we can arrange arbitrarily slow convergence of (x_2^n); see the proof of the last theorem.

The consistent case. Our journey is over: the consistent case of Theorem 5.7.5 is the von Neumann/Halperin result (Fact 1.1.1) we started with. A vast number of papers on this result have been appearing; we feel the reader is in good hands when we refer him or her once more to Deutsch's survey article [**28**]. Two special cases have received much attention:

- two subspaces with positive angle. Sharper upper bounds on the rate of convergence can be found in [**3, 27, 58**].
- intersecting hyperplanes. Here one obtains the well-known method of Kaczmarz [**57**] (see also [**49**]) for solving systems of linear equations.

For more general algorithmic schemes, see [**8**] and the references therein.

5.8. Related algorithms.

Random projections. Suppose r is a *random mapping* for $\{1, \dots, N\}$, i.e. a surjective mapping from $\mathbb{N}$ to $\{1, \dots, N\}$ that assumes every value infinitely often. Then generate a *random sequence* (x_n) by

$$x_0 \in H \text{ arbitrary}, \; x_{n+1} := P_{r(n+1)} x_n, \; \text{for all } n \geq 0.$$

This *method of random projections* is more general than the method of cyclic projections; indeed, if $r(\cdot) \equiv [\cdot]$, then the random sequence (x_n) and the orbit generated by x_0 (in the sense of Section 1) coincide.
It is not surprising that little is known on random projections for the inconsistent case with three or more sets, because the preceding material for the more restrictive case of cyclic projections has left some important questions open.

The two set case. Each projection P_i is *idempotent*, i.e. $P_i^2 = P_i$; thus the random sequence (x_n) is essentially the orbit (for cyclic projections) with starting point x_0.

The consistent case. The random sequence (x_n) converges weakly to some point in $\bigcap_{i=1}^N C_i$, when each C_i is a closed subspace (Amemiya and Ando [**2**]), or $N = 3$ (Dye and Reich [**37**]), or the sets C_i have a common "inner point" (Dye and Reich [**37**], Youla [**82**]; see also Bruck [**18**], Dye [**34**], Dye et al. [**35**], and Dye and Reich [**36**]). It is not known whether or not every random sequence converges weakly to some point in $\bigcap_{i=1}^N C_i$.

Norm convergence of the random sequence (x_n) to some point in $\bigcap_i C_i$ follows when some C_i is boundedly compact (Bauschke and Borwein [**8**, Example 4.6]; see also Aharoni and Censor [**1**], Bruck [**18**], Elsner et al. [**39**], Flåm and Zowe [**41**], and Tseng [**79**]) or when the N-tuple $(C_1, \dots, C_N)$ is *innately boundedly regular*, i.e. $(C_j)_{j \in J}$ is boundedly regular, for every nonempty subset J of $\{1, \dots, N\}$ (Bauschke [**7**]).

All of the above authors have established more general results.

Weighted projections. The method of (equally) *weighted projections* is given by

$$x_0 \in H \text{ arbitrary}, \; x_{n+1} := \textstyle\sum_{i=1}^N \frac{1}{N} P_i x_n, \; \text{for all } n \geq 0.$$

At first glance, this looks quite different from the method of cyclic projections; however Pierra's product space formalization ([**70**]; see also [**41, 55**]) allows an interpretation as the method of cyclic projections in a suitable product space:

Let $\mathbf{H} := \prod_{i=1}^N (H, \frac{1}{N}\langle \cdot, \cdot \rangle)$, $\mathbf{C}_1 := \{\mathbf{x} = (x_i)_{i=1}^N \in \mathbf{H} : x_1 = \cdots = x_N\}$, and $\mathbf{C}_2 := \prod_{i=1}^N C_i$.
The projection of $\mathbf{x} = (x_i)_{i=1}^N$ onto $\mathbf{C}_1$, $\mathbf{C}_2$ is given by

$$P_{\mathbf{C}_1}\mathbf{x} = \left(\textstyle\sum_{i=1}^N \frac{1}{N} x_i, \dots, \sum_{i=1}^N \frac{1}{N} x_i\right), \; P_{\mathbf{C}_2}\mathbf{x} = (P_1 x_1, \dots, P_N x_N).$$

Moreover, by [**10, 9**],

$$\begin{aligned} \mathbf{F}_1 &:= \operatorname{Fix}(P_{\mathbf{C}_1} P_{\mathbf{C}_2}) \\ &= \{(x, \dots, x) \in \mathbf{H} : x \in \operatorname{Fix}(\tfrac{P_1 + \cdots + P_N}{N})\} \\ &= \{(x, \dots, x) \in \mathbf{H} : x \in \operatorname{arginf}_{x \in H} \textstyle\sum_{i=1}^N d^2(x, C_i)\}, \end{aligned}$$

and

$$\mathbf{F}_2 := \operatorname{Fix}(P_{\mathbf{C}_2} P_{\mathbf{C}_1}) = \operatorname{arginf}_{(x_1, \dots, x_N) \in \mathbf{C}_2} \textstyle\sum_{i,j} \|x_i - x_j\|^2.$$

If we now consider the method of cyclic projections for $\mathbf{C}_1$, $\mathbf{C}_2$ in $\mathbf{H}$ with starting point $\mathbf{x}^0 = (x_0, \dots, x_0)$, then

$$\mathbf{x}_2^n = (x_{n+1}, \dots, x_{n+1}), \text{ for all } n \geq 0.$$

Hence all results on the two set case can be "translated" to this particular setting; see, for instance, [**9**, Section 6].
Also, Reich proved norm convergence for weighted projections if each set C_i is symmetric; see [**72**] for this result and more.

Flexible weighted relaxed projections. The method of *flexible weighted relaxed projections* is a generalization of the two previously discussed methods: Given $x_0 \in H$, generate a sequence (x_n) by

$$x_{n+1} := \left(\sum_{i=1}^N \lambda_i^{(n)} \left[(1 - \alpha_i^{(n)}) I + \alpha_i^{(n)} P_i \right] \right) x_n, \text{ for all } n \geq 0.$$

The $\alpha_i^{(n)} \in [0,2]$ are called *relaxation parameters* and the $\lambda_i^{(n)}$ are nonnegative *weights* with $\sum_{i=1}^N \lambda_i^{(n)} = 1$; they might well depend on n and i.

Under additional assumptions on the relaxation parameters and on the weights, numerous (linear, norm, and weak) convergence results have been obtained for the consistent case; see [**8**] and the references therein for further information.

Moreover, we would like to mention that at least some of the results on the inconsistent case given in the previous sections generalize to the method of *relaxed projections*, which is the method of flexible weighted projections with $\alpha_i^{(n)} \equiv \alpha_i \in]0,2[$ and $\lambda_{[n]}^{(n-1)} \equiv 1$.

Dykstra's algorithm. Dykstra's algorithm is closely related to the method of cyclic projections; in fact, if each C_i is a closed affine subspace, then Dykstra's algorithm is essentially the method of cyclic projections. It is defined as follows:

Fix an arbitrary $x_0 \in H$ and set $e_{-(N-1)} := e_{-(N-2)} := \cdots := e_0 := 0$. Then define two sequences (x_n), (e_n) by

$$x_{n+1} := P_{[n+1]}(x_n + e_{n+1-N}), e_{n+1} := (I - P_{[n+1]})(x_n + e_{n+1-N}), \text{ for all } n \geq 0.$$

CONJECTURE 5.8.1. (on Dykstra's algorithm) If all F_i are empty, then $\lim_n \|x_n\| = +\infty$. Otherwise, all F_i are nonempty and

$$(x_n)_{[n]=i} \text{ converges in norm to } P_{F_i} x_0, \text{ for every } i.$$

The following fact not only supports this conjecture but also demonstrates the power of Dykstra's algorithm:

FACT 5.8.2. The previous conjecture is true for: (i) the two set case; (ii) the consistent case; and (iii) the affine subspace case.

PROOF. (i) is Bauschke and Borwein's [**10**, Theorem 3.8], (ii) is Boyle and Dykstra's [**15**, Theorem 2]. (iii): Denote the orbit (for the method of cyclic projections) with starting point x_0 by (x_i^n). It is not hard to check that $x_i^n := x_{(n-1)N+i}$, for all n and every i. The result now follows from Corollary 5.7.6. $\square$

For further information on Dykstra's algorithm, the reader is referred to [**15, 10, 30, 32, 54**]

Acknowledgments. It is our pleasure to thank Frank Deutsch and Jon Vanderwerff for numerous helpful discussions and remarks. We also wish to thank Yair Censor, Simeon Reich, and an anonymous referee for careful reading and for helpful suggestions.

References

[1] R. AHARONI and Y. CENSOR. Block-iterative projection methods for parallel computation of solutions to convex feasibility problems. *Linear Algebra and its Applications*, 120:165–175, 1989.

[2] I. AMEMIYA and T. ANDO. Convergence of random products of contractions in Hilbert space. *Acta scientiarum mathematicarum (Szeged)*, 26:239–244, 1965.

[3] N. ARONSZAJN. Theory of reproducing kernels. *Transactions of the American Mathematical Society*, 68:337–404, May 1950.

[4] J.-P. AUBIN and I. EKELAND. *Applied Nonlinear Analysis.* Wiley-Interscience, New York, 1984.

[5] J.B. BAILLON and R.E. BRUCK. Ergodic theorems and the asymptotic behavior of contraction semigroups. In K.K. Tan, editor, *Fixed Point Theory and Applications*, pages 12–26, Singapore, 1992. World Scientific Publ. Proceedings of the second international conference held in Halifax, Nova Scotia, Canada, June 9–14, 1991.

[6] J.B. BAILLON, R.E. BRUCK, and S. REICH. On the asymptotic behaviour of nonexpansive mappings and semigroups in Banach spaces. *Houston Journal of Mathematics*, 4(1):1–9, 1978.

[7] H.H. BAUSCHKE. A norm convergence result on random products of relaxed projections in Hilbert space. *Transactions of the American Mathematical Society*, 347(4):1365–1374, April 1995.

[8] H.H. BAUSCHKE and J.M. BORWEIN. On projection algorithms for solving convex feasibility problems. *SIAM Review.* To appear.

[9] H.H. BAUSCHKE and J.M. BORWEIN. On the convergence of von Neumann's alternating projection algorithm for two sets. *Set-Valued Analysis*, 1(2):185–212, 1993.

[10] H.H. BAUSCHKE and J.M. BORWEIN. Dykstra's alternating projection algorithm for two sets. *Journal of Approximation Theory*, 79(3):418–443, December 1994.

[11] J.M. BORWEIN and S. FITZPATRICK. Mosco convergence and the Kadec property. *Proceedings of the American Mathematical Society*, 106(3):843–851, July 1989.

[12] J.M. BORWEIN and A.S. LEWIS, 1993. Unpublished manuscript.

[13] J.M. BORWEIN and D.T. YOST. Absolute norms on vector lattices. *Proceedings of the Edinburgh Mathematical Society*, 27:215–222, 1984.

[14] J.E. BOYD and J.J. LITTLE. Complementary data fusion for limited-angle tomography. In *Proceedings of the IEEE Conference on Computer Vision and Pattern Recognition*, pages 288–294, 1994.

[15] J.P. BOYLE and R.L. DYKSTRA. A method for finding projections onto the intersection of convex sets in Hilbert spaces. In R.L Dykstra, T. Robertson, and F.T. Wright, editors, *Advances in Order Restricted Statistical Inference*, pages 28–47. Springer-Verlag, 1985. Lecture Notes in Statistics; vol 37. Proceedings, Iowa City.

[16] L.M. BREGMAN. The method of successive projection for finding a common point of convex sets. *Soviet Mathematics Doklady*, 6:688–692, 1965.

[17] F.E. BROWDER. Nonexpansive nonlinear operators in a Banach space. *Proceedings of the National Academy of Science, USA*, 54:1041–1044, 1965.

[18] R.E. BRUCK. Random products of contractions in metric and Banach spaces. *Journal of Mathematical Analysis and Applications*, 88:319–332, 1982.

[19] R.E. BRUCK and S. REICH. Nonexpansive projections and resolvents of accretive operators in Banach spaces. *Houston Journal of Mathematics*, 3(4):459–470, 1977.

[20] C. CASTAING and M. VALADIER. *Convex Analysis and Measurable Multifunctions*, volume 580 of *Lecture notes in mathematics.* Springer-Verlag, Berlin, 1977.

[21] Y. CENSOR. On variable block algebraic reconstruction techniques. In G.T. Herman, A.K. Louis, and F. Natterer, editors, *Mathematical Methods in Tomography*, pages 133–140. Springer, 1990. Proceedings of a Conference held in Oberwolfach, Germany, June 11-15. LNM Vol 1497.

[22] Y. CENSOR, P.P.B. EGGERMONT, and D. GORDAN. Strong underrelaxation in Kaczmarz's method for inconsistent systems. *Numerische Mathematik*, 41:83–92, 1983.

[23] W. CHENEY and A.A. GOLDSTEIN. Proximity maps for convex sets. *Proceedings of the American Mathematical Society*, 10:448–450, 1959.

[24] P.L. COMBETTES. The foundations of set theoretic estimation. *Proceedings of the IEEE*, 81(2):182–208, February 1993.

[25] P.L. COMBETTES. *The Convex Feasibility Problem in Image Recovery*, volume 95 of *Advances in Imaging and Electron Physics*, pages 155–270. Academic Press, Inc., 1996.

[26] C.A. DESOER and B.H. WHALEN. A note on pseudoinverses. *SIAM Journal*, 11(2):442–447, June 1963.

[27] F. DEUTSCH. Rate of convergence of the method of alternating projections. In B. Brosowski and F. Deutsch, editors, *Parametric optimization and approximation*, pages 96–107. Birkhäuser, 1983. International Series of Numerical Mathematics Vol 72.

[28] F. DEUTSCH. The method of alternating orthogonal projections. In S.P Singh, editor, *Approximation Theory, Spline functions and Applications*, pages 105–121. Kluwer Academic Publ., 1992. Proceedings of a Conference held in the Hotel villa del Mare, Maratea, Italy between April 28, 1991 and May 9, 1991.

[29] F. DEUTSCH. The angle between subspaces of a Hilbert space. In S.P. Singh, editor, *Approximation Theory, Wavelets and Applications*, pages 107–130. Kluwer Academic Publishers, 1995.

[30] F. DEUTSCH. Dykstra's cyclic projections algorithm: the rate of convergence. In S.P. Singh, editor, *Approximation Theory, Wavelets and Applications*, pages 87–94. Kluwer Academic Publishers, 1995.

[31] F. DEUTSCH and H. HUNDAL. The rate of convergence for the method of alternating projections II. *Journal of Mathematical Analysis and Applications.* to appear.

[32] F. DEUTSCH and H. HUNDAL. The rate of convergence of Dykstra's cyclic projections algorithm: the polyhedral case. *Numerical Functional Analysis and Optimization*, 15(5-6):537–565, 1994.

[33] R. DEVILLE, G. GODEFROY, and V. ZIZLER. *Smoothness and Renormings in Banach Spaces*, volume 64 of *Monographs and Surveys in Pure and Applied Mathematics.* Pitnam, 1993.

[34] J.M. DYE. Convergence of random products of compact contractions in Hilbert space. *Integral Equations and Operator Theory*, 12:12–22, 1989.

[35] J.M. DYE, T. KUCZUMOW, P.-K. LIN, and S. REICH. Random products of nonexpansive mappings in spaces with the Opial property. In B.-L. Lin and W.B. Johnson, editors, *Banach Spaces*, pages 87–93, Providence, Rhode Island, 1993. American Mathematical Society. Proceedings of an International Workshop on Banach Space Theory, held January 6–17, 1992. Contemporary mathematics; v. 144.

[36] J.M. DYE and S. REICH. Random products of nonexpansive mappings. In A. Ioffe, M. Marcus, and S. Reich, editors, *Optimization and Nonlinear Analysis*, pages 106–118, Harlow, England, 1992. Longman Scientific & Technical. Proceedings of a Binational Workshop on Optimization and Nonlinear Analysis, held at Technion City, Haifa, 21–27 March 1990. Pitman Research Notes in Mathematics Series, Vol 244.

[37] J.M. DYE and S. REICH. Unrestricted iterations of nonexpansive mappings in Hilbert space. *Nonlinear Analysis*, 18(2):199–207, 1992.

[38] P.P.B. EGGERMONT, G.T. HERMAN, and A. LENT. Iterative algorithms for large partitioned linear systems, with applications to image reconstruction. *Linear Algebra and its Applications*, 40:37–67, 1981.

[39] L. ELSNER, I. KOLTRACHT, and M. NEUMANN. Convergence of sequential and asynchronous nonlinear paracontractions. *Numerische Mathematik*, 62:305–319, 1992.

[40] Ky FAN. On systems of linear inequalities. In H.W. Kuhn and A.W. Tucker, editors, *Linear inequalities and related systems*, pages 99–156. Princeton University Press, 1956. Annals of Mathematics Studies Number 38.

[41] S.D. FLÅM and J. ZOWE. Relaxed outer projections, weighted averages and convex feasibility. *BIT*, 30:289–300, 1990.

[42] C. FRANCHETTI and W. LIGHT. On the von Neumann alternating algorithm in Hilbert space. *Journal of Mathematical Analysis and Applications*, 114:305–314, 1986.

[43] K. FRIEDRICHS. On certain inequalities and characteristic value problems for analytic functions and for functions of two variables. *Transactions of the American Mathematical Society*, 41:321–364, 1937.

[44] N. GAFFKE and R. MATHAR. A cyclic projection algorithm via duality. *Metrika*, 36:29–54, 1989.

[45] K. GOEBEL and W.A. KIRK. *Topics in metric fixed point theory.* Cambridge University Press, 1990. Cambridge studies in advanced mathematics 28.

[46] K. GOEBEL and S. REICH. *Uniform convexity, hyperbolic geometry, and nonexpansive mappings*, volume 83 of *Monographs and textbooks in pure and applied mathematics.* Marcel Dekker, New York, 1984.

[47] D. GÖHDE. Zum Prinzip der kontraktiven Abbildung. *Mathematische Nachrichten*, 30:251–258, 1965.

[48] M. GOLDBURG and R.J. MARKS II. Signal synthesis in the presence of an inconsistent set of constraints. *IEEE Transactions on Circuits and Systems*, CAS-32(7):647–663, July 1985.

[49] R. GORDON, R. BENDER, and G.T. HERMAN. Algebraic reconstruction techniques (ART) for three-dimensional electron microscopy and X-ray photography. *Journal of theoretical Biology*, 29:471–481, 1970.

[50] C.W. GROETSCH. *Generalized inverses of linear operators.* Marcel Dekker, Inc., New York, 1977. Monographs and textbooks in pure and applied mathematics, Vol 37.

[51] L.G. GUBIN, B.T. POLYAK, and E.V. RAIK. The method of projections for finding the common point of convex sets. *U.S.S.R. Computational Mathematics and Mathematical Physics*, 7(6):1–24, 1967.

[52] I. HALPERIN. The product of projection operators. *Acta scientiarum mathematicarum (Szeged)*, 23:96–99, 1962.

[53] G.T. HERMAN, A. LENT, and P.H. LUTZ. Relaxation methods for image reconstruction. *Communications of the Association for Computing Machinery*, 21(2):152–158, February 1978.

[54] H. HUNDAL and F. DEUTSCH. Two generalizations of Dykstra's cyclic projection algorithm, 1995. Preprint.

[55] A.N. IUSEM and A.R. DE PIERRO. On the convergence of Han's method for convex programming with quadratic objective. *Mathematical Programming*, 52:265–284, 1991.

[56] G.J.O. JAMESON. *Topology and Normed Spaces.* Chapman and Hall, 1974.

[57] S. KACZMARZ. Angenäherte Auflösung von Systemen linearer Gleichungen. *Bulletin internationel de l'Académie Polonaise des Sciences et des Lettres. Classe des Sciences mathématiques et naturelles. Séries A: Sciences mathématiques*, pages 355–357, 1937. Cracovie, Imprimerie de l'Université.

[58] S. KAYALAR and H.L. WEINERT. Error bounds for the method of alternating projections. *Mathematics of Control, Signals, and Systems*, 1:43–59, 1988.

[59] W.A. KIRK. A fixed point theorem for mappings which do not increase distances. *American Mathematical Monthly*, 72:1004–1006, 1965.

[60] P. KOSMOL. Über die sukzessive Wahl des kürzesten Weges. In Opitz and Rauhut, editors, *Ökonomie und Mathematik*, pages 35–42, Berlin, 1987. Springer-Verlag.

[61] P. KOSMOL and X. ZHOU. The limit points of affine iterations. *Numerical Functional Analysis and Optimization*, 11(3&4):403–409, 1990.

[62] P. KOSMOL and X.-L. ZHOU. The product of affine orthogonal projections. *Journal of Approximation Theory*, 64:351–355, 1991.

[63] V. LAKSHMIKANTHAM and D. TRIGIANTE. *Theory of Difference Equations*, volume 181 of *Mathematics in Science and Engineering.* Academic Press, San Diego, 1987.

[64] J.P. LASALLE. *The stability and control of discrete processes*, volume 62 of *Applied Mathematical Sciences.* Springer-Verlag, New York, 1986.

[65] J. MANDEL. Convergence of the cyclic relaxation method for linear inequalities. *Mathematical Programming*, 30:218–228, 1984.

[66] M. MARTELLI. *Discrete dynamical sytems and chaos*, volume 62 of *Pitman Monographs and Surveys in Pure and Applied Mathematics.* Longman Scientific & Technical, Harlow, England, 1992.

[67] E.H. MOORE. Abstract. *Bulletin of the American Mathematical Society*, 26:394–395, 1920.

[68] A. PAZY. Asymptotic behavior of contractions in Hilbert spaces. *Israel Journal of Mathematics*, 9:235–240, 1971.

[69] R. PENROSE. A generalized inverse for matrices. *Proceedings of the Cambridge Philosophical Society*, 51(3):406–413, July 1955.

[70] G. PIERRA. Decomposition through formalization in a product space. *Mathematical Programming*, 28:96–115, 1984.

[71] A.R. DE PIERRO and A.N. IUSEM. On the asymptotic behaviour of some alternate smoothing series expansion iterative methods. *Linear Algebra and its Applications*, 130:3–24, 1990.

[72] S. REICH. A limit theorem for projections. *Linear and Multilinear Algebra*, 13:281–290, 1983.

[73] S.M. ROBINSON. Some continuity properties of polyhedral multifunctions. *Mathematical Programming Study*, 14:206–214, 1981.

[74] R.T. ROCKAFELLAR. Monotone operators and the proximal point algorithm. *SIAM Journal on Control and Optimization*, 14(5):877–898, August 1976.

[75] M.I. SEZAN. An overview of convex projections theory and its applications to image recovery problems. *Ultramicroscopy*, 40:55–67, 1992.

[76] K.T. SMITH, D.C. SOLMON, and S.L. WAGNER. Practical and mathematical aspects of the problem of reconstructing objects from radiographs. *Bulletin of the American Mathematical Society*, 83(6):1227–1270, November 1977.

[77] K. TANABE. Projection method for solving a linear system of linear equations and its applications. *Numerische Mathematik*, 17:203–214, 1971.

[78] M.R. TRUMMER. Rekonstruktion von Bildern aus ihren Projektionen. Diplomarbeit, Seminar für Angewandte Mathematik, ETH Zürich, 1979.

[79] P. TSENG. On the convergence of the products of firmly nonexpansive mappings. *SIAM Journal on Optimization*, 2(3):425–434, August 1992.

[80] M.A. VIERGEVER. Introduction to discrete reconstruction methods in medical imaging. In M.A. Viergever and A. Todd-Pokropek, editors, *Mathematics and Computer Science in Medical Imaging*, pages 43–65, Berlin, 1988. Springer-Verlag. NATO Advanced Science Institute Series F: Computer and Systems Sciences Vol. 39. Proceedings, held in Il Ciocco, Italy, September 21 – October 4, 1986.

[81] J. von NEUMANN. *Functional Operators*, volume II. The Geometry of Orthogonal Spaces. Princeton University Press, 1950. Annals of mathematics studies Vol. 22. Reprint of mimeographed lecture notes first distributed in 1933.

[82] D.C. YOULA. On deterministic convergence of iterations of relaxed projection operators. *Journal of Visual Communication and Image Representation*, 1(1):12–20, September 1990.

[83] D.C. YOULA and V. VELASCO. Extensions of a result on the synthesis of signals in the presence of inconsistent constraints. *IEEE Transactions on Circuits and Systems*, CAS-33(4):465–468, April 1986.

[84] E.H. ZARANTONELLO. Projections on convex sets in Hilbert space and spectral theory. In E.H. Zarantonello, editor, *Contributions to Nonlinear Functional Analysis*, pages 237–424, New York, 1971. Academic Press. University of Wisconsin. Mathematics Research Center; Publication No. 27.

CENTRE FOR EXPERIMENTAL AND CONSTRUCTIVE MATHEMATICS, DEPARTMENT OF MATHEMATICS AND STATISTICS, SIMON FRASER UNIVERSITY, BURNABY, BRITISH COLUMBIA V5A 1S6, CANADA
Current address: Department of Combinatorics and Optimization, Faculty of Mathematics, University of Waterloo, Waterloo, Ontario N2L 3G1, CANADA
E-mail address: bauschke@cecm.sfu.ca

CENTRE FOR EXPERIMENTAL AND CONSTRUCTIVE MATHEMATICS, DEPARTMENT OF MATHEMATICS AND STATISTICS, SIMON FRASER UNIVERSITY, BURNABY, BRITISH COLUMBIA V5A 1S6, CANADA
E-mail address: jborwein@cecm.sfu.ca

DEPARTMENT OF COMBINATORICS AND OPTIMIZATION, FACULTY OF MATHEMATICS, UNIVERSITY OF WATERLOO, WATERLOO, ONTARIO N2L 3G1, CANADA
E-mail address: aslewis@orion.uwaterloo.ca

Contemporary Mathematics
Volume **204**, 1997

Newton's method with modified functions

Adi Ben-Israel

ABSTRACT. Applying the Newton method to a modified function

$$f(x)\,(x-\theta)^{\alpha} \qquad \text{where } \theta,\ \alpha \text{ are suitable parameters ,}$$

results in the iteration $\quad x_{k+1} \ := \ x_k - \dfrac{(x_k-\theta)\,f(x_k)}{(x_k-\theta)\,f'(x_k)+\alpha f(x_k)}$,

whose convergence is related to the convexity of f relative to the family of functions

$$\mathcal{F}_{\theta,\alpha} \ := \ \left\{ \frac{a+b\,(x-\theta)}{(x-\theta)^{\alpha}} \ : \ a,\ b \in \mathbb{R} \right\}$$

We study useful selections of the parameters α and θ, as well as the case where these are updated at each iteration.

1. Introduction

The **Newton method**

$$x_{k+1} \ := \ x_k - \frac{f(x_k)}{f'(x_k)}\ , \qquad k = 0,1,2,\dots \tag{1}$$

may perform better, near a zero ζ of f, if applied to a modified function $\widehat{f}$ with the same zero. Examples:

method	iteration	obtained by applying (1) to:
A 2nd order method for zeros with multiplicity m (known) [**10**, Chapter 8]	$x_{k+1} \ := \ x_k - m\,\dfrac{f(x_k)}{f'(x_k)}$	$\widehat{f}(x) \ := \ f^{1/m}(x)$
A 2nd order method for zeros of any multiplicity [**15**, Example 2–5]	$x_{k+1} \ := \ x_k - \dfrac{f(x_k)}{f'(x_k) - \dfrac{f''(x_k)}{f'(x_k)}\,f(x_k)}$	$\widehat{f}(x) \ := \ \dfrac{f(x)}{f'(x)}$
The **Halley** method: A 3rd order method ([**1**], [**15**], [**13**] and references therein)	$x_{k+1} \ := \ x_k - \dfrac{f(x_k)}{f'(x_k) - \dfrac{f''(x_k)}{2f'(x_k)}\,f(x_k)}$	$\widehat{f}(x) \ := \ \dfrac{f(x)}{\sqrt{f'(x)}}$

1991 *Mathematics Subject Classification.* Primary 65H05; Secondary 49M15, 52A41.
Key words and phrases. Newton method, Halley method, generalized convexity.
Research done at the Department of Mathematics, Indian Institute of Technology, Kanpur, and supported by a Fulbright Grant and the US Educational Foundation in India.

The above modified functions are special cases of

$$(2) \qquad \widehat{f}(x) := e^{-\int a(x)\,dx} f(x) , \text{ with a suitable integrand } a(x) ,$$

for which the Newton method gives

$$(3) \qquad x_{k+1} := x_k - \frac{f(x_k)}{f'(x_k) - a(x_k) f(x_k)} , \quad k = 0, 1, 2, \ldots$$

The order of (3) is determined by the first nonzero (continuous) derivative of its iteration function

$$\Phi_{(3)}(x) := x - \frac{f(x)}{f'(x) - a(x) f(x)}$$

at the fixed point $\zeta = \Phi_{(3)}(\zeta)$, see e.g. [**15**, Theorem 2.2]. Differentiating $\Phi_{(3)}$ at ζ and substituting $f(\zeta) = 0$ we get

$$\Phi'_{(3)}(\zeta) = 0 , \quad \Phi''_{(3)}(\zeta) = \frac{f''(\zeta) - 2\, a(\zeta)\, f'(\zeta)}{f'(\zeta)} ,$$

showing that in general the order is 2 (as expected, since (3) is a Newton method). The order is 3 for the selection

$$(4) \qquad a(x) := \frac{f''(x)}{2\, f'(x)}$$

which renders $\Phi''_{(3)}(\zeta) = 0$. Indeed, substituting (4) in (3) we get the Halley method

$$(5) \qquad x_{k+1} := x_k - \frac{f(x_k)}{f'(x_k) - \dfrac{f''(x_k)}{2 f'(x_k)} f(x_k)} , \quad k = 0, 1, 2, \ldots$$

Sometimes it is advantageous to modify f as a composition, (or transformation of variables),

$$\widehat{f}(x) := f(g(x)) , \quad \text{see e.g. [9]} ,$$

rather than the multiplicative form (2).

In this paper we consider a special case of (2)

$$(6) \qquad \widehat{f}(x) := (x - \theta)^{\alpha} f(x) \text{ , with suitable parameters } \theta \text{ and } \alpha ,$$

corresponding to the selection of $a(x)$ as

$$(7) \qquad a(x) := -\frac{\alpha}{x - \theta} .$$

Applying the Newton method to (6) (i.e. substituting (7) in (3)), we get

$$(8) \qquad x_{k+1} := x_k - \frac{(x_k - \theta)\, f(x_k)}{(x_k - \theta)\, f'(x_k) + \alpha f(x_k)} , \quad k = 0, 1, 2, \ldots$$

The parameter θ may be adjusted in each iteration, in particular,

$$(9) \qquad \theta_k := x_{k-1} , \quad \text{the last iterate,}$$

in which case (8) becomes

$$(10) \qquad x_{k+1} := x_k - \frac{(x_k - x_{k-1})\, f(x_k)}{(x_k - x_{k-1})\, f'(x_k) + \alpha f(x_k)} , \quad k = 0, 1, 2, \ldots$$

A geometric interpretation of (8) is given in Theorem 1. First we require the following notation: two differentiable functions f and g are called **tangent** at a point x_k if $f(x_k) = g(x_k)$ and $f'(x_k) = g'(x_k)$, a fact denoted by $f \overset{x_k}{\sim} g$. Clearly

$$(11) \qquad f(x) \overset{x_k}{\sim} g(x) \qquad \text{if and only if} \qquad (f(x)\,h(x)) \overset{x_k}{\sim} (g(x)\,h(x))$$

whenever h is differentiable, and $h(x_k) \neq 0$.

Given θ and α, consider the function

$$(12) \qquad F(x\,|\,\theta,\alpha) \;:=\; \frac{a + b\,(x-\theta)}{(x-\theta)^\alpha}$$

which is tangent to f at x_k (the coefficients a and b are determined by the tangency $f(x) \overset{x_k}{\sim} F(x\,|\,\theta,\alpha)$). Note that $x = \theta$ is a zero of $F(x\,|\,\theta,\alpha)$ if $\alpha < 0$, and is a pole if $\alpha > 0$.

If $\theta \neq x_k$ it follows from (11) that

$$(13) \quad f(x) \overset{x_k}{\sim} \frac{a + b\,(x-\theta)}{(x-\theta)^\alpha} \qquad \text{if and only if} \qquad (x-\theta)^\alpha f(x) \overset{x_k}{\sim} a + b\,(x-\theta)\,.$$

The RHS of (13) states that the affine function

$$(14) \qquad \ell(x) \;=\; a + b\,(x-\theta)$$

is tangent to $(x-\theta)^\alpha f(x)$ at x_k. Moreover, the function (12) and the affine function (14) have a common zero at

$$(15) \qquad x \;=\; \theta - \frac{a}{b}$$

provided b is nonzero. We summarize:

THEOREM 1. *Let f be differentiable, let θ, α be fixed, and let $x_k \neq \theta$ be a point where $(x_k-\theta)\,f'(x_k) + \alpha f(x_k) \neq 0$. The function*

$$F(x\,|\,\theta,\alpha) \;:=\; \frac{a + b\,(x-\theta)}{(x-\theta)^\alpha} \qquad (12)$$

tangent to f at x_k has a zero at

$$x_{k+1} \;=\; x_k - \frac{(x_k-\theta)\,f(x_k)}{(x_k-\theta)\,f'(x_k) + \alpha f(x_k)} \qquad (8)$$

which is the zero of the affine function

$$\ell(x) \;=\; a + b\,(x-\theta) \qquad (14)$$

tangent at x_k to

$$\widehat{f}(x) \;:=\; (x-\theta)^\alpha f(x)\,. \qquad (6)$$

Theorem 1 states that the iteration (8) is equivalent to finding the zero of the function (12) which is tangent to $f(x)$ at x_k .

The iterative method (8) uses the function (12) to interpolate f and f' at the single point x_k, thereby fixing the two free coefficients a and b (the parameters α and θ are fixed at the outset). In comparison, the rational interpolation methods (see e.g. [**8**], [**7**]) use a rational function, such as $(x-a)/\sum_{j=0}^{m} b_j\,x^j$ to interpolate f (or f and some of its derivatives) in as many points as needed to determine the coefficients. This was generalized to nonpolynomial interpolation in [**2**]. However, the iterative methods (8) and (10) are not covered by the theory in [**8**], [**7**] or [**2**], even if α is a positive integer (making (12) rational). In particular, the iterative method (10) uses information in an asymmetric way: f and f' at x_k, nothing at

x_{k-1} (except x_{k-1} itself).

Outline of this paper:

Four "bad" examples, presenting situations where Newton method is inadequate, are given in § 2.

The iterative method (8) is related to a notion of generalized convexity discussed in § 3. Let $\mathcal{F}_{\theta,\alpha}$ be the family of functions (12), and let f be strictly $\mathcal{F}_{\theta,\alpha}$–convex. Then Theorem 2 shows the convergence of (8) to be monotone.

Useful selections of the parameters θ and α are indicated in § 4.

In § 5 we discuss the case where the parameters θ and α are updated in each iteration, in particular a quasi–Halley method (38) with order 2.41, using at each iteration the current values of f and f', and a previous (available) value of f'.

In § 6, the Newton and Halley steps, emanating from the same point, are compared in Theorem 4. Similarly, the Halley and quasi–Halley steps are compared in Theorem 5.

2. Bad examples for the Newton method

Examples 1–4 below illustrate what can go wrong with the Newton method. First we recall typical conditions for the convergence of (1), written as $x_{k+1} - x_k = -u_k$ where

$$u_k := \frac{f(x_k)}{f'(x_k)}, \quad k = 0, 1, 2, \dots \tag{16}$$

is the kth **Newton step**. Let x_0 be an initial point, and let J_0 be the interval with endpoints x_0 and $x_0 - 2\,u_0$, in which f is assumed twice–differentiable. The conditions

$$\sup_{x \in J_0} |f''(x)| = M \tag{17a}$$

$$|f'(x_0)| \geq 2\,|u_0|\,M \tag{17b}$$

are sufficient for the existence of a unique zero ζ of f in J_0, and for the convergence of (1) to that solution, see [**10**, Theorem 7.1]. If ζ is a simple root, the convergence is quadratic

$$|x_{k+1} - \zeta| \leq \left|\frac{f''(\zeta)}{2\,f'(\zeta)}\right| |x_k - \zeta|^2 . \tag{18}$$

Also, the conditions (17) hold throughout the iterations:

$$|f''(x_k)| \leq M \tag{19a}$$

$$|f'(x_k)| \geq 2\,|u_k|\,M , \quad k = 1, 2, \dots \tag{19b}$$

EXAMPLE 1 (Repulsion). The function

$$f_1(x) := x^{1/3} \tag{20}$$

has a unique zero at 0. The Newton method diverges for any nonzero x_0: the iteration (1) gives $$x_{k+1} = -2\,x_k .$$

EXAMPLE 2 (Linear convergence for multiple zeros). The function

$$f_2(x) := x^p$$

has a multiple zero at 0 if $p \geq 2$. The Newton method gives $x_{k+1} = \frac{p-1}{p} x_k$, i.e. linear convergence. For $p = 2$, the conditions (17) hold at all x_0, but convergence is linear since the root $\zeta = 0$ is multiple.

EXAMPLE 3 (Large step). Let

$$f_3(x) := e^{1-x} - 1 \tag{21}$$

If $x_0 = 1 + \ln a$ where $a > 0$ is large, then the first step $u_0 = -(a-1)$ is large and negative, as is $x_1 = \ln a - a$. Many consecutive steps are ≈ 1. For example,

$$x_0 = 10,\ x_1 = -8092.08,\ x_2 = -8091.08,\ x_3 = -8090.08, \ldots$$

and thousands of iterations are required to approach the root $\zeta = 1$.

EXAMPLE 4 (Wrong direction). The function

$$f_4(x) := x\,e^{-x} \tag{22}$$

has a unique zero at 0. The derivative of f_4 is zero for $x = 1$, and negative for $x > 1$. For any initial $x_0 > 1$ the Newton iterates move away from the zero. For example: $x_0 = 2,\ x_1 = 4,\ x_2 = 5.3333,\ x_3 = 6.5641,\ x_4 = 7.74382, \ldots$

3. Convexity and monotone convergence

The iterative method (8) is related to a notion of generalized convexity discussed below.

All functions in this section are twice continuously differentiable in a real interval I. The function f is **supported** by g at x_0 if

$$f(x_0) = g(x_0) \quad \text{and} \quad f(x) \geq g(x) \quad \text{for all } x \in I,$$

and **supported strictly** if the above inequality is strict for all $x \neq x_0$.

Let $\mathcal{F}$ be a family of functions $: I \to \mathbb{R}$. The function f is called [**strictly**] $\mathcal{F}$**–convex** if at each point in I it is [strictly] supported by a member of $\mathcal{F}$, see [**3**], [**4**].

We use the family of functions (12)

$$\mathcal{F}_{\theta,\alpha} := \left\{ \frac{a + b\,(x-\theta)}{(x-\theta)^{\alpha}} \ : \ a,\, b \in \mathbb{R} \right\} \tag{23}$$

where θ and α are given parameters. In particular, for $\alpha = 0$ and any θ, the **affine functions**

$$\mathcal{F}_{\theta,0} := \{a + b\,(x-\theta) \ : \ a,\, b \in \mathbb{R}\}. \tag{24}$$

Since convex functions are supported by affine functions, $\mathcal{F}_{\theta,0}$–convexity is the same as ordinary convexity.

LEMMA 1. *Let I be a real interval, let $f : I \to \mathbb{R}$ be a finite function, and let $\theta, \alpha \in \mathbb{R}$ be given, with $\theta \notin I$. If $\left\{ \begin{array}{l} (x-\theta)^{\alpha} > 0 \\ (x-\theta)^{\alpha} < 0 \end{array} \right.$ $\forall x \in I$, then f is $\mathcal{F}_{\theta,\alpha}$–convex in I if and only if $(x-\theta)^{\alpha} f(x)$ is $\left\{ \begin{array}{l} \text{convex} \\ \text{concave} \end{array} \right.$ in I.*

PROOF. Let $\mathcal{F}$ be a family of functions $: I \to \mathbb{R}$, and p a function positive in I. Then for any function $f : I \to \mathbb{R}$,

$$f \text{ is } \mathcal{F}\text{–convex if and only if the product function } (p\,f) \text{ is } (p\,\mathcal{F})\text{–convex}$$

where $(p\mathcal{F})$ is the family $\{pF : F \in \mathcal{F}\}$. The lemma is then proved by the observations: (a) $\mathcal{F}_{\theta,0} = (x-\theta)^\alpha \mathcal{F}_{\theta,\alpha}$,
(b) $\mathcal{F}_{\theta,0}$–convexity is ordinary convexity, and
(c) f is convex iff $-f$ is concave. □

In Lemma 1 the point θ is assumed outside the interval I, so that

$$(x-\theta)^\alpha \neq 0 \, , \quad \forall x \in I \, . \tag{25}$$

If θ is updated at each iteration, according to (9), we update I accordingly by deleting the interval between θ and x_k, to assure that (25) holds throughout the iterations.

Lemma 1 can be explained using the 2nd derivative characterization of convexity. The family (23) satisfies the 2nd order differential equation

$$y'' = -\frac{2\alpha}{x-\theta}\, y' - \frac{\alpha\,(\alpha-1)}{(x-\theta)^2}\, y \, . \tag{26}$$

It follows from [**11**] and [**3**] that the $\mathcal{F}_{\theta,\alpha}$–convexity of f is characterized by the corresponding 2nd order differential inequality

$$f''(x) \geq -\frac{2\alpha}{x-\theta}\, f'(x) - \frac{\alpha\,(\alpha-1)}{(x-\theta)^2}\, f(x) \, . \tag{27}$$

Consider the 2nd derivative of the modified function $\widehat{f}(x) := (x-\theta)^\alpha f(x)$

$$\widehat{f}''(x) = (x-\theta)^\alpha f''(x) + 2\,\alpha\,(x-\theta)^{\alpha-1} f'(x) + \alpha\,(\alpha-1)\,(x-\theta)^{\alpha-2} f(x) \, .$$

If $(x-\theta)^\alpha > 0$ holds in I then $\widehat{f}'' \geq 0$ ($\widehat{f}$ is convex) iff the inequality (27) holds, i.e. f is $\mathcal{F}_{\theta,\alpha}$–convex. Similarly, if $(x-\theta)^\alpha < 0$ in I then (27) is equivalent to $\widehat{f}'' \leq 0$.

Convexity is related to the monotone convergence of Newton's iterations

$$x_{k+1} := x_k - \frac{f(x_k)}{f'(x_k)} \, , \quad k = 0, 1, 2, \ldots \tag{1}$$

If f is strictly convex and $f(x_k) > 0$ then $f(x_j) > 0$ for all $j > k$, i.e. f is positive at all successive points. If $f(x_0) < 0$ then $f(x_1) > 0$ and thereafter f is positive at all iterations. An analogous result holds for the iteration (8).

THEOREM 2. *Let f be strictly $\mathcal{F}_{\theta,\alpha}$–convex in an interval I, $\theta \notin I$, and let I include x_0 and all iterates generated by*

$$x_{k+1} = x_k - \frac{(x_k-\theta)\, f(x_k)}{(x_k-\theta)\, f'(x_k) + \alpha f(x_k)} \, , \quad k = 0, 1, 2, \ldots \tag{8}$$

(a) If $f(x_0) > 0$ then f is positive for all successive points generated by (8).
(b) If $f(x_0) < 0$ then $f(x_1) > 0$ and thereafter f is positive at all iterations.

PROOF. Follows from Theorem 1 which shows that (8) is equivalent to the Newton method applied to the modified function $(x-\theta)^\alpha f(x)$, and Lemma 1. □

For other applications of generalized convexity in Newton's method see [**4**, § 9], [**5**].

4. The method (8) with constant α

Consider the selection of the parameter α in (8) or (10). If monotone convergence is desired, the key to the selection is provided by the inequality (27).

For example, consider the case $x > \theta$ and $\alpha \leq 1$. If $f(x) > 0$ and $f'(x) > 0$ it follows that the RHS of (27) has the same sign as α. Therefore, if the second derivative f'' is bounded, there is an $\alpha < 0$ satisfying (27). A positive α makes sense, in this case, only if $f'' > 0$, i.e. if f is strictly convex.

In §§ 4.1–4.2 we consider the special cases $\alpha = -1$ [$\alpha = 1$] as representative of negative [positive] α.

4.1. $\alpha = -1$. Here the family (23) consists of the quadratic functions with a zero at θ,

$$\mathcal{F}_{\theta,-1} = \{a\,(x-\theta) + b\,(x-\theta)^2 \,:\, a,\, b \in \mathbb{R}\}\,. \tag{28}$$

A function $F \in \mathcal{F}_{\theta,-1}$ supporting f at x_k has its second zero at

$$x_{k+1} = x_k - \frac{(x_k - \theta)\, f(x_k)}{(x_k - \theta)\, f'(x_k) - f(x_k)}\,, \tag{29}$$

which is (8) with $\alpha = -1$. The modified function $\widehat{f}$ of (6) is

$$\widehat{f}(x) = \frac{f(x)}{x - \theta} \tag{30}$$

its pole at θ acts as a barrier, repulsing the iterates of (8).

The method (29) becomes, for the selection (9),

$$x_{k+1} = x_k - \frac{(x_k - x_{k-1})\, f(x_k)}{(x_k - x_{k-1})\, f'(x_k) - f(x_k)}\,, \quad k = 0, 1, 2, \ldots \tag{31}$$

requiring two initial points x_0 and x_{-1}. Since x_{-1} is the initial barrier, the initial point x_0 should be between the sought zero ζ and x_{-1}. Thus to apply (31) it is required to know on which side of x_0 lies ζ.

As long as the barrier is "away" from the sought zero, the step length $|x_{k+1} - x_k|$ of (31) is shorter than a corresponding step length of the Newton method (1).

EXAMPLE 5. We apply the method (31) to the function

$$f_1(x) := x^{1/3} \tag{20}$$

of Example 1. The table below gives a selected iterate x_k for various combinations of x_0 and x_{-1}.

x_{-1}	x_0	iterate
1.01	1	$x_{10} = 0.901845$
		$x_{100} = 0.193404$
1.1	1	$x_{20} = -2.31673\ 10^{-6}$
1.2	1	$x_{10} = -1.13652\ 10^{5}$
1.5	1	$x_3 = 0$

The table shows tricky dependence on x_{-1}. One expects short steps and slow convergence if x_{-1} is very close to x_0. In general, reducing $|x_{-1} - x_0|$ does not retain convergence, as shown by the last two entries in the table.

EXAMPLE 6. We apply the method (31) to the function

$$f_3(x) := e^{1-x} - 1 \tag{21}$$

of Example 3. In contrast with the Newton method, the method (31) converges fast for all $1 < x_0$, as long as the initial barrier x_{-1} is to the right of x_0. Examples:

x_{-1}	x_0	x_1	x_2	x_3	x_4	x_5
20	10	0.0123266	0.681959	0.875605	0.948526	0.978248
6	4	2.1897	1.17941	1.01489	1.00114	1.00008

EXAMPLE 7. We apply the method (31) to the function

$$f_4(x) := x\, e^{-x} \tag{22}$$

of Example 4. The table below gives a selected iterate x_k for various combinations of x_0 and x_{-1}.

x_{-1}	x_0	iterate
2.1	2	$x_{10} = 0.733696$
		$x_{100} = 2.25589\ 10^{-15}$
2.5	2	$x_7 = 9.93368\ 10^{-7}$
3	2	$x_2 = 0$
3.5	2	$x_5 = -2.20516$
		$x_{300} = -0.0473647$

EXAMPLE 8. ([**6**, p. 178, Example 4]) We use (31) to find a root of

$$f(x) := e^{-x} - \sin x$$

x_{-1}	x_0	x_1	x_2	x_3	x_4	x_5
0.5	0.6	0.586979	0.588741	0.588504	0.588536	0.588532

This is the smallest root. There are infinitely many roots lying close to π, 2π, 3π, ... which can be computed recursively by (31) if the initial barrier x_{-1} is taken between the initial x_0 and the last found root. For example, to compute the root near 2π we can use (31) with $x_{-1} = 4$ and $x_0 = 5$,

x_{-1}	x_0	x_1	x_2	x_3	x_4	x_5
4	5	5.76880	6.09502	6.21599	6.25999	6.27596

In contrast, for the same initial $x_0 = 5$ the Newton method converges to the root closest to 3π,

x_0	x_1	x_2	x_3	x_4	x_5
5	8.32528	10.2880	9.11860	9.43463	9.42469

i.e. the nearest root 6.27596 is skipped.

4.2. $\alpha = 1$. In this case the family $\mathcal{F}_{\theta,\alpha}$ of (23) consists of the hyperbolas with pole in θ

$$\mathcal{F}_{\theta,1} = \left\{ \frac{a}{x-\theta} + b \,:\, a,\, b \in \mathbb{R} \right\} . \tag{32}$$

The parameter θ should be sufficiently far from x_k, for stability reasons. For $\alpha = 1$ the method (8) gives

$$x_{k+1} = x_k - \frac{(x_k - \theta)\, f(x_k)}{(x_k - \theta)\, f'(x_k) + f(x_k)} , \quad k = 0, 1, 2, \ldots \tag{33}$$

and the inequality (27) becomes,

$$f''(x) \geq -\frac{2}{x-\theta} f'(x) .$$

For $x < \theta$ it follows that $\alpha = 1$ is a good choice if f is strictly convex and $|f'|$ is sufficiently small, for example, near a multiple root.

5. The quasi–Halley method

In this section we consider a version of (10), where also α is updated at each iteration.

The derivatives of the iteration function of (8)

$$\Phi_{(8)}(x) := x - \frac{(x-\theta)\, f(x)}{(x-\theta)\, f'(x) + \alpha\, f(x)} \tag{34}$$

at a fixed point $\zeta \neq \theta$, where necessarily $f(\zeta) = 0$, are

$$\Phi'_{(8)}(\zeta) = 0 , \tag{35a}$$

$$\Phi''_{(8)}(\zeta) = \frac{2\,\alpha\, f'(\zeta) + (\zeta - \theta)\, f''(\zeta)}{(\zeta - \theta)\, f'(\zeta)} . \tag{35b}$$

The parameter α may be chosen so as to make $|\Phi''_{(8)}(\zeta)|$ in (35b) as small as possible. The ideal choice (making $|\Phi''_{(8)}(\zeta)| = 0$) is

$$\alpha = -\frac{(\zeta - \theta)\, f''(\zeta)}{2\, f'(\zeta)} \tag{36}$$

If f'' is continuous and θ is close to ζ, then (36) can be approximated as

$$\alpha \approx -\frac{f'(\zeta) - f'(\theta)}{2\, f'(\zeta)}$$

which can be implemented, using (9) for θ, by the selection

$$\alpha_k := -\frac{f'(x_k) - f'(x_{k-1})}{2\, f'(x_k)}\ . \tag{37}$$

Substituting (9) and (37) in (8) gives the **quasi–Halley method**

$$x_{k+1} := x_k - \frac{f(x_k)}{f'(x_k) - \dfrac{f'(x_k) - f'(x_{k-1})}{2\,(x_k - x_{k-1})\, f'(x_k)}\, f(x_k)}\ , \quad k = 0, 1, 2, \ldots \tag{38}$$

The first iteration uses x_0 and an additional point x_{-1} . Alternatively, the quasi–Halley method (38) can be obtained from (3) by approximating (4) at x_k as

$$a(x_k) = \frac{f''(x_k)}{2\, f'(x_k)} \approx \frac{f'(x_k) - f'(x_{k-1})}{2\,(x_k - x_{k-1})\, f'(x_k)}\ .$$

The quasi–Halley method requires only the current values of f and f', and a previous (available) value of f', while the Halley method uses the current f, f' and f''. The order of the quasi–Halley method is ≥ 2.41 (see Theorem 3), as compared with order 3 for the Halley method.

The quasi–Halley method (38) is expected to perform, on the average, worse than the Halley method (5). Indeed, for $f(x) = x^{1/3}$ the table below shows convergence of (5) and divergence of (38). However, the two methods perform similarly for the other examples in the table, and in general for sufficiently smooth functions. In particular, both methods diverge for $x\, e^{-x}$ and $x_0 \geq 2$.

This similarity is explained in Theorem 5 which gives a local comparison of the Halley method and the quasi–Halley method.

function	method	initial point(s)	x_1	x_2	x_3	x_4	x_5
$x^{1/3}$	Halley	$x_0 = 1$	-0.5	0.25	-0.125	0.0625	-0.0312
	quasi–Halley	$x_{-1} = 1.1,\ x_0 = 1$	-0.559693	1.4699	-0.755481	2.01688	-0.995203
$x\, e^{-x}$	Halley	$x_0 = 2$	4	6.4	8.69177	10.9142	13.0937
	quasi–Halley	$x_{-1} = 2.1,\ x_0 = 2$	4.09816	6.82057	4.60104	6.27622	25.7395
$x^3 - 10$	Halley	$x_0 = 2$	2.15384	2.15443	2.15443	2.15443	2.15443
	quasi–Halley	$x_{-1} = 2.5,\ x_0 = 2$	2.15238	2.15443	2.15443	2.15443	2.15443
$e^{1-x} - 1$	Halley	$x_0 = 10$	8.00049	6.00413	4.03079	2.21501	1.00018
	quasi–Halley	$x_{-1} = 11,\ x_0 = 10$	6.83728	0.357585	0.849529	1	1
$x^4 + 2x^2$	Halley	$x_0 = 1$	0.4	0.135137	0.045055	0.015018	0.005006
	quasi–Halley	$x_{-1} = 1.1,\ x_0 = 1$	0.370739	0.009415	0.002829	0.000943	0.000314

The quasi–Halley method is usually better than the Halley method in the case of multiple roots (see e.g. the last example in the above table). Indeed, the quasi–Halley method "remembers" the last iterate, and is therefore less sensitive to the multiplicity of the root.

We illustrate next the Halley and quasi–Halley methods for complex roots.

EXAMPLE 9. ([**6**, p. 177, Example 3]) Consider the complex polynomial

$$f(z) := z^5 + (7-2i)\,z^4 + (20-12i)\,z^3 + (20-28i)\,z^2 + (19-12i)\,z + (13-26i)$$

One of the five roots is found by the Halley and quasi–Halley methods as follows.

iterate	Halley method $x_0 = 3i$	quasi–Halley method $x_{-1} = 1,\ x_0 = 3i$
x_1	$-0.499312 + 2.19129\,i$	$-0.343620 + 2.52897\,i$
x_2	$-0.987763 + 1.89479\,i$	$-1.01552 + 1.84006\,i$
x_3	$-1.00026 + 1.99934\,i$	$-1.02212 + 1.97408\,i$
x_4	$-1 + 2\,i$	$-0.999892 + 2\,i$
x_5		$-1 + 2\,i$

For simple roots of sufficiently smooth functions, the order of convergence of the quasi–Halley method is 2.41:

THEOREM 3. *If f'' is Lipschitz continuous near a root ζ, if $f'(\zeta) \neq 0$, and if the iterates* (38) *converge to ζ, then as $k \to \infty$,*

$$|x_{k+1} - \zeta| = O\left(|x_k - \zeta|^{1+\sqrt{2}}\right) \tag{39}$$

PROOF. The iteration function of (38) at x_k is

$$\Phi_{(38)}(x) = x - \frac{f(x)}{f'(x) - \left(\dfrac{f'(x) - f'(x_{k-1})}{2(x - x_{k-1})f'(x)}\right) f(x)}$$

and its second derivative at the zero ζ is

$$\Phi''_{(38)}(\zeta) = \frac{f''(\zeta)}{f'(\zeta)} - \frac{f'(\zeta) - f'(x_{k-1})}{(\zeta - x_{k-1})f'(\zeta)} = \frac{f''(\zeta) - f''(\xi)}{f'(\zeta)}$$

for some ξ between x_{k-1} and ζ. Since f'' is Lipschitz continuous

$$|f''(\zeta) - f''(\xi)| \le L\,|\zeta - \xi|\,, \quad \text{for some } L\,.$$

$$\therefore\ |\Phi''_{(38)}(\zeta)| \le \frac{L\,|\zeta - \xi|}{|f'(\zeta)|} < \frac{L\,|x_{k-1} - \zeta|}{|f'(\zeta)|}\,. \tag{40}$$

$$\therefore\ |x_{k+1} - \zeta| \approx \frac{|\Phi''_{(38)}(\zeta)|}{2}\,|x_k - \zeta|^2 \approx \frac{L}{2\,|f'(\zeta)|}\,|x_{k-1} - \zeta|\,|x_k - \zeta|^2\,.$$

$$\therefore\ |x_{k+1} - \zeta| = O\left(|x_{k-1} - \zeta|^{\gamma^2}\right) = O\left(|x_{k-1} - \zeta|^{2\gamma+1}\right),$$

where the order of convergence, γ, satisfies the quadratic equation

$$\gamma^2 - 2\gamma - 1 = 0\,.$$

□

Other first–derivative methods, with order $1 + \sqrt{2}$, are known, see e.g. Method 9a in [**15**, p. 234]. Since the second inequality in (40) is strict, the quasi-Halley method may in fact have a higher order, i.e. its order γ satisfies

$$1 + \sqrt{2} \le \gamma < 3\,. \tag{41}$$

The possibility that $1 + \sqrt{2} < \gamma$ is supported by numerical experience, and the results of the next section, showing that (for sufficiently smooth functions) the quasi-Halley method is virtually indistinguishable from the Halley method, near a root to which both converge.

6. Comparison of steps

Iterative methods can be compared locally by comparing their steps at given points. The steps can be compared in terms of **length**, as we do here, or by their effect on the function value, see [**2**].

The proofs in this section are tedious, hence omitted.

We first compare the steps of the Newton and Halley methods, assuming both steps emanate from the same point x_k, arrived at by Newton's method. This corresponds to a hypothetical situation where at an iterate x_k of Newton's method we have an option of continuing (and making a Newton step) or switching to the Halley method (5), making a **Halley step**

$$h_k := \frac{f(x_k)}{f'(x_k) - \dfrac{f''(x_k)}{2f'(x_k)}\, f(x_k)}\,, \quad k = 0, 1, 2, \ldots \tag{42}$$

To simplify the writing we denote by f_k the function f evaluated at x_k. Similarly, f'_k and f''_k denote the derivatives f', f'' evaluated at x_k. The steps to be compared are

$$u_k := \frac{f_k}{f'_k} \quad \text{and} \quad h_k := \frac{f_k}{f'_k - \dfrac{f''_k}{2f'_k}\, f_k}\,.$$

The next lemma gives a condition for the Newton step u_k and the Halley step h_k to have the same sign.

LEMMA 2. *The steps u_k and h_k have the same sign iff*

$$|f'_k|^2 > \frac{f''_k f_k}{2}\,, \tag{43}$$

in which case

$$|h_k| \geq |u_k| \quad \text{if} \quad f_k f''_k \geq 0\,,$$
$$|h_k| < |u_k| \quad \text{if} \quad f_k f''_k < 0\,.$$

It is reasonable to assume that conditions (19) hold at the point x_k, which is arrived at by the Newton method, and where the steps u_k and h_k are compared. Under these conditions, we have the following comparison of the Newton and Halley steps.

THEOREM 4. *If conditions* (19) *hold, then the Newton step u_k and the Halley step h_k have the same sign, and are related by*

$$\frac{2}{3}\,|u_k| \leq |h_k| \leq \frac{4}{3}\,|u_k|\,. \tag{44}$$

We next compare the Halley step h_k of (42) and the quasi–Halley step

$$q_k := \frac{f(x_k)}{f'(x_k) - \dfrac{f'(x_k) - f'(x_{k-1})}{2\,(x_k - x_{k-1})\, f'(x_k)}\, f(x_k)}\,, \quad k = 0, 1, 2, \ldots \tag{45}$$

evaluated at the same point x_k arrived at by the Newton method.

THEOREM 5. *Let f have continuous third derivative in the interval J_0, and let*

$$\sup_{x \in J_0} |f'''(x)| = N\,.$$

If conditions (19) *hold, then the Halley step h_k and the quasi–Halley step q_k are related by*

$$|h_k - q_k| \;\leq\; \frac{N}{2\,|f_k'|}\,|u_{k-1}|^3\,. \tag{46}$$

Theorem 5 gives a comparison of the Halley step h_k and the quasi–Halley step q_k in terms of the underlying Newton step. If the point x_k is sufficiently close to a root, so that the last Newton step u_{k-1} is small, then the steps h_k and q_k are close within $O(|u_{k-1}|^3)$.

If the two steps h_k and q_k are compared at a point x_k arrived at by the Halley method, we can show that

$$|h_k - q_k| \;=\; O(|h_{k-1}|^3)\,,$$

as can be expected from Theorem 5 and the comparison between the Newton and Halley steps in Theorem 4.

Acknowledgments

Thanks to Ronny Ben-Tal, Aaron Melman, David Shanno, Prabha Sharma, Ron Stern, Henry Wolkowicz and the referee, for help and constructive suggestions. The numerical examples were computed using DERIVE, [**12**].

References

[1] G. Alefeld, *On the convergence of Halley's method*, Amer. Math. Monthly **88**(1981), 530–536.

[2] J. Barzilai and A. Ben-Tal, *Nonpolynomial and inverse interpolation for line search: Synthesis and convergence rates*, SIAM J. Numer. Anal. **19**(1982), 1263–1277.

[3] A. Ben-Tal and A. Ben-Israel, *A generalization of convex functions via support properties*, J. Austral. Math. Soc. **21A**(1976), 341–361.

[4] A. Ben-Tal and A. Ben-Israel, *$\mathcal{F}$–convex functions: Properties and applications*, pp. 301–334 in [**14**]

[5] S.L. Brumelle and M.L. Puterman, *Newton's method for W–convex operators*, pp. 399–414 in [**14**]

[6] C.-E. Fröberg, *Numerical Mathematics: Theory and Computer Applications*, The Benjamin/Cummings Publishing Co., 1985

[7] P. Jarratt, *A rational iteration function for solving equations*, Computer J. **9**(1966), 304–307.

[8] P. Jarratt and D. Nudds, *The use of rational functions in the iterative solution of equations on a digital computer*, Computer J. **8**(1965), 62–65.

[9] A. Melman, *Numerical solution of a secular equation*, Numer. Math. **69**(1995), 483–493.

[10] A.M. Ostrowski, *Solution of Equations in Euclidean and Banach Spaces*, 3rd Edition, Academic Press, 1973

[11] M.M. Peixoto, *Generalized convex functions and second order differential inequalities*, Bull. Amer. Math. Soc. **55**(1949), 563–572.

[12] A. Rich, J. Rich and D. Stoutemyer, DERIVE *User Manual*, Soft Warehouse, Inc., 3660 Waialae Avenue, Suite 304, Honolulu, Hawaii, 96816-3236.

[13] T.R. Scavo and J.B. Thoo, *On the geometry of Halley's method*, Amer. Math. Monthly **102**(1995), 417–426.

[14] S. Schaible and W.T. Ziemba (editors), *Generalized Concavity in Optimization and Economics*, Academic Press, 1981

[15] J.F. Traub, *Iterative Methods for the Solution of Equations*, Prentice–Hall, 1964

RUTCOR–RUTGERS CENTER FOR OPERATIONS RESEARCH, RUTGERS UNIVERSITY, NEW BRUNSWICK, NJ 08903-5062, USA

E-mail address: `bisrael@rutcor.rutgers.edu`

Contemporary Mathematics
Volume **204**, 1997

Numerical Stability of Methods for Solving Augmented Systems

Åke Björck

ABSTRACT. We consider questions of scaling and accuracy of methods for solving the two coupled symmetric linear systems $Hy + Ax = b$, $A^T y = c$, where $A \in R^{m \times n}$, $m \geq n$, and $H \in R^{m \times m}$ is symmetric positive definite. Such systems, known in optimization as KKT systems, are symmetric indefinite and form the kernel in many optimization algorithms. An often used method for solving KKT systems is to compute an LDL^T factorization of the system matrix, where L is lower triangular and D is block-diagonal with 1×1 and symmetric 2×2 pivots. The pivots are chosen according to a scheme by Bunch and Kaufman.

For the standard case, when $H = I$, we review results from Björck [1991] which show that for the above method to be numerically stable a scaling of the $(1,1)$ block is needed. We give a closed expression for the optimal scaling factor, and show that for this scaling the errors in x and y are of the same size as from a norm-wise backward stable method. For the case when $H = D^{-2}$ is diagonal we recommend a diagonal scaling such that rows of A are equilibrated before a block scaling is performed as in the standard case.

For sparse problems the scaling has to be a compromise between stability and sparsity, since the optimal scaling usually leads to unacceptable fill-in in the factorization. Accuracy of the computed solution can then usually be restored by fixed precision iterative refinement. A different scheme, due to Saunders, which uses SYMMLQ with a preconditioner derived from the LDL^T factorization is also described.

1. Introduction

In interior point methods for optimization problems symmetric linear systems of the form

$$\begin{pmatrix} H & A \\ A^T & 0 \end{pmatrix} \begin{pmatrix} y \\ x \end{pmatrix} = \begin{pmatrix} b \\ c \end{pmatrix} \tag{1.1}$$

occur, where $A \in \mathbf{R}^{m \times n}$, $m \geq n$, and $H \in \mathbf{R}^{m \times m}$ is symmetric and positive semi-definite. In this context equation (1.1) is called a Karush-Kuhn-Tucker (KKT) system. We remark that in practice H and A are large and sparse matrices.

The system (1.1) is nonsingular if $\operatorname{rank}(A) = n$ and H is positive definite. In this case it gives the conditions for the solution of the following two optimization problems.

1991 *Mathematics Subject Classification.* Primary 65F05, 65G05.

1. *Generalized linear least squares problem* (GLLS)

$$\min_x (Ax-b)^T H^{-1}(Ax-b) + 2c^T x, \tag{1.2}$$

where H is the covariance matrix of the error vector.

2. *Equality constrained quadratic optimization* (ECQO)

$$\min_y \frac{1}{2} y^T H y - b^T y, \quad A^T y = c. \tag{1.3}$$

We refer to (1.1) as the augmented system formulation of these two problems. For problem 1 (GLLS), (1.1) is obtained by differentiating the objective function of (1.2) to give $A^T H^{-1}(b - Ax) = c$, and setting y to be the weighted least squares residual $y = H^{-1}(b - Ax)$. For problem 2 (ECQO), (1.1) is obtained by differentiating the Lagrangian $\frac{1}{2}y^T H y - 2b^T y + x^T(A^T y - c)$, and equating to zero. Here x is the vector of Lagrange multipliers.

If H is positive definite the augmented matrix

$$M = \begin{pmatrix} H & A \\ A^T & 0 \end{pmatrix} \tag{1.4}$$

can be shown to have m positive and n negative eigenvalues. Hence M is symmetric but indefinite. The system can be solved by using symmetric Gaussian elimination to compute a factorization $M = LDL^T$, where L is triangular and D is block-diagonal with blocks of dimension 1 or 2 that may be indefinite. The most common pivoting strategy for symmetric indefinite systems is that of Bunch and Kaufman [**5**]. This pivoting strategy has been extended to the sparse case by Duff et al. [**12**] and is implemented in the widely used Harwell subroutine MA27, see Duff and Reid [**11**].

In Section 2 we describe the Bunch-Kaufman pivoting strategy and discuss its stability properties for general symmetric indefinite systems. In Section 3 we show that the choice of pivots will depend on the relative scaling of the blocks in the augmented system matrix (1.4), and discuss the consequences. Using results from [**4**] for the case when $H = I$, we then give a closed expression for the "optimal" scaling parameter. The scaling problem for (1.4) with a general positive definite matrix H is then discussed. A combination of row equilibration and block scaling is recommended.

In the sparse case the optimum scaling for stability often leads to severe fill-in, and can usually not be used in practice. In Section 4 we describe two different schemes to refine computed solutions. Finally, a survey of numerical results from different studies is given in Section 5.

2. The Bunch-Kaufman pivoting strategy

A stable scheme for computing the LDL^T factorization of a symmetric indefinite matrix was first given by Bunch and Parlett [**6**]. We here describe a modified pivoting strategy due to Bunch and Kaufman [**5**], which is more efficient. For simplicity of notations we restrict our attention to the first stage of the elimination. All later stages proceed similarly. First determine the off-diagonal element of largest magnitude in the first row $\lambda = |m_{1r}| = \max_{2\le j\le p}|m_{1j}|$. If

$$(2.1) \qquad |m_{11}| \geq \rho |m_{1r}|, \qquad \rho = (\sqrt{17}+1)/8 \approx 0.6404,$$

then m_{11} is used as a pivot. If not, determine the largest off-diagonal element in column r, $\sigma = \max_{1 \leq i \leq p} |m_{ir}|$. If $|m_{11}| \geq \rho\lambda^2/\sigma$, then again take m_{11} as pivot, else if $|m_{rr}| \geq \rho\sigma$, take m_{rr} as pivot. Otherwise, take the 2×2 pivot

$$\begin{pmatrix} m_{11} & m_{1r} \\ m_{1r} & m_{rr} \end{pmatrix}.$$

Note that at most two rows need to be searched in each step, and at most p^2 comparisons are needed in all. It can be further shown that if M is positive definite then only 1×1 pivots will be chosen.

With the choice of ρ given in (2.1) the element growth is bounded by the factor

$$(2.2) \qquad g_p = \frac{\max_{i,j,k} |m_{ij}^{(k)}|}{\max_{i,j} |m_{ij}|} \leq (1 + 1/\rho)^{p-1} < (2.57)^{p-1},$$

where p is the dimension of the system. This bound is slightly larger than the bound 2^{p-1}, which holds for Gaussian elimination with partial pivoting. but, in practice, element growth is usually small.

As Higham [**18**] recently has pointed out, backward stability does not directly follow from the above result, since large growth can occur in the intermediate stages of solution of the 2×2 systems. He gives a strict proof of normwise backward stability for the Bunch-Kaufman pivoting, provided that the 2×2 systems are solved by Gaussian elimination with partial pivoting. Hence for *general* symmetric indefinite systems this method is stable in a similar sense as Gaussian elimination with partial pivoting.

However, when the Bunch-Kaufman algorithm is used to solve the system (1.4) we would like a backward error analysis to take account of its special structure. Ideally we would like to require that no perturbations to the (2,2) block of the system matrix occur. Also the relative scaling between x and y should not influence the accuracy. These points show that a separate analysis is needed.

3. Scaling and stability

The round-off introduced will not, in general, respect the structure of the augmented system (1.1). Therefore, we *cannot conclude* from the backward stability of the Bunch-Kaufman method that the computed solution to the system (1.1) will be the exact solution to a problem with data $H + \Delta H$, $A + \Delta A$, where

$$\|\Delta H\| \leq cu\|H\|, \qquad \|\Delta A\| \leq cu\|A\|,$$

c is a constant, and u the machine precision. This is easily shown by considering the effect of the relative scaling of the matrices H and A. Introducing the scaled vector $\alpha^{-1}y$, the augmented system can be written

$$(3.1) \qquad \begin{pmatrix} \alpha H & A \\ A^T & 0 \end{pmatrix} \begin{pmatrix} \alpha^{-1}y \\ x \end{pmatrix} = \begin{pmatrix} b \\ \alpha^{-1}c \end{pmatrix} \quad \Leftrightarrow \quad M_\alpha z_\alpha = d_\alpha.$$

(We will refer to this as an α-scaling.) The choice of pivots in the Bunch-Kaufman scheme will depend on α and hence may affect the accuracy. If H is positive definite then for sufficiently large values of α the first m pivots will be chosen from the block αH, and the resulting reduced system is

$$-A^T H^{-1} A x = c - A^T H^{-1} b, \tag{3.2}$$

where $-A^T H^{-1} A$ is the Schur complement. For the case $H = I$ and $c = 0$ these equations are the normal equations, and so this example suffices to show that the Bunch–Kaufman choice of pivots will not always give a backward stable method for (1.1).

Using smaller values of α will introduce 2×2 pivots of the form

$$\begin{pmatrix} \alpha h_{kk} & a_{rk} \\ a_{rk} & 0 \end{pmatrix}, \quad (k < r)$$

called *tile pivots* in [**16**], and will improve the stability. Such pivots may also circumvent the catastrophic fill-in which occurs in (3.2) when A has some full rows.

The code MA27 was never intended for systems like (1.1) which have many zeros on the diagonal. The new code MA47 for sparse LDL^T factorization in the Harwell Subroutine Library [**1**] has been designed to take the structure of system of the form (1.1) into account. The new pivoting strategy in MA47 is discussed by Duff et al. in [**8**], and the design of the code in [**10**].

3.1. Optimal scaling for the standard case. The above discussion raises the question of the *optimal choice of α for stability*. In this subsection we consider the special case when $H = I$. and survey results from [**4**]. For this case, which will be referred to as the standard case, the problems (1.2) and (1.3) simplify to

$$\min_x \|Ax - b\|_2^2 + 2c^T x, \tag{3.3}$$

$$\min_y \|y - b\|_2, \qquad A^T y = c. \tag{3.4}$$

Setting $c = 0$ in (3.3) gives the standard linear least squares problem. Setting $b = 0$ in (3.4) gives the problem of finding the minimum 2-norm solution of a linear underdetermined system.

It has been shown, see Björck [**4**], that the $m + n$ eigenvalues λ of M_α are

$$\lambda = \begin{cases} \frac{\alpha}{2} \pm \sqrt{\frac{\alpha^2}{4} + \sigma_i^2}, & \text{i=1,2, \dots ,n;} \\ \alpha. \end{cases} \tag{3.5}$$

If $\operatorname{rank}(A) = r \le n$, then the eigenvalue α has multiplicity $(m - r)$, and 0 is an eigenvalue of multiplicity $(n - r)$. Since M_α is symmetric its singular values equal the absolute values of its eigenvalues. From this it can be deduced that, for nonsingular A, the condition number $\kappa_2(M_\alpha) = \sigma_{\max}/\sigma_{\min}$ is minimized for

$$\alpha = \tilde{\alpha} = \frac{1}{\sqrt{2}} \sigma_n(A), \qquad \kappa_2(M_{\tilde{\alpha}}) \approx \sqrt{2} \kappa_2(A). \tag{3.6}$$

Because of the above result, $\tilde{\alpha}$ (or σ_n) has been suggested as the optimal scaling factor in the augmented system method.

A more refined blockwise error analysis has been given in Björck [**4**]. This uses the explicit form of the inverse of M_α,

$$M_\alpha^{-1} = \begin{pmatrix} \frac{1}{\alpha} P_{\mathcal{N}(A^T)} & (A^\dagger)^T \\ A^\dagger & -\alpha (A^T A)^{-1} \end{pmatrix}, \tag{3.7}$$

where $P_{\mathcal{N}(A^T)} = I - AA^\dagger$ is the orthogonal projector onto the nullspace of A^T. It is assumed that $0 \leq \alpha \leq \|A\|_2 = \sigma_1(A)$, so that the element of maximum magnitude in M_α is bounded by $\sigma_1(A)$. Upper bounds for the round-off errors in the computed solutions $\bar{y}$, $\bar{x}$, of the form

$$\begin{pmatrix} \|\bar{y} - y\|_2 \\ \|\bar{x} - x\|_2 \end{pmatrix} \leq c(m+n)g_{m+n}uf(\alpha) \begin{pmatrix} \sigma_1(A) \\ \kappa_2(A) \end{pmatrix}, \tag{3.8}$$

are derived, where g_{m+n} is the growth factor, u the machine precision, and

$$f(\alpha) = \left(1 + \frac{\alpha}{\sigma_n}\right)\left(\frac{1}{\alpha}\|y\|_2 + \|x\|_2\right). \tag{3.9}$$

If $x \neq 0$ then $f(\alpha)$ is minimized when

$$\alpha = \alpha_{\text{opt}} = \left(\frac{\sigma_n \|y\|_2}{\|x\|_2}\right)^{1/2}, \tag{3.10}$$

and

$$f(\alpha_{\text{opt}}) = \left(1 + \frac{\alpha_{\text{opt}}}{\sigma_n}\right)^2 \|x\|_2 = \left(1 + \frac{\sigma_n}{\alpha_{\text{opt}}}\right)^2 \sigma_n^{-1}\|y\|_2. \tag{3.11}$$

An *acceptable-error-stable* algorithm is defined to be one which gives a solution whose error is never significantly worse than the error bound obtained from a tight perturbation analysis. Consider a perturbation of the problem (3.3), such that

$$\|\delta A\|_2 \leq cu\|{\dagger}A\|_2.$$

Then it can be shown that tight upper bounds for the resulting perturbations in the solutions are

$$\begin{pmatrix} \|\delta y\|_2 \\ \|\delta x\|_2 \end{pmatrix} \leq cu\left(\|x\|_2 + \frac{1}{\sigma_n}\|y\|_2\right) \begin{pmatrix} \sigma_1(A) \\ \kappa_2(A) \end{pmatrix}. \tag{3.12}$$

Using

$$f(\sigma_n) = 2\left(\|x\|_2 + \frac{1}{\sigma_n}\|y\|_2\right) \tag{3.13}$$

in (3.8) shows that the error bounds obtained with $\alpha = \sigma_n$, and therefore also with α_{opt}, are similar to those for a backward stable algorithm. Hence the augmented system method is acceptable-error-stable for these choices of α. For sparse least squares problems taking $\alpha = \sigma_n$ will often give rise to unacceptable fill-in, see [**2**].

The optimal value α_{opt} depends on σ_n, $\|y\|$, and $\|x\|$. Using a condition estimator α_{opt} can be estimated a posteriori. It has been pointed out by G. H. Golub (private communication) that α_{opt} could be estimated a priori by using a few steps of a Lanczos-type method.

3.2. Scaling for weighted problems. We now consider the general case when $H \neq I$. If H is positive definite and R its Cholesky factor we have $H = R^T R$, where R is upper triangular and nonsingular. In this case the system (3.1) can be transformed into the problem of standard form

$$\begin{pmatrix} \alpha I & R^{-T}A \\ A^T R^{-1} & 0 \end{pmatrix} \begin{pmatrix} \alpha^{-1}z \\ x \end{pmatrix} = \begin{pmatrix} R^{-T}b \\ \alpha^{-1}c \end{pmatrix}, \tag{3.14}$$

where $z = Ry$. However, this transformation to standard form cannot be recommended in general.

In the following we restrict ourselves to the case when $H = D_w^{-2} > 0$ is a diagonal matrix. Then (3.14) becomes

$$\begin{pmatrix} \alpha I & D_w A \\ A^T D_w & 0 \end{pmatrix} \begin{pmatrix} \alpha^{-1}(D_w^{-1} y) \\ x \end{pmatrix} = \begin{pmatrix} D_w b \\ \alpha^{-1} c \end{pmatrix}. \tag{3.15}$$

Consider a case when A is well-conditioned but D_w is ill-conditioned. Then $\kappa_2(D_w A)$ will be large even though the problem of computing x and y may be well-conditioned. In such a case computing the LDL^T factorization of (3.15) may lead to a loss of stability.

For stability, it is preferable to keep the system in the form

$$\begin{pmatrix} \alpha D_w^{-2} & A \\ A^T & 0 \end{pmatrix} \begin{pmatrix} \alpha^{-1} y \\ x \end{pmatrix} = \begin{pmatrix} b \\ \alpha^{-1} c \end{pmatrix}, \tag{3.16}$$

This also has the advantage that by a further diagonal scaling D_A we can make A *row equilibrated*,

$$\max_{1 \le j \le n} |a_{ij}| = 1, \qquad i = 1, \ldots, m.$$

It is well that such a scaling will come within a factor of $m^{1/2}$ of minimizing $\kappa(D_A^{-1} A)$, see van der Sluis [**21**]. After row equilibration we get a system of the form

$$\begin{pmatrix} \alpha D^{-2} & \tilde{A} \\ \tilde{A}^T & 0 \end{pmatrix} \begin{pmatrix} \alpha^{-1} y \\ x \end{pmatrix} = \begin{pmatrix} \tilde{b} \\ \alpha^{-1} c \end{pmatrix}, \tag{3.17}$$

where

$$\tilde{A} = D_A^{-1} A, \qquad \tilde{b} = D_A^{-1} b, \qquad D = D_w D_A.$$

In order to generalize the analysis in Section 3.1 and determine an optimal value of α in (3.17) we need the inverse of the system matrix. In the case that $D > 0$, the inverse can be written

$$M_\alpha^{-1} = \begin{pmatrix} \frac{1}{\alpha} D P_{\mathcal{N}(\tilde{A}^T D)} D & D(\tilde{A}^T D)^\dagger \\ (D\tilde{A})^\dagger D & -\alpha(\tilde{A}^T D^2 \tilde{A})^{-1} \end{pmatrix}, \tag{3.18}$$

which generalizes (3.7). Note that if $D \subset \mathcal{D}_+$, the class of positive diagonal matrices, then $\|(D\tilde{A})^\dagger D\|$ is bounded, but not $\|(\tilde{A}^T D^2 \tilde{A})^{-1}\|$, see Stewart [**22**]. However, if $c = 0$ we can still obtain a solution in this case, since then the last block column in (3.18) is not needed.

Proceeding as in the standard case and minimizing the error bound for $\bar{x}$, we obtain the optimal α

$$\alpha_{\text{opt}} = \left(\frac{\|(D\tilde{A})^\dagger D\|_2 \|y\|_2}{\|(\tilde{A}^T D^2 \tilde{A})^{-1}\|_2 \|x\|_2} \right)^{1/2}. \tag{3.19}$$

Similarly, minimizing the error bound for $\bar{y}$ instead of that for $\bar{x}$, we get

$$\alpha_{\text{opt}} = \left(\frac{\|D P_{\mathcal{N}(\tilde{A}^T D)} D\|_2 \|y\|_2}{\|(D\tilde{A})^\dagger D\|_2 \|x\|_2} \right)^{1/2}. \tag{3.20}$$

In general, these two values of α are not equal and we conclude that no simple result exists for weighted problems.

4. Refinement of solutions

We have seen that a problem which arises when using the augmented system method is that a value $\alpha_{\rm opt}$ such that is stable can only be estimated a posteriori. A perhaps more serious difficulty is that, for sparse problems, a small value of α will often generate too much fill-in during factorization. Hence the choice of α must be a compromise between preserving sparsity and stability. If an unsuitable value for α has been used, it may then be necessary to refactorize M_α using a smaller value of α.

A refactorization can often be avoided if iterative refinement of the solution using single precision residuals is used to restore stability; see Björck [**3**]. This idea goes back at least to Wilkinson 1977, who in an unpublished report remarked that " ... when $\bar{x}$ has been determined by a direct method of some *poorer* numerical stability than Gaussian elimination with pivoting [...] the use of $\delta^{(1)}$ as an actual correction should yield substantial dividends [...] and may be of great value in the solution of sparse systems when pivoting requirements have been relaxed." Iterative refinement was also used by Arioli, Duff, and de Rijk in their augmented system approach to sparse least-squares problems [**2**].

In fixed precision iterative refinement for the augmented system we proceed as follows:

$$\begin{aligned}
&\text{Solve } M_\alpha z_0 = d;\\
&\text{for } s = 0, 1, 2, \dots,\\
&\qquad r_s = d - A z_s;\\
&\qquad \text{solve } M_\alpha \delta z_s = r_s;\\
&\qquad \text{update } z_{s+1} = x_s + \delta z_s;\\
&\text{end}
\end{aligned}$$

It can be shown by a simplified error analysis where roundoff in the residual $d - M\bar{z}$ is neglected, that initially the error norm will be reduced by a factor

$$\rho \le cu\kappa(M_\alpha).$$

in each iteration.

We now describe a different way to use a Bunch-Kaufman factorization to compute a refined solution. Gill et al. [**17**] have shown how to derive a preconditioner from the Bunch-Kaufman factorization that can be used with the Paige-Saunders iterative method SYMMLQ (see [**20**]). SYMMLQ is a Lanczos-type algorithm, which can be used to solve a symmetric linear system $M_\alpha z = d$ even when M_α is indefinite. It can also be used with a positive definite preconditioner $K = C^T C$, see [**17**] by implicitly applying SYMMLQ to the system

$$C^{-1} M C^{-T} w = C^{-1} d, \qquad z = C^{-T} w. \tag{4.1}$$

Suppose that we have the Bunch-Kaufman factorization $M_\alpha = P^T L D L^T P$ where P is a permutation matrix. Using an idea of Saunders (see [**17**]) we let

$$D = Q \text{diag}\,(\lambda_j) Q^T, \qquad \overline{D} = Q \text{diag}\,(|\lambda_j|) Q^T,$$

where λ_j are the eigenvalues of D and Q is orthogonal. Note that $\overline{D}$ can be computed simply from D as follows. Every 1×1 pivot α in D is replaced by $|\alpha|$.

Every 2×2 pivot

$$\begin{pmatrix} \alpha & \beta \\ \beta & \gamma \end{pmatrix} = \begin{pmatrix} c & s \\ -s & c \end{pmatrix} \begin{pmatrix} \lambda_1 & \\ & \lambda_2 \end{pmatrix} \begin{pmatrix} c & -s \\ s & c \end{pmatrix}$$

in D is replaced by

$$\begin{pmatrix} \bar{\alpha} & \bar{\beta} \\ \bar{\beta} & \bar{\gamma} \end{pmatrix} = \begin{pmatrix} c & s \\ -s & c \end{pmatrix} \begin{pmatrix} |\lambda_1| & \\ & |\lambda_2| \end{pmatrix} \begin{pmatrix} c & -s \\ s & c \end{pmatrix}.$$

Hence we only have to solve a set of 2×2 eigenproblems. Then if we let $C = P^T L \overline{D}^{1/2}$, $\overline{D} = \operatorname{diag}(|\lambda_j|)$, it is easily verified that

$$\overline{M}_a = C^{-1} M_\alpha C^{-T} = \operatorname{diag}(\pm 1).$$

From the convergence properties of SYMMLQ it follows that SYMMLQ applied to the preconditioned system (4.1), with the *exact* preconditioner, will converge in at most two iterations. Therefore we can expect rapid convergence and an accurate solution when using the preconditioner derived from the computed factorization.

5. Survey of numerical results

Matstoms [**19**] did some MATLAB experiments studying the influence of iterative refinement. He used the matrix WELL1033 from the Harwell-Boeing collection of sparse matrix test problems [**9**], with the first $n-1$ equations multiplied by $16^{-5} \approx 9.5 \cdot 10^{-7}$. The optimal value of α given by (3.10) is about $1.5 \cdot 10^{-8}$. In practice full accuracy was obtained for $\alpha \le 10^{-5}$. With one (two) refinement steps full accuracy was obtained for $\alpha \le 10^{-4}$ ($\alpha \le 10^{-3}$), where the larger values of α required much less storage and computation.

Duff [**7**] reports on experiments with scaling of weighted systems. He also considered a different scaling strategy, where the matrix (1.4) was scaled using the Harwell Subroutine Library scaling routine MC30, before applying the α-scaling. He used a test problem where A is the matrix FFFFF800 from the test set of LP problems distributed by Gay [**15**]. His test results for various ill-conditioned weight matrices $H = D_w$, show that row equilibration followed by α-scaling performed almost as well as MC30 and α scaling. In some cases the errors in the solution were reduced by a factor of 10^{-13} compared to no scaling.

Fourer and Mehrotra [**13**] have made a detailed study of a refinement of the Bunch-Parlett procedure of [**8**]. They use an α scaling without the initial row equalibration recommended above. A further parameter δ is used in the pivot selection to balance sparsity and stability criteria. They give comparisons with the normal equation approach for a primal-dual interior point LP code. On a large variety of test problems the LDL^T factorizations involved an extra cost of about 40% over the normal equations, but could be much faster in a minority of cases.

6. Conclusions

We have shown that when solving KKT systems using LDL^T factorizations the relative scaling between the H and A matrices, and row equlibration of A are crucial for the stability of the Bunch–Kaufman scheme. For large sparse matrices this scaling may lead to unacceptable fill-in in the L-factor, and, in practice, it is therefore essential to complement the solution by one of the refinement procedures discussed.

There are several complicating aspects of solving KKT systems which have not been taken into account here. For example, in interior point methods a *sequence of* KKT systems with varying diagonal H matrices have to be solved. Further, these systems typically have extremely large (and small) elements in some rows of the matrix H. A recent attempt to analyze this much more complex situation is described in [**14**].

Acknowledgement

The author gratefully acknowledges comments from a referee which greatly improved the presentation.

References

[1] Anonnymous. *Harwell Subroutine Library. A Catalogue of Subroutines (Release 11)*, Theoretical Studies Department, AEA Industrial Technology, Harwell, Didcot, U.K., 1993.

[2] M. Arioli, I. S. Duff, and P.P.M. de Rijk, *On the augmented system approach to sparse least-squares problems*, Numer. Math., **55** (1989), 667–684.

[3] Å. Björck. *Iterative refinement and reliable computing*, in Reliable Numerical Computation, M. G. Cox and S. J. Hammarling, eds., Clarendon Press, Oxford, UK, 1990, 249–266.

[4] Å. Björck. *Pivoting and stability in the augmented system method*, in D. F. Griffiths and G. A. Watson, editors, Numerical Analysis 1991: Proceedings of the 14th Dundee Conference, June 1991, Pitman Research Notes in Mathematics 260, 1–16. Longman Scientific and Technical, 1992.

[5] J. R. Bunch and L. Kaufman. *Some stable methods for calculating inertia and solving symmetric linear systems*, Math. Comp., **31** (1977), 162–179.

[6] J. R. Bunch and B. Parlett, *Direct methods for solving symmetric indefinite systems of linear equations*, SIAM J. Numer. Anal., **8** (1971), 639–655.

[7] I. S. Duff, *The solution of augmented systems*, in D. F. Griffiths and G. A. Watson, eds., Numerical Analysis 1993: Proceedings of the 15th Dundee Conference, June 1993, Pitman Research Notes in Mathematics. Longman Scientific and Technical, 1994.

[8] I. S. Duff, N. I. M. Gould, J. K. Reid, J. A. Scott, and K. Turner, *The factorization of sparse symmetric indefinite matrices*, IMA J. Numer. Anal., **11** (1991), 181–204.

[9] I. S. Duff, R. G. Grimes, and J. G. Lewis, Sparse matrix test problems, ACM Trans. Math. Software, **15** (1989),1–14.

[10] I. S. Duff and J. K. Reid, *MA47, a Fortran code for direct solution of indefinite sparse symmetric linear systems*, Tech. Report. RAL-95-001, Rutherford Appleton Laboratory, Didcot, U.K., 1995.

[11] I. S. Duff and J. K. Reid, *The multifrontal solution of indefinite sparse symmetric linear systems*, ACM Trans. Math. Software, **9** (1983), 302–325.

[12] I. S. Duff, N. Munksgaard, H. B. Nielsen, and J. K. Reid, *Direct solution of sets of linear equations whose matrix is sparse symmetric and indefinite*, J. Inst. Maths. Applics., **23** (1979), 235–250.

[13] R. Fourer and S. Mehrotra, *Solving symmetric indefinite systems in an interior-point method for linear programming*, Technical Report 92-01, Dept. of Industrial Engineering and management Sciences, Northwestern University, Evanston, IL, USA, 1992.

[14] A. Forsgren, P. E. Gill, and J. R. Shinnerl, *Stability of symmetric ill-conditioned systems arising in interior point methods for constrained optimization*, Tech. Report. TRITA-MAT-1994-24, Royal Institute of Technology, Stockholm, Sweden, 1994.

[15] D. M. Gay. *Electronic mail distribution of linear programming test problems. Math. Programming Society, COAL Newsletter*, 13:10–12, 1985.

[16] P. E. Gill, W. Murray, M. A. Saunders, and M. H. Wright, *A Schur-complement method for sparse quadratic programming*, in Reliable Numerical Computation, M. G. Cox and S. J. Hammarling, eds., Clarendon Press, Oxford, UK, 1990, 113–138.

[17] P. E. Gill, W. Murray, D. Ponceleón, and M. A. Saunders, *Preconditioners for indefinite systems arising in optimization*, SIAM. J. Matrix Anal. Appl., **13** (1992), 292–311.

[18] N. J. Higham. *Stability of the diagonal pivoting method with partial pivoting*, Technical Report 265, July 1995, University of Manchester. Department of Mathematics, Manchester, England, 1995.

[19] P. Matstoms, *Sparse QR factorization in MATLAB*, ACM Trans. Math. Software, **20** (1994), 136–159.

[20] C. C. Paige and M. A. Saunders, *Solution of sparse indefinite systems of linear equations*, SIAM. J. Numer. Anal., **12** (1975), 617–629.

[21] A. van der Sluis, *Condition numbers and equilibration of matrices*, Numer. Math., **14** (1969), 14–23.

[22] G. W. Stewart, *On scaled projections and pseudoinverses*, Linear Algebra Appl., **112** (1989), 189–193.

Department of Mathematics, Linköping University, S-581 83 Linköping, Sweden.
E-mail address: akbjo@math.liu.se

Contemporary Mathematics
Volume **204**, 1997

LOCAL MODULI OF CONVEXITY AND THEIR APPLICATION TO FINDING ALMOST COMMON FIXED POINTS OF MEASURABLE FAMILIES OF OPERATORS

DAN BUTNARIU AND ALFREDO N. IUSEM

ABSTRACT. The aim of this work is twofold: First, to present the basic properties of the local moduli of convexity associated to functions on Banach spaces and second, to apply these results in order to find almost common fixed points of measurable families of operators. The local moduli of convexity measure the convexity of the function around specific points. Significant nonlinear operators on Banach spaces, although not necessarily nonexpansive, relate to the underlying topology via functions with positive local moduli of convexity in a manner which is similar to that in which nonexpansive operators on Hilbert spaces relate to the underlying topology via the square of the norm. The local moduli of convexity of the interconnecting function prove to be useful tools for dealing with such operators. We explicitly determine the local moduli of convexity for the function $f(x) = \|x\|^p$ on $\mathcal{L}^p([a,b])$ and on ℓ^p, which is the interconnecting function for significant classes of operators encountered in practice.

1. INTRODUCTION

In this paper we study the properties of the local moduli of convexity associated to convex functions defined on a Banach space B and we show how they can be applied to establishing convergence of iterative procedures for finding almost common fixed points of measurable families of operators. To be precise, let $f : B \to (-\infty, +\infty]$ be a convex function whose domain $\mathcal{D} = Dom(f)$ has nonempty interior $\mathcal{D}^\circ$. For each $x \in \mathcal{D}^\circ$ and for any $d \in B$, the right-hand sided derivative in the

Date: Submitted: April 1996; Revised: July 1996 .
1991 *Mathematics Subject Classification.* Primary 46B07, 49M30; Secondary: 47H07, 52A41.
Key words and phrases. local modulus of convexity, uniformly convex function, totally convex function, almost common fixed point.

direction d given by

$$f^{\circ}(x,d) := \lim_{t\searrow 0} \frac{f(x+td)-f(x)}{t}, \tag{1}$$

exists, and $f^{\circ}(x,\cdot)$ is a sublinear functional on B such that, for every $y \in B$, we have (see, for instance, [12, Lemma 1.5])

$$D_f(y,x) := f(y) - f(x) - f^{\circ}(x, y-x) \geq 0. \tag{2}$$

The *local modulus of convexity of* f *at the point* $x \in \mathcal{D}^{\circ}$, introduced in [4], is the function $\nu_f(x,\cdot) : [0,+\infty) \to [0,+\infty]$ defined by [1]

$$\nu_f(x,t) = \inf\{D_f(y,x) : y \in \mathcal{D},\ \|y-x\| = t\}. \tag{3}$$

Obviously, $\nu_f(x,0) = 0$ and it may happen that $\nu_f(x,t) = 0$, for any $t \in [0,+\infty)$. The function f is called *totally convex at* x if $\nu_f(x,t) > 0$ whenever $t > 0$.

The basic properties of the local moduli of convexity are presented in Section 2 below. Section 3 deals with the inter-connections between the concept of totally convex function and the concepts of strictly and uniformly convex function: Totally convex functions are strictly convex and uniformly convex functions are totally convex. Totally convex functions may fail to be uniformly convex but they still have some of the properties which make uniformly convex functions of interest in fields ranging from functional analysis to operations research (see [14], [2], [6], [4] and [13]). In Section 4 we prove that total convexity is a common property of such functions like the negentropy on $\mathbb{R}^n$ and $\|\cdot\|^p$ on $\mathcal{L}^p([a,b])$ and ℓ^p when $p \in (1,+\infty)$. In fact, we give an explicit formula (see Theorem 4.1) for computing the local moduli of convexity of the function $f(x) = \|x\|^p$ on the spaces $\mathcal{L}^p([a,b])$ and ℓ^p with $p \in (1,2)$ on which, as shown in [14], this function is not uniformly convex. Our interest in local moduli of convexity and totally convex functions comes from the usefulness of these concepts when dealing with a class of recursive procedures of solving stochastic convex feasibility problems which includes as particular cases optimization problems (cf. [3]), integral equations and best approximation problems (see [4] and the references therein), stochastic systems of inequalities ([5]), etc. The stochastic convex feasibility problem is a particular instance of the problem of finding almost common fixed points of measurable families of operators presented in Section 5. A commonly used method of approximating almost common fixed points of measurable families

[1] With the usual convention that the infimum of an empty set of real numbers is $+\infty$.

of operators consists of iteratively averaging the actual values of the given operators as shown in (44) below. The behavior of this method for firmly nonexpansive operators in Hilbert spaces was studied before (see [10], [3], [5] and the references therein). The results in [4] and [6] suggest that the same method can be applied to operators which are not, necessarily, nonexpansive and on spaces which are not Hilbertian. However, the proofs of convergence given in these works invoke Bregman projection operators and take advantage of the special features of these functions. In what follows we show how one can exploit the properties of the local moduli of convexity associated to totally convex functions in order to ensure existence and weak convergence of the iterative averaging procedure in a more general context (see Theorem 5.3 and its corollaries). Our proofs make clear that having a large pool of totally convex functions with respect to which the requirements of our convergence results can be verified may help enlarge the area of applicability of the iterative averaging method far beyond the particular cases listed in Section 5. This naturally leads to the question of whether, and under what conditions, given a Banach space B, a totally convex function $f : B \to (-\infty, +\infty]$ does exist. We do not have an answer to this question outside the context of the examples given in Section 4. It seems to us that the answer is intimately related to the geometry of B itself and, in particular, to the smoothness of its norm (see [9]).

2. Properties of the local moduli of convexity

We start our study of the local moduli of convexity by proving the following extension of Proposition 2.4 in [4].

Proposition 2.1. *If* $x \in \mathcal{D}^\circ$, *then*

(i) The domain of $\nu_f(x,\cdot)$ *is an interval* $[0,\tau_f)$ *or* $[0,\tau_f]$ *with* $\tau_f \in (0,+\infty]$; τ_f *is finite if and only if* $\mathcal{D}$ *is bounded.*

(ii) If $c \in [1,+\infty)$ *and* $t \geq 0$, *then* $\nu_f(x,ct) \geq c\nu_f(x,t)$.

(iii) The function $\nu_f(x,\cdot)$ *is superadditive, that is, for any* $s,t \in [0,+\infty)$, *we have*

$$\nu_f(x,s+t) \geq \nu_f(x,s) + \nu_f(x,t).$$

(iv) The function $\nu_f(x,\cdot)$ *is nondecreasing; it is strictly increasing if and only if* f *is totally convex at* x.

Proof. (*i*) Since $x \in \mathcal{D}^\circ$, there exists a closed ball of center x and some radius $r > 0$ which is included in $\mathcal{D}^\circ$. Therefore, the interval $[0,r]$ is contained in the domain of $\nu_f(x,\cdot)$.

Suppose that $\nu_f(x,t) < +\infty$. According to (3), there exists a point $y_t \in \mathcal{D}$ such that $\|y_t - x\| = t$. The set $\mathcal{D}$ is convex because it is the domain of a convex function. Hence, the segment $[x, y_t]$ is included in $\mathcal{D}$. This implies that for any $s \in [0,t]$ there exists a point $y_s \in \mathcal{D}$ such that $\|y_s - x\| = s$. By consequence, $[0,t]$ is contained in the domain of $\nu_f(x,\cdot)$ whenever $\nu_f(x,t) < +\infty$. This shows that the domain of $\nu_f(x,\cdot)$ is convex, that is, it is an interval.

(ii) If $c = 1$ or if $t = 0$ or if $\nu_f(x,ct) = +\infty$, then the result is obvious. Otherwise, let ε be a positive real number. According to (3), there exists a point $u \in \mathcal{D}$ such that $\|u - x\| = ct$ and

$$\nu_f(x,ct) + \varepsilon > D_f(u,x) = f(u) - f(x) - f^\circ(x, u-x). \tag{4}$$

For every $\alpha \in (0,1)$, denote $u_\alpha = \alpha u + (1-\alpha)x$. Let $\beta = c^{-1}$ and observe that

$$\|u_\beta - x\| = \beta \|u - x\| = t.$$

Note that, for any $\alpha \in (0,1)$,

$$\frac{\alpha}{\beta} u_\beta + (1 - \frac{\alpha}{\beta})x = \frac{\alpha}{\beta}[\beta u + (1-\beta)x] + (1 - \frac{\alpha}{\beta})x = u_\alpha. \tag{5}$$

The function

$$\alpha \to \frac{f(x + \alpha(u-x)) - f(x)}{\alpha}$$

is nondecreasing on $(0,1)$ because f is convex (see, for instance, [12, p. 2]). Therefore, according to (1) and (4) we have

$$\nu_f(x,ct) + \varepsilon > f(u) - f(x) - \frac{f(x + \alpha(u-x)) - f(x)}{\alpha},$$

for all $\alpha \in (0,1)$. As a consequence,

$$\nu_f(x,ct) + \varepsilon > \frac{\alpha f(u) + (1-\alpha) f(x) - f(x + \alpha(u-x))}{\alpha} =$$

$$\frac{\alpha f(u) + (1-\alpha) f(x) - \frac{\alpha}{\beta} f(u_\beta) - (1 - \frac{\alpha}{\beta}) f(x)}{\alpha} +$$

$$\frac{\frac{\alpha}{\beta} f(u_\beta) + (1 - \frac{\alpha}{\beta}) f(x) - f(u_\alpha)}{\alpha} =$$

$$\frac{\beta f(u) + (1-\beta) f(x) - f(u_\beta)}{\beta} +$$

$$\frac{\frac{\alpha}{\beta} f(u_\beta) + (1 - \frac{\alpha}{\beta}) f(x) - f(\frac{\alpha}{\beta} u_\beta + (1 - \frac{\alpha}{\beta})x)}{\alpha},$$

where the last equality results from (5). The first term of the last sum is nonnegative because f is convex. Thus,

$$\nu_f(x, ct) + \varepsilon > \frac{\frac{\alpha}{\beta} f(u_\beta) + (1 - \frac{\alpha}{\beta}) f(x) - f(\frac{\alpha}{\beta} u_\beta + (1 - \frac{\alpha}{\beta}) x)}{\alpha}$$

$$= \frac{1}{\beta} \left[f(u_\beta) - f(x) - \frac{f(x + \frac{\alpha}{\beta}(u_\beta - x)) - f(x)}{\frac{\alpha}{\beta}} \right].$$

Letting $\alpha \searrow 0$ and taking into account (4) and (1) we deduce that

$$\nu_f(x, ct) + \varepsilon > cD_f(u_\beta, x) \geq c\nu_f(x, t).$$

Since ε is an arbitrary positive real number, this proves (ii).

(iii) Let s and t be positive real numbers. Then, according to (ii), we have

$$\nu_f(x, s+t) = \nu_f(x, \frac{s+t}{s} s) \geq \frac{s+t}{s} \nu_f(x, s),$$

and

$$\nu_f(x, s+t) = \nu_f(x, \frac{s+t}{t} t) \geq \frac{s+t}{t} \nu_f(x, t).$$

Thus, it results that

$$s\nu_f(x, s+t) \geq (s+t)\nu_f(x, s),$$

and

$$t\nu_f(x, s+t) \geq (s+t)\nu_f(x, t).$$

Summing up these two inequalities the superadditivity of $\nu_f(x, \cdot)$ follows.

(iv) Suppose that $0 < s < t$. Then,

$$\nu_f(x, t) = \nu_f(x, \frac{t}{s} s) \geq \frac{t}{s} \nu_f(x, s) \geq \nu_f(x, s), \tag{6}$$

where the first inequality follows from (ii). Thus, $\nu_f(x, \cdot)$ is nondecreasing. If f is totally convex, then the last inequality in (6) is strict and this shows that the function $\nu_f(x, \cdot)$ is strictly increasing on $(0, +\infty)$. The converse is obvious. Hence, (iv) is proven. □

It is asked in [6] whether the local moduli of convexity of a continuous function f are continuous on their domains. A *partial* answer in given by the next result.

Proposition 2.2. *If $x \in \mathcal{D}^\circ$, then*

(i) The function $\nu_f(x, \cdot)$ is continuous from the right at $t = 0$;

(ii) If f is continuous on $\mathcal{D}$, then $\nu_f(x, \cdot)$ is continuous from the right on $[0, \tau_f)$ when $\mathcal{D}$ is open;

(iii) If B has finite dimension, if $\mathcal{D}$ is closed, zzzzand if f is continuous on $\mathcal{D}$, then $\nu_f(x,\cdot)$ is continuous from the left on its domain;

(iv) If B has finite dimension and $\mathcal{D}=B$, then $\nu_f(x,\cdot)$ is continuous on $(0,\tau_f)$.

Proof. (*i*) Let $\{t_k\}_{k\in\mathbb{N}}$ be a sequence in $(0,1)$ converging nonincreasingly to 0. Applying Proposition 2.1(*ii*) we deduce

$$\nu_f(x,1) \geq \nu_f(x,\sqrt{t_k}) = \nu_f(x,\frac{t_k}{\sqrt{t_k}}) \geq \frac{1}{\sqrt{t_k}}\nu_f(x,t_k).$$

Hence,

$$0 = \nu_f(x,0) \leq \lim_{k\to\infty} \nu_f(x,t_k) \leq \nu_f(x,1) \lim_{k\to\infty} \sqrt{t_k} = 0.$$

(*ii*) Let $0 < s < t < \tau_f$. Fix $\varepsilon > 0$. According to (3), there exists a point $y_\varepsilon \in \mathcal{D}$ such that $\|y_\varepsilon - x\| = s$ and

$$\nu_f(x,s) + \frac{\varepsilon}{4} > D_f(y_\varepsilon,x).$$

According to Proposition 2.1, we have

$$0 \leq \nu_f(x,t) - \nu_f(x,s) < \nu_f(x,t) - D_f(y_\varepsilon,x) + \frac{\varepsilon}{4}.$$

The function $D_f(\cdot,x)$ is continuous on $\mathcal{D}$ (see (2)) because f is continuous and, consequently, so is $f^\circ(x,\cdot)$ (cf. [12, Corollary 1.7]). Therefore, there exists a number $\delta(\varepsilon) > 0$ such that, for any $z \in B$ with $\|z - y_\varepsilon\| < \delta(\varepsilon)$, we have $z \in \mathcal{D}$ and

$$|D_f(z,x) - D_f(y_\varepsilon,x)| < \frac{\varepsilon}{4}.$$

If $0 < t - s < \delta(\varepsilon)$, then the vector

$$y'_\varepsilon = \frac{t}{s}y_\varepsilon + (1 - \frac{t}{s})x$$

satisfies $\|y'_\varepsilon - y_\varepsilon\| = t - s < \delta(\varepsilon)$ and $\|y'_\varepsilon - x\| = t$. Hence $y'_\varepsilon \in \mathcal{D}$ and, thus,

$$0 \leq \nu_f(x,t) - \nu_f(x,s) < D_f(y'_\varepsilon,x) - D_f(y_\varepsilon,x) + \frac{\varepsilon}{4} < \frac{\varepsilon}{4} + \frac{\varepsilon}{4} < \varepsilon.$$

This shows that $\nu_f(x,\cdot)$ is continuous from the right.

(*iii*) Fix t in the domain of $\nu_f(x,\cdot)$. Let $\{t_k\}_{k\in\mathbb{N}}$ be a sequence which converges nondecreasingly to t. Since B is finite dimensional, the sets $\{y \in \mathcal{D} : \|y - x\| = t_k\}$ are compact because they are bounded and closed. The function $D_f(\cdot,x)$ is continuous on $\mathcal{D}$. By consequence, for each nonnegative integer k, there exists a vector $y^k \in \mathcal{D}$ such that

$\|y^k - x\| = t_k$ and $\nu_f(x, t_k) = D_f(y^k, x)$. The sequence $\{y^k\}_{k\in\mathbb{N}}$ is bounded. Hence, it has a convergent subsequence $\{y^{p_k}\}_{k\in\mathbb{N}}$. Let $y^* = \lim_{k\to\infty} y^{p_k}$. Then, according to Proposition 2.1, the limits below exist and we have

$$\nu_f(x,t) \geq \lim_{k\to\infty} \nu_f(x,t_k) = \lim_{k\to\infty} \nu_f(x,t_{p_k}) =$$

$$\lim_{k\to\infty} D_f(y^{p_k},x) = D_f(y^*,x) \geq \nu_f(x,t),$$

where the last inequality holds because $y^* \in \mathcal{D}$ and

$$\|y^* - x\| = \lim_{k\to\infty} \|y^{p_k} - x\| = t.$$

Hence, $\nu_f(x,t) = \lim_{k\to\infty} \nu_f(x,t_k)$.

(iv) is a consequence of (ii) and (iii). □

3. Totally convex functions

The aim of this section is to establish connections between the notion of total convexity and the notions of strict and uniform convexity. The function f is called *totally convex* if it is totally convex at every point $x \in \mathcal{D}^\circ$. According to [14], the function f is called *uniformly convex* if the function $\delta_f : [0,+\infty) \to [0,+\infty)$ defined below is positive on $(0,+\infty)$:

$$\delta_f(t) = \inf_{\substack{x,y\in\mathcal{D} \\ \|x-y\|=t \\ s\in(0,1)}} \left\{\frac{sf(x) + (1-s)f(y) - f\left[sx+(1-s)y\right]}{s(1-s)}\right\}. \tag{7}$$

The function δ_f is called *the modulus of convexity of* f.

It was shown in [4] that there are functions on $\mathbb{R}^n$ which are totally convex without being uniformly convex. Also, it is known (see [14]) that, if $p \in (1,2)$, then the function $f(x) = \|x\|^p$ on the space $B = \mathcal{L}^p([a,b])$ or $B = \ell^p$ *is not* uniformly convex. However, this function is totally convex as it is shown in the next section. These prove that in finite, as well as in infinite, dimensional Banach spaces totally convex functions may not be uniformly convex.

Strictly convex real functions defined on infinite dimensional spaces are not necessarily totally convex. An example in this sense is given in [6, Section 2.6]. In spite of these facts, the following results (proved in [6] under some additional conditions concerning the space B and the function f) hold:

Proposition 3.1. *(i) If f is totally convex, then f is strictly convex on $\mathcal{D}^\circ$.*

(ii) If B is finite dimensional, $\mathcal{D}$ is closed, and f is continuous and strictly convex on $\mathcal{D}$, then f is totally convex.

(iii) If B is finite dimensional and $\mathcal{D} = B$, then f is strictly convex if and only if f is totally convex.

(iv) If f is uniformly convex, then f is totally convex and $\delta_f(t) \leq \nu_f(x,t)$, for any $t \in [0,+\infty)$.

Proof. (i) Recall (see, for instance, [6, Lemma 2.3]) that f is strictly convex on $\mathcal{D}^\circ$ if and only if, for any pair $x, y \in \mathcal{D}^\circ$ such that $x \neq y$, we have

$$f^\circ(x, x-y) > f^\circ(y, x-y). \tag{8}$$

Suppose, by contradiction, that f is totally convex but not strictly convex on $\mathcal{D}^\circ$. Then, there exists a pair $x^0, y^0 \in \mathcal{D}^\circ$ such that $x^0 \neq y^0$ and

$$f^\circ(x^0, x^0 - y^0) \leq f^\circ(y^0, x^0 - y^0). \tag{9}$$

Since f is convex, we also have

$$f^\circ(x^0, x^0 - y^0) \geq f(x^0) - f(y^0) \geq f^\circ(y^0, x^0 - y^0).$$

This and (9) imply

$$f^\circ(x^0, x^0 - y^0) = f^\circ(y^0, x^0 - y^0) = f(x^0) - f(y^0). \tag{10}$$

Hence,

$$D_f(x^0, y^0) = f(x^0) - f(y^0) - f^\circ(y^0, x^0 - y^0) = 0.$$

This implies $\nu_f(y^0, \|y^0 - x^0\|) = 0$ which cannot hold unless $\|y^0 - x^0\| = 0$ because f is totally convex at y^0. Thus, we obtain $y^0 = x^0$ and this is a contradiction.

(ii) Let t be a positive real number such that $\nu_f(x,t)$ is finite. Then, the set $\{y \in \mathcal{D} : \|y - x\| = t\}$ is compact and $D_f(\cdot, x)$ is continuous on this set. Therefore, there exists a point $y^* \in \mathcal{D}$ such that $\nu_f(x,t) = D_f(y^*, x)$. Note that, for any $\tau \in (0,1)$,

$$D_f(y^*, x) = f(y^*) - f(x) - f^\circ(x, y^* - x) \geq$$

$$f(y^*) - f(x) - \tfrac{f(x+\tau(y^*-x)) - f(x)}{\tau} > 0,$$

where the first inequality holds because f is convex and, thus, the function $\tau \to \frac{f(x+\tau(y^*-x))-f(x)}{\tau}$ is nonincreasing on $(0,1)$ and the second inequality follows from the strict convexity of f. Hence, $\nu_f(x,t) = D_f(y^*, x) > 0$.

(iii) results from (i) and (ii).

(iv) From [14, Remark 2] it follows that $\delta_f(t) \leq \nu_f(x,t)$, for any $t \geq 0$. Hence, $\nu_f(x,t) > 0$ whenever $\delta_f(t) > 0$. □

An interesting question in applications is whether, or under what conditions, totally convex functions on a Banach space do exist. It seems that the answer to this question is strongly related to the nature of the Banach space itself and reflects the degree of smoothness of the space. Sometimes, having a specific totally convex function on a given Banach space B is not enough for making computational procedures like those discussed in [6] efficient and one has to look for specially designed totally convex functions. In this respect, the following rule of generating totally convex functions from given ones may be of use.

Proposition 3.2. *Let* $f_1, ..., f_m : B \to (-\infty, +\infty]$ *be totally convex functions with the domains* $\mathcal{D}_1, ..., \mathcal{D}_m$, *respectively, such that* $Int(\bigcap_{i=1}^m \mathcal{D}_i) \neq \emptyset$. *Then, for any* m *nonnegative real numbers* $c_1, ..., c_m$ *such that* $\sum_{i=1}^m c_i > 0$, *the function* $h := \sum_{i=1}^m c_i f_i$ *is totally convex and, for any* $x \in Int(\bigcap_{i=1}^m \mathcal{D}_i) \neq \emptyset$ *and for all* $t \in [0, +\infty)$,

$$\nu_h(x,t) \geq \sum_{i=1}^m c_i \nu_{f_i}(x,t). \tag{11}$$

Proof. Apply (3) for h instead of f and use the superadditivity of the infimum taking into account that

$$D_h = \sum_{i=1}^m c_i D_{f_i}.$$

In this way (11) results and implies that h is totally convex. □

4. Local moduli of convexity of particular functions

In this section we identify several totally convex functions which are useful in practical applications. First, observe that if B is a Hilbert space, then the function $f(x) = \|x\|^2$ defined on B is uniformly convex with $\delta_f(t) = t^2 = \nu_f(x,t)$, for all $x \in B$ and $t \geq 0$. Hence, according to Proposition 3.1(iv), the square of the norm of a Hilbert space is totally convex. It was shown in [14, Theorem 4] that, if $p > 2$ and if $B = \mathcal{L}^p([a,b])$ or $B = \ell^p$, then the function $f(x) = \|x\|^p$ is uniformly convex and, thus, this function is totally convex. If $B = \mathbb{R}^n$, then the negentropy, that is the function $f : \mathbb{R}^n \to (-\infty, +\infty]$ defined by[2]

$$f(x) = \begin{cases} \sum_{i=1}^n x_i \ln x_i & if\ x_1, ..., x_n \geq 0, \\ +\infty & otherwise, \end{cases} \tag{12}$$

is totally convex (cf. [4]).

The following result shows that, in spite of not being uniformly convex, the function $f(x) = \|x\|^p$ on the spaces $B = \mathcal{L}^p([a, b])$ and $B = \ell^p$ with $p \in (1, 2)$ is still totally convex.

Theorem 4.1. *If $p \in (1, 2)$, then, in each of the following cases:*
(i) $B = \mathbb{R}^n$ provided with the norm $\|x\| = \left(\sum_{i=1}^n |x_i|^p\right)^{1/p}$;
(ii) $B = \ell^p$;
(iii) $B = \mathcal{L}^p([a, b])$;
the function $f(x) = \|x\|^p$ is totally convex and, for any $x \in B$ and for all $t \in [0, +\infty)$, we have

$$\nu_f(x, t) = (\|x\| + t)^p - \|x\|^p - p\|x\|^{p-1} t. \tag{13}$$

The proof consists of several stages. We start by showing the following:

Lemma 4.2. *The function $\varphi_1 : \mathbb{R} \to \mathbb{R}$ defined by*

$$\varphi_1(t) = \begin{cases} \frac{|t+1|^p - |t-1|^p}{t} & if\ t \neq 0, \\ 2p & if\ t = 0, \end{cases} \tag{14}$$

is symmetric and, for any $t \in \mathbb{R}$, we have

$$\varphi_1(t) \leq 2p. \tag{15}$$

Proof. The symmetry of φ_1 is obvious. Therefore, it is sufficient to show that (15) holds for any $t > 0$. To this end, observe that

$$\lim_{t \searrow 0} \varphi_1(t) = \lim_{t \searrow 0} p\left[(t+1)^{p-1} + (1-t)^{p-1}\right] = 2p.$$

Thus, if we show that the function φ_1 is nonincreasing on $(0, +\infty)$, then we have $\varphi_1(t) \leq \varphi_1(0) = 2p$, for any $t > 0$. In order to show that φ_1 is nonincreasing on $(0, +\infty)$, note that

$$\varphi_1'(t) = \frac{1}{t^2}\left\{[(p-1)t - 1](t+1)^{p-1} - [(p-1)t + 1](t-1)^{p-1}\right\}$$

if $t > 1$ and

$$\varphi_1'(t) = \frac{1}{t^2}\left\{[(p-1)t - 1](t+1)^{p-1} + [(p-1)t + 1](t-1)^{p-1}\right\}$$

[2]Here we make the usual convention that $0 \ln 0 = 0$.

if $t < 1$; and that

$$\lim_{t \searrow 1} \varphi_1'(1) = 2^{p-1}(p-2) = \lim_{t \nearrow 1} \varphi_1'(1).$$

These imply that φ_1 is differentiable at $t = 1$ and $\varphi_1'(1) = 2^{p-1}(p-2) < 0$. If $t > 1$, then

$$1 - \frac{2}{(p-1)t+1} \leq \left(1 - \frac{2}{t+1}\right)^{p-1}$$

and, consequently, $\varphi_1'(t) < 0$. Suppose that $t \in (0,1)$. In this situation, in order to show that $\varphi_1'(t) \leq 0$ it is enough to prove that

$$\left(1 + \frac{2}{\frac{1}{(p-1)t} - 1}\right)^{\frac{1}{p-1}} \leq 1 + \frac{2}{\frac{1}{t} - 1}. \tag{16}$$

Denote

$$r = \frac{1}{2}\left[\frac{1}{(p-1)t} - 1\right]$$

and

$$s = \frac{1}{2}\left(\frac{1}{t} - 1\right).$$

Clearly, $s < r$. Define the function $\xi : (0, +\infty) \to (0, +\infty)$ by

$$\xi(q) = \left(1 + \frac{1}{q}\right)^{q + \frac{1}{2}}$$

and observe that, to establish (16), it is sufficient to prove that $\xi(s) \geq \xi(r)$ and that this results if we show that the function ξ is nonincreasing, that is $\xi'(q) \leq 0$, for all $q \in (0, +\infty)$. After computing $\xi'(q)$ and denoting $u = 1 + 1/t$, one can easily see that this amounts to proving that

$$u^2 - 2u \ln u - 1 \geq 0, \tag{17}$$

whenever $u > 1$. Let $\theta(u)$ be the left hand side of (17). We have $\theta'(u) \geq 0$, whenever $u \geq 1$. Hence, $\theta(u) \geq \theta(1) = 0$, for all $u \geq 1$, i.e. the inequality (17) holds and Lemma 4.2 is proved. □

Define the function $\varphi_2 : \mathbb{R} \to \mathbb{R}$ by

$$\varphi_2(t) = |1 + t|^p - pt.$$

From Lemma 4.2 it follows that, whenever $t \geq 0$,

$$\varphi_2(t) \leq \varphi_2(-t). \tag{18}$$

This fact and the next result will be used below.

Lemma 4.3. *(a) The function* $\varphi_3 : \mathbb{R} \to \mathbb{R}$ *defined by*

$$\varphi_3(t) = \begin{cases} |1+t|^{p-2}(1+t) - |t|^{p-2}t & \text{if } t \neq 0, \\ 1 & \text{if } t = 0, \end{cases}$$

is strictly decreasing on $(0, +\infty)$.

(b) The function $\varphi_4 : \mathbb{R}_+^2 \to \mathbb{R}$ *defined by*

$$\varphi_4(s,t) = (s+t)^p - s^p - ps^{p-1}t, \tag{19}$$

has the property that, for each $t > 0$, *the function* $\varphi_4(\cdot, t)$ *is strictly decreasing on* $[0, +\infty)$.

Proof. Observe that $\varphi_3'(t) < 0$, whenever $t > 0$, and this proves (a). For proving (b), note that, for $s, t > 0$, we have

$$\frac{\partial \varphi_4}{\partial s}(s,t) = ps^{p-1}\left[\left(\frac{t}{s}+1\right)^{p-1} - 1 - (p-1)\frac{t}{s}\right]. \tag{20}$$

For any $r \geq 0$, let $\varsigma(r) := (r+1)^{p-1}$. We have

$$\varsigma'(r) = (p-1)(r+1)^{p-2} \text{ and } \varsigma''(r) = (p-1)(p-2)(r+1)^{p-2} < 0,$$

showing that ς is strictly concave on $[0, +\infty)$ and, thus,

$$\varsigma(r) - \varsigma(0) < r\varsigma'(0),$$

for any $r > 0$. Hence, if $r > 0$, we have

$$(1+r)^{p-1} - 1 < (p-1)r,$$

and, letting here $r = \frac{t}{s}$, one obtains that the right hand side of (20) is negative. The proof of Lemma 4.3 is complete. □

Now, we prove the theorem case by case. To this end, recall (see [2]) that in all cases (i), (ii) and (iii), for any $x, y \in B$, we have

$$D_f(x,y) = \|x\|^p - \|y\|^p - p\left\langle |y|^{p-2}y, x-y \right\rangle. \tag{21}$$

Fix $y \in B$ and $t \in (0, +\infty)$. In order to determine the local modulus of convexity $\nu_f(y,t)$ of $f(\cdot) = \|\cdot\|^p$ it is sufficient to compute the number $D_f(x,y)$, where $x \in B$ is an optimal solution of the optimization problem

$$\min D_f(x,y) \text{ such that } \|y - x\| = t. \tag{22}$$

Our objective is to solve this optimization problem in each specific case of the theorem.

Case (i). Let $B = \mathbb{R}^n$ and, for any $z \in B$, let $\|z\| = (\sum_{i=1}^n |z_i|^p)^{1/p}$. Denote

$$g_n(z) = \sum_{i=1}^{n} (|z_i + y_i|^p - |y_i|^p - p\,|y_i|^{p-2}\, y_i z_i). \tag{23}$$

If $z \in B$ is an optimal solution of the optimization problem

$$\min g_n(z) \textit{ such that } \sum_{i=1}^{n} |z_i|^p = t^p, \tag{24}$$

then the vector $x = y + z$ is an optimal solution of the problem (22). In order to solve the optimization problem (24) we associate to any $z \in B$ the set

$$J(z) = \{j : 1 \le j \le n \textit{ and } z_j \ne 0\}$$

and we use the results summarized in the next

Lemma 4.4. *For any $z \in B$ with $\|z\| = t$, there exists a vector $\widehat{z} \in B$ such that $\|\widehat{z}\| = t$, $J(z) = J(\widehat{z})$, $\frac{y_j}{\widehat{z}_j} \ge 0$, whenever $j \in J(\widehat{z})$, and*

$$g_n(\widehat{z}) \le g_n(z). \tag{25}$$

Proof.

$$L(z) = \left\{ j : j \in J(z) \textit{ and } \frac{y_j}{z_j} < 0 \right\},$$

and, for any $i = 1, ..., n$, put

$$\widehat{z}_i = \begin{cases} -z_i & \textit{if } i \in L(z), \\ z_i & \textit{if } i \notin L(z). \end{cases} \tag{26}$$

Clearly, $\|\widehat{z}\| = \|z\| = t$, $J(z) = J(\widehat{z})$ and $\frac{y_j}{\widehat{z}_j} \ge 0$, whenever $j \in J(\widehat{z})$. Observe that, for any $u \in B$,

$$g_n(u) = \sum_{i=1}^{n} g_n^i(u), \tag{27}$$

where

$$g_n^i(u) := |y_i + u_i|^p - |y_i|^p - p\,|y_i|^{p-2}\, y_i u_i.$$

According to (26), if $j \notin L(z)$, then

$$g_n^j(z) = |z_j|^p \left(\left| 1 + \frac{y_j}{z_j} \right|^p - \left| \frac{y_j}{z_j} \right|^p - p \left| \frac{y_j}{z_j} \right|^{p-2} \frac{y_j}{z_j} \right) = g_n^j(\widehat{z}).$$

If $y_j = 0$, then $g_n^j(z) = |z_j|^p = g_n^j(\widehat{z})$. Hence, for proving (25) it is sufficient to show that

$$g_n(z) - g_n(\widehat{z}) = \sum_{j \in L(z)} (g_n^j(z) - g_n^j(\widehat{z})) \geq 0. \tag{28}$$

Let $j \in L(z)$ and denote $\tau_j = \widehat{z}_j / y_j$. Since $\tau_j > 0$ and $z_j / y_j = -\tau_j$, we deduce

$$g_n^j(z) - g_n^j(\widehat{z}) = |y_j|^p \left[\varphi_2(-\tau_j) - \varphi_2(\tau_j)\right] \geq 0,$$

because of (18). This completes the proof of Lemma 4.4. □

Let S be the set of optimal solutions of the problem (24). This set is nonempty because the objective function of the optimization problem (24) is continuous and the feasibility set is compact. Lemma 4.4 ensures that, if $z \in S$, then there exists $\widehat{z} \in S$ such that $\frac{y_j}{\widehat{z}_j} \geq 0$ for any $j \in J(\widehat{z})$. Denote

$$\widehat{S} = \left\{ \widehat{z} \in S : \frac{y_j}{\widehat{z}_j} \geq 0 \text{ for any } j \in J(\widehat{z}) \right\}.$$

If $\widehat{z} \in \widehat{S}$, then it must satisfy the Kuhn-Tucker condition, that is, there exists a real number λ such that

$$p \left|\widehat{z}_j + y_j\right|^{p-2} (z_j + y_j) - p \left|y_j\right|^{p-2} y_j = \lambda p \left|\widehat{z}_j\right|^{p-2} \widehat{z}_j,$$

for all $j = 1, ..., n$. This implies that, for any $j \in J(\widehat{z})$,

$$\lambda = \left|1 + \frac{y_j}{\widehat{z}_j}\right|^{p-2} (1 + \frac{y_j}{\widehat{z}_j}) - \left|\frac{y_j}{\widehat{z}_j}\right|^{p-2} \frac{y_j}{\widehat{z}_j} - \left|\frac{y_j}{\widehat{z}_j}\right|^{p-2} \frac{y_j}{\widehat{z}_j} = \varphi_3(\frac{y_j}{\widehat{z}_j}).$$

In other words, for each $j \in J(\widehat{z})$, the number $y_j / \widehat{z}_j$ is a nonnegative solution of the equation

$$\varphi_3(s) = \lambda. \tag{29}$$

According to Lemma 4.4, the function φ_3 is strictly increasing on $[0, +\infty)$. Therefore, the equation (29) has only one solution $\rho \in [0, +\infty)$. Hence,

$$j \in J(\widehat{z}) \Longrightarrow \frac{y_j}{\widehat{z}_j} = \rho. \tag{30}$$

Consequently, for each $j = 1, \ldots, n$, we either have $\widehat{z}_j = 0$ or $\rho \widehat{z}_j = y_j$. We distinguish two possible situations: (I) $\rho = 0$ and (II) $\rho \neq 0$.

Lemma 4.5. *If* $\rho = 0$, *then* $g_n(\widehat{z}) = t^p$ *and* $y = 0$.

Proof. Since $\rho = 0$, for any $j \in J(\widehat{z})$, we have $y_j = \rho\widehat{z}_j = 0$, by (30). Hence, for each $j = 1, ..., n$, we either have $\widehat{z}_j = 0$ or $y_j = 0$. Therefore,

$$g_n(\widehat{z}) = \sum\nolimits_{j:\widehat{z}_j=0} g_n^j(\widehat{z}) + \sum\nolimits_{j:\widehat{z}_j\neq 0} g_n^j(\widehat{z}) = \tag{31}$$
$$\sum\nolimits_{j:\widehat{z}_j=0} (|y_j|^p - |y_j|^p) + \sum\nolimits_{j:\widehat{z}_j\neq 0} |\widehat{z}_j|^p = \|\widehat{z}\|^p = t^p.$$

Suppose, by contradiction, that $\rho = 0$ and $y \neq 0$. Then the vector $\overline{z} \in B$ with the coordinates $\overline{z}_j = ty_j / \|y\|$, if $\widehat{z}_j = 0$, and $\overline{z}_j = 0$, otherwise, has $\|\overline{z}\| = t$ and

$$g_n(\overline{z}) = \sum_{j \notin J(\widehat{z})} g_n^j(\overline{z}) =$$

$$\sum_{j \notin J(\widehat{z})} \frac{|y_j|^p}{\|y\|^p} \left[(t + \|y\|)^p - \|y\|^p - p\|y\|^{p-1} t\right] =$$

$$\varphi_4(\|y\|, t) < \varphi_4(0, t) = t^p = g(\widehat{z}),$$

where the last inequality follows from the strict monotonicity of $\varphi_4(\cdot, t)$ (see Lemma 4.3) and the last equality results from (31). This contradicts the optimality of $\widehat{z}$ as a solution of the problem (24). Hence, $y = 0$. □

Lemma 4.5 shows that, if $\rho = 0$, then $y = 0$ and, thus, the vector $x := y + \widehat{z} = \widehat{z}$ is an optimal solution of the problem (22), that is,

$$\nu_f(0, t) = D_f(x, 0) = g_n(\widehat{z}) = t^p \tag{32}$$

and this proves (13) in the situation (I) mentioned above. Note that (32) holds whenever $y = 0$, because, in this case, any feasible solution of the problem (24) is optimal. Therefore, in order to completely prove (13) in Case (i), it remains to show that (13) holds when $y \neq 0$ and $\rho \neq 0$.

Lemma 4.6. *If* $\rho \neq 0$ *and* $y \neq 0$, *then* $\widehat{z} = \frac{t}{\|y\|}y$, *and the vector*

$$x = \left(\frac{t}{\|y\|} + 1\right) y$$

is an optimal solution of the problem (22).

Proof. According to (30), we have that $\widehat{z}_j = \frac{1}{\rho}y_j$, for any $j \in J(\widehat{z})$. Therefore,

$$t^p = \|\widehat{z}\|^p = \sum_{j \in J(\widehat{z})} |\widehat{z}_j|^p = \rho^{-p}\beta, \tag{33}$$

where

$$\beta := \sum_{j \in J(\widehat{z})} |y_j|^p .$$

According to (33) and (27),

$$g_n(\widehat{z}) = \sum_{j \in J(\widehat{z})} g_n^j(\widehat{z}) = \varphi_4(\beta^{1/p}, t). \tag{34}$$

We claim that $J(\widehat{z}) \supseteq J(y)$. In order to prove that, suppose, by contradiction, that there exists an element $l \in J(y)$ such that $\widehat{z}_l = 0$. Let $u \in B$ be the vector whose coordinates are $u_j = ty_j/\gamma$, if $j \in J(\widehat{z}) \cup \{l\}$, and $u_j = 0$, otherwise, where $\gamma := \beta + |y_l|^p$. It can be easily seen that $\|u\| = t$ and

$$g_n(u) = \sum_{j \in J(\widehat{z}) \cup \{l\}} g_n^j(\widehat{z}) = \varphi_4(\gamma^{1/p}, t).$$

Hence, by (34), we deduce

$$g_n(\widehat{z}) - g_n(u) = \varphi_4(\beta^{1/p}, t) - \varphi_4(\gamma^{1/p}, t),$$

where $\gamma > \beta$ because $y_l \neq 0$. According to Lemma 4.3(b), this means that $g_n(\widehat{z}) > g_n(u)$, which contradicts the assumption that $\widehat{z} \in \widehat{S}$. Hence, $J(\widehat{z}) \supseteq J(y)$. This, combined with (30), gives $J(\widehat{z}) = J(y)$. Consequently, $\beta = \|y\|^p$. Now, an application of (33) yields $\rho = \beta/t$ and, thus,

$$\widehat{z} = \frac{t}{\|y\|} y.$$

This shows that the vector

$$x = y + \widehat{z} = \left(\frac{t}{\|y\|} + 1 \right) y$$

is an optimal solution of the problem (22). □

Applying Lemma 4.6 shows that, when $y \neq 0$ and $\rho \neq 0$, we have

$$\nu_f(y, t) = D_f(x, y) = g_n(\widehat{z}) = \varphi_4(\beta^{1/p}, t) = \varphi_4(\|y\|, t).$$

This implies that (13) holds in this situation too. Thus, the proof of Theorem 4.1 in Case (i) is complete.

Case (ii). Suppose that $B = \ell^p$ is provided with its usual norm. For any $z = (z_1, z_2, ..., z_k, ...) \in B$ and for any positive integer n, let $z^{[n]} = (z_1, z_2, ..., z_n)$. Define $g : B \to \mathbb{R}$ by

$$g(z) = \sum_{j=1}^{\infty} \left(|z_j + y_j|^p - |y_j|^p - p |y_j|^{p-2} y_j z_j \right).$$

Note that, if $\widehat{z} \in B$ is an optimal solution of the problem

$$\min g(z) \text{ such that } \|z\| = t, \tag{35}$$

then $x = \widehat{z} + y$ is an optimal solution of the problem (22) and

$$\nu_f(y,t) = D_f(x,y) = g(\widehat{z}). \tag{36}$$

If $y = 0$, then $g(z) = \|z\|^p$ and, thus, any feasible solution of (35) is optimal. By consequence, if $y = 0$, then application of (36) yields (13). Thus, in order to complete the proof of (13) in Case (*ii*), it is sufficient to deal with the situation in which $y \neq 0$.

Lemma 4.7. *If* $y \neq 0$, *then* $z^* := \frac{t}{\|y\|} y$ *is an optimal solution of the problem* (35).

Proof. Note that $\|z^*\| = t$. Fix an arbitrary element $z \in B$ such that $\|z\| = t$ and denote $t_n = \|z^{[n]}\|$. For each positive integer n, consider the problem (24)) with $y^{[n]}$ instead of y and t_n instead of t. It follows from Lemma 4.5 and Lemma 4.6 that this problem has an optimal solution $\widehat{z}(n) \in \mathbb{R}^n$ and, for each positive integer n such that $y^{[n]} \neq 0$, we have $\widehat{z}(n) = \frac{t_n}{\|y^{[n]}\|} y^{[n]}$ and, thus,

$$g_n(\widehat{z}(n)) = (\|y^n\| + t_n)^p - \|y^n\|^p - p\,\|y^n\|^{p-1} t_n.$$

Since $y \neq 0$, there exists a positive integer n_0 such that, for each integer $n \geq n_0$, we have $y^{[n]} \neq 0$ and, therefore

$$g_n(\widehat{z}(n)) \leq g_n(z^{[n]}).$$

Letting here $n \to \infty$ we obtain $g(z^*) \leq g(z)$. Since z was an arbitrary element of B with $\|z\| = t$, it results that z^* is an optimal solution of the problem (35). □

Combining Lemma 4.7 and (36), we obtain that (13) holds when $y \neq 0$ and this completes the proof of Theorem 4.1 in Case (*ii*).

Case (*iii*). Suppose that $B = \mathcal{L}^p([a,b])$ is provided with its usual norm. Define the function $g_* : B \to \mathbb{R}$ by

$$g_*(z) = \int_a^b \left(|z(s) + y(s)|^p - |y(s)|^p - p\,|y(s)|^{p-2}\, y(s)z(s)\right) ds.$$

Note that, if $z^* \in B$ is an optimal solution of the problem

$$\min g_*(z) \text{ such that } \|z\| = t, \tag{37}$$

then $x = z^* + y$ is a solution of the problem (22) and thus, $\nu_f(y,t) = g_*(z^*)$. If $y = 0$, then $g_*(z) = \|z\|^p = t^p$ for any $z \in B$ with $\|z\| = t$. Therefore, if $y = 0$, then (13) holds. It remains to show that (13) is true when $y \neq 0$.

Lemma 4.8. *If* $y \neq 0$, *then* $z^* := \frac{t}{\|y\|} y$ *is an optimal solution of the problem given by* (37).

Proof. To each element $w \in B$ we associate the sequence of step functions $\{w^k\}_{k \in \mathbb{N}}$ consisting of the elements

$$w^k = \sum_{i=1}^{k} c_{k,i} 1_{[a_{k,i}, a_{k,i+1})},$$

where

$$a_{k,i} = a + \frac{(i-1)(b-a)}{k}, \quad (1 \leq i \leq k).$$

Recall that this sequence converges to w almost everywhere uniformly on the interval $[a, b]$. For each positive integer n, we denote by $\widetilde{w}^n$ the vector in $\mathbb{R}^n$ with the coordinates

$$\widetilde{w}_j^n = c_{n,j} \alpha_n^{1/p}, \quad (1 \leq j \leq n),$$

where $\alpha_n = (b-a)/n$. Provide $\mathbb{R}^n$ with the norm $\|u\| = (\sum_{i=1}^n |u_i|^p)^{1/p}$. Then, the norm $\|\widetilde{w}^n\|$ of $\widetilde{w}^n$ in $\mathbb{R}^n$ is exactly the norm of w^n in B. Suppose that $\|w\| = t$. Consider the optimization problem (24) with $\widetilde{y}^n$ instead of y and $t_n := \|\widetilde{w}^n\|$ instead of t. Clearly, $\lim_{n \to \infty} t_n = t$. Since $y \neq 0$, there exists a positive integer n_0 such that $\widetilde{y}^n \neq 0$, for any $n \geq n_0$. Thus, for any $n \geq n_0$, the optimal solution of (24) is the vector $\frac{t_n}{\|\widetilde{y}^n\|} \widetilde{y}^n$ (cf. Lemma 4.6) and we have

$$(\|\widetilde{y}^n\| + t_n)^p - \|\widetilde{y}^n\|^p - p\,\|\widetilde{y}^n\|^{p-1} t_n = g_n(\tfrac{t_n}{\|\widetilde{y}^n\|} \widetilde{y}^n) \leq$$

$$g_n(\widetilde{w}^n) = \int_a^b \left[|w^n(s) + t_n|^p - |w^n(s)|^p - p\,|w^n(s)|^{p-1} t_n \right] ds.$$

Therefore, the following limits exist and we have

$$\lim_{n \to \infty} g_n(\frac{t_n}{\|\widetilde{y}^n\|} \widetilde{y}^n) \leq \lim_{n \to \infty} g_n(\widetilde{w}^n) = g_*(w). \tag{38}$$

For each positive integer $n \geq n_0$, let z^n be the step function defined by

$$z^n(s) = \frac{t_n}{\|\widetilde{y}^n\|} y^n.$$

We have

$$\left| \frac{t}{\|y\|} y(s) - z^n(s) \right| \leq \left| \frac{t}{\|y\|} y(s) - \frac{t_n}{\|\widetilde{y}^n\|} y(s) \right| + \left| \frac{t_n}{\|\widetilde{y}^n\|} y(s) - z^n(s) \right| =$$

$$\left| \frac{t}{\|y\|} - \frac{t_n}{\|\widetilde{y}^n\|} \right| |y(s)| + |y(s) - y^n(s)| \frac{t_n}{\|\widetilde{y}^n\|},$$

where each term of the last sum converges to zero as $n \to \infty$ almost everywhere on $[a, b]$. This shows that the sequence of step functions $\{z^n\}_{n \geq n_0}$ converges to $z^* := \frac{t}{\|y\|} y$ almost everywhere on $[a, b]$. Note that, for each $n \geq n_0$,

$$g_n\left(\frac{t_n}{\|\widetilde{y}^n\|}\widetilde{y}^n\right) = g_*(z^n).$$

This and (38) imply $g_*(z^*) \leq g_*(w)$. Since $w \in B$ is an arbitrarily chosen function such that $\|w\| = t$, we deduce that z^* is an optimal solution of the problem (37). □

Application of Lemma 4.8 yields $\nu_f(y, t) = g_*(z^*)$ and this implies (13) in this case. The proof of Theorem 4.1 is complete.

A significant consequence of Theorem 4.1 is the next result:

Corollary 4.9. *If $p > 1$, then, in each of the three cases enumerated in Theorem 4.1, the function $f(x) = \|x\|^p$ is a Bregman function, that is, it satisfies the following conditions:*

(i) The function f is Fréchet differentiable and strictly convex;

(ii) For each $x \in B$ and for any real number $\alpha \geq 0$ the level sets

$$L_\alpha^f(x) = \{y \in B : D_f(y, x) \leq \alpha\}$$

and

$$R_\alpha^f(x) = \{y \in B : D_f(x, y) \leq \alpha\}$$

are bounded;

(iii) If the sequence $\{x^k\}_{k\in\mathbb{N}}$ is bounded in B and if $\{y^k\}_{k\in\mathbb{N}} \subset B$ is such that

$$\lim_{k\to\infty} D_f(y^k, x^k) = 0, \tag{39}$$

then

$$\lim_{k\to\infty} \|x^k - y^k\| = 0.$$

Proof. Our argument is based on the fact that, if $p \in (1, 2)$, then the function f is totally convex as follows from Theorem 4.1. If $p \geq 2$, then, as observed above, the function f is uniformly convex and, thus, it is also totally convex (cf. Proposition 3.1(iv)).

(i) The fact that f is Fréchet differentiable is well-known (for instance, it follows from [9, Corollary 1, p. 57]). Since f is totally convex, Proposition 3.1(i) implies that f is strictly convex.

(ii) The fact that the set $L_\alpha^f(x)$ is bounded follows from the total convexity of f combined with [6, Lemma 2.5(ii)]. Suppose, by contradiction, that $R_\alpha^f(x)$ includes an unbounded sequence $\{y^k\}_{k\in\mathbb{N}}$. We distinguish two possible cases.

Case 1. Assume that $p \geq 2$. Then, according to Proposition 3.1(iv), for any nonnegative integer k, we have

$$\delta_f(\|y^k - x\|) \leq \nu_f(y^k, \|y^k - x\|) \leq D_f(x, y^k) \leq \alpha.$$

It follows from [14, Theorem 4] that

$$\delta_f(t) = \frac{1}{2^{p-1}} t^p. \tag{40}$$

Thus, for any $k \in \mathbb{N}$,

$$\frac{1}{2^{p-1}} \|y^k - x\|^p \leq \alpha$$

and letting $k \to \infty$ we get a contradiction.

Case 2. Assume that $p \in (1, 2)$ Without loss of generality we may assume that, for all k, we have $\|y^k\| \neq 0$. According to (13), for any nonnegative integer k, we have

$$\left(\|y^k\| + \|y^k - x\|\right)^p - \|y^k\|^p - p \|y^k\|^{p-1} \|y^k - x\| =$$
$$\nu_f(y^k, \|y^k - x\|) \leq D_f(x, y^k) \leq \alpha.$$

This implies

$$\|y^k\|^p[(1 + t_k)^p - 1 - pt_k] \leq \alpha, \tag{41}$$

where the sequence

$$t_k := \frac{\|y^k - x\|}{\|y^k\|}$$

satisfies

$$\left| \frac{\|x\|}{\|y^k\|} - 1 \right| \leq t_k \leq 1 + \frac{\|x\|}{\|y^k\|}.$$

The function

$$\psi(t) = (1 + t)^p - pt - 1 \tag{42}$$

is strictly increasing on $[0, +\infty)$ and $\psi(0) = 0$. Observe that the left hand side of (41) is exactly $\|y^k\|^p \psi(t_k)$. Therefore, the inequality (41) implies

$$\alpha \geq \overline{\lim}_{k\to\infty} \|y^k\|^p \psi(t_k) \geq \overline{\lim}_{k\to\infty} \|y^k\|^p \psi\left(\left|1 - \frac{\|x\|}{\|y^k\|}\right|\right) = +\infty$$

and this is a contradiction.

(iii) Suppose that the sequence $\{x^k\}_{k\in\mathbb{N}}$ is bounded in B and that (39) holds. Then, if $p \geq 2$, we have

$$0 \leq \frac{1}{2^{p-1}} \underline{\lim}_{\ k\to\infty} \|y^k - x^k\|^p \leq \frac{1}{2^{p-1}} \overline{\lim}_{k\to\infty} \|y^k - x^k\|^p$$

$$\leq \lim_{k\to\infty} D_f(x^k, y^k) = 0,$$

because of (40) and of Proposition 3.1(iv). This implies $\lim_{k\to\infty}\|y^k - x^k\| = 0$ in this case. Now, assume that $p \in (1,2)$. Then, an application of (39) yields

$$\lim_{k\to\infty} \nu_f(x^k, \|y^k - x^k\|) = 0. \tag{43}$$

Note that, according to (13),

$$\nu_f(x^k, \|y^k - x^k\|) = \varphi_4(\|x^k\|, \|y^k - x^k\|),$$

where φ_4 is the function defined by (19). Let M be a positive upper-bound of the bounded sequence $\{\|x^k\|\}_{k\in\mathbb{N}}$. Since $\varphi_4(\cdot, \|y^k - x^k\|)$ is decreasing (cf. Lemma 4.3 above) we have

$$\varphi_4(M, \|y^k - x^k\|) \leq \varphi_4(\|x^k\|, \|y^k - x^k\|) = \nu_f(x^k, \|y^k - x^k\|),$$

for all nonnegative integers k. Hence,

$$M^p\left[\left(1 + \frac{\|y^k - x^k\|}{M}\right)^p - 1 - p\frac{\|y^k - x^k\|}{M}\right] =$$

$$\left(M + \|y^k - x^k\|\right)^p - M^p - pM^{p-1}\|y^k - x^k\| =$$

$$\varphi_4(\|x^k\|, \|y^k - x^k\|) \leq \nu_f(x^k, \|y^k - x^k\|).$$

This implies

$$0 \leq M^p \lim_{k\to\infty} \psi\left(\frac{\|y^k - x^k\|}{M}\right) \leq \lim_{k\to\infty} \nu_f(x^k, \|y^k - x^k\|) = 0,$$

where ψ is the function defined by (42). Since ψ is strictly increasing and continuous on $[0, +\infty)$, this yields $\lim_{k\to\infty}\|y^k - x^k\| = 0$ and the proof is complete. □

5. Computing almost common fixed points

In this section we apply the results concerning the local moduli of convexity presented above in order to prove existence and convergence of an iterative algorithm of computing almost common fixed points of measurable families of operators. Precisely, the problem we consider is the following:

Let $(\Omega, \mathcal{A}, \mu)$ be a probability space, let B be a reflexive Banach space and let $C \subseteq B$ be a nonempty closed convex subset of B. Suppose that the family of operators $T_\omega : C \to C$, $\omega \in \Omega$, is measurable, that is, for each $x \in C$, the function $T_\bullet(x) : \omega \to T_\omega(x)$ is measurable[3]. The question is whether and under what conditions, the sequences recursively

[3] *In this work measurability and integrability of functions from Ω to B is in the sense of Bochner.*

generated according to the rule

$$\text{Chose } x^0 \in C \text{ and let } x^{k+1} := \int_\Omega T_\omega(x^k)d\mu(\omega), \; k \in \mathbb{N}, \tag{44}$$

exist and converge (weakly or strongly) to almost common fixed points of the operators T_ω no mater how the initial point x^0 is chosen in C. Recall that an *almost common fixed point* of the measurable family of operators T_ω, $\omega \in \Omega$, is a point $x^* \in C$ such that

$$\mu\left(\{\omega \in \Omega : T_\omega(x^*) = x^*\}\right) = 1. \tag{45}$$

We denote by $Afix(T_\bullet)$ the set of almost common fixed points of the given family of operators and we presume that $Afix(T_\bullet) \neq \emptyset$.

Our convergence analysis of the procedure (44) is done under the assumption that there exists a totally convex function $f : B \to (-\infty, +\infty]$ with closed domain $\mathcal{D} := Dom(f)$, which is continuous on $\mathcal{D}$, Fréchet differentiable on $\mathcal{D}^\circ := Int(\mathcal{D})$ and satisfies the following conditions of compatibility with the given problem:

(A) The set C is included in $\mathcal{D}^\circ$ and, for any $x \in C$, the function $\omega \to D_f(T_\omega(x), x)$ is integrable;

(B) If Ω is infinite, then either C is bounded or, for each $x \in C$, the function $\nu_f(x, \cdot)$ is continuous on its domain;

(C) If $z \in Afix(T_\bullet)$, then the function $D_f(z, \cdot)$ is convex and, for each $x \in C$,

$$D_f(z, T_\omega(x)) \leq D_f(z, x) - D_f(T_\omega(x), x), \; a.e. \tag{46}$$

(D) For any $x \in C$ and for each real number $\alpha \geq 0$, the level set

$$R^f_\alpha(x) = \{y \in C : D_f(x, y) \leq \alpha\} \tag{47}$$

is bounded.

(E) If $\omega \in \Omega$, then the function $x \to D_f(T_\omega(x), x) : C \to [0, +\infty)$ is sequentially weakly lower semicontinuous.

The following result shows that, under these conditions, the sequence $\{x^k\}_{k\in\mathbb{N}}$ generated according to the rule (44) exists no matter how the initial point $x^0 \in C$ is chosen and, thus, the question asked above concerning the convergence of such sequences makes sense.

Lemma 5.1. *If the conditions (A) and (B) are satisfied, then the function $\mathbf{T} : C \to C$ given by*

$$\mathbf{T}(x) = \int_\Omega T_\omega(x)d\mu(\omega), \tag{48}$$

is well-defined, that is, for any $x \in C$, the function $T_\bullet(x)$ is integrable and its integral belongs to C.

Proof. Fix an arbitrary point $x \in C$. If Ω is finite, then the result is obvious. Suppose that Ω is infinite. If the function $T_\bullet(x)$ is integrable, then its integral is the limit of a sequence of convex combination of the form $\sum_{i=1}^{r} \mu(A_i)T_{\omega_i}(x)$, where $A_1, ..., A_r$ is a measurable partition of Ω and, for each $i \in \{1, ..., r\}$, we have $\omega_i \in A_i$. Therefore, if $T_\bullet(x)$ is integrable, then its integral $\mathbf{T}(x)$ belongs to C. If C is bounded, then the integrability of $T_\bullet(x)$ follows from the measurability of $T_\bullet(x)$ and the fact that

$$\int_\Omega \|T_\omega(x)\| \, d\mu(\omega) \le M,$$

where M is an upperbound of the set C. Suppose that C is unbounded. Then, the set $\mathcal{D}$ is unbounded because, according to (A), we have $C \subseteq \mathcal{D}$, This implies that the domain of the function $\eta(t) := \nu_f(x,t)$ is $[0, +\infty)$. Since this function is continuous (cf. (B)) and strictly increasing (cf. Proposition 2.1), it is invertible and its inverse function, η^{-1}, is continuous too. According to Proposition 2.1(ii), we have that, for any $c \ge 1$ and for each $t \in [0, +\infty)$,

$$\eta^{-1}(ct) \le c\eta^{-1}(t). \tag{49}$$

Also, since

$$\eta(\|T_\omega(x) - x\|) = \nu_f(x, \|T_\omega(x) - x\|) \le D_f(T_\omega(x), x), \tag{50}$$

we deduce

$$\|T_\omega(x) - x\| \le \eta^{-1}(D_f(T_\omega(x), x)),$$

for any $\omega \in \Omega$. The functions on both sides of the last inequality are measurable and nonnegative. Thus, the following integrals exist and we have

$$\int_\Omega \|T_\omega(x) - x\| \, d\mu(\omega) \le \int_\Omega \eta^{-1}(D_f(T_\omega(x), x)) d\mu(\omega). \tag{51}$$

Since

$$\int_\Omega \|T_\omega(x)\| \, d\mu(\omega) \le \|x\| + \int_\Omega \|T_\omega(x) - x\| \, d\mu(\omega), \tag{52}$$

if we show that

$$I := \int_\Omega \eta^{-1}(D_f(T_\omega(x), x)) d\mu(\omega) < +\infty, \tag{53}$$

then the integrability of $T_\bullet(x)$ follows from (51) and (52). In order to prove (53), observe that the convex function f is continuous at $x \in \mathcal{D}^\circ$

and, therefore, it is locally Lipschitz at this point, that is, there exist two positive real numbers r and L such that

$$|f(x) - f(y)| \leq L \|x - y\|, \tag{54}$$

whenever $\|x - y\| < r$. Let

$$\Gamma = \{\omega \in \Omega : \|x - T_\omega(x)\| < r\}$$

and $\Lambda = \Omega \backslash \Gamma$. If $\omega \in \Gamma$, then

$$D_f(T_\omega(x), x) \leq |f(T_\omega(x)) - f(x)| + |f^\circ(x, T_\omega(x) - x)|$$

$$\leq 2L \|T_\omega(x) - x\| \leq 2Lr,$$

because of (54) and of [8, Proposition 2.1.1]. This implies that, for any $\omega \in \Gamma$,

$$\eta^{-1}(D_f(T_\omega(x), x)) \leq \eta^{-1}(2Lr).$$

Consequently,

$$I_1 := \int_\Gamma \eta^{-1}(D_f(T_\omega(x), x)) d\mu(\omega) \leq \eta^{-1}(2Lr) < +\infty. \tag{55}$$

For any $\omega \in \Lambda$, define

$$c(\omega) = \frac{D_f(T_\omega(x), x)}{\eta(\|T_\omega(x) - x\|)}.$$

According to (50) we have that $c(\omega) \geq 1$. Therefore,

$$I_2 := \int_\Lambda \eta^{-1}(D_f(T_\omega(x), x)) d\mu(\omega) = \int_\Lambda \eta^{-1}(c(\omega)\eta(\|T_\omega(x) - x\|)) d\mu(\omega)$$

$$\leq \int_\Lambda c(\omega)\eta^{-1}(\eta(\|T_\omega(x) - x\|)) d\mu(\omega) = \int_\Lambda c(\omega) \|T_\omega(x) - x\| \, d\mu(\omega),$$

where the inequality results from (49). According to Proposition 2.1(*ii*), if $\omega \in \Lambda$, we also have

$$\eta(\|T_\omega(x) - x\|) = \eta\left(\frac{\|T_\omega(x) - x\|}{r} r\right) \geq \frac{\|T_\omega(x) - x\|}{r} \eta(r),$$

that is,

$$c(\omega) \leq \frac{D_f(T_\omega(x), x)}{\frac{\|T_\omega(x) - x\|}{r} \eta(r)}.$$

This implies

$$I_2 \leq \int_\Lambda \frac{D_f(T_\omega(x), x)}{\frac{\|T_\omega(x) - x\|}{r} \eta(r)} \|T_\omega(x) - x\| \, d\mu(\omega) =$$

$$\frac{r}{\eta(r)} \int_\Lambda D_f(T_\omega(x), x) d\mu(\omega) < +\infty,$$

because of the integrability of the function $\omega \to D_f(T_\omega(x), x)$ (see $(A1)$). Since $I = I_1 + I_2$ and since I_1 is finite (cf. (55)), the last inequality implies (53) and the proof is complete. □

The conditions (A) and (B) not only guarantee existence of the sequences $\{x^k\}_{k\in\mathbb{N}}$ generated according to the rule (44), but also ensure well-definedness of the function $\Upsilon_f : C \to [0, +\infty)$ given by

$$\Upsilon_f(x) = \int_\Omega D_f(T_\omega(x), x) d\mu(\omega). \tag{56}$$

This function is relevant in our context because of the following result.

Lemma 5.2. *(i) Suppose that the condition (A) and (B) hold and that $x^* \in C$. Then x^* is an almost common fixed point of the family of operators T_ω, $\omega \in \Omega$, if and only if $\Upsilon_f(x^*) = 0$.*

(ii) If the conditions (A), (B) and (C) are satisfied, then, for any $z \in Afix(T_\bullet)$ and for any $x \in C$, the function $\omega \to D_f(z, T_\omega(x))$ is integrable and we have

$$D_f(z, \mathbf{T}(x)) \leq \int_\Omega D_f(z, T_\omega(x)) d\mu(\omega) \leq D_f(z, x) - \Upsilon_f(x). \tag{57}$$

Proof. (i) The "if" part is obvious. If $\Upsilon_f(x^*) = 0$, then, for almost all $\omega \in \Omega$,

$$0 \leq \nu_f(x^*, \|T_\omega(x^*) - x^*\|) \leq D_f(T_\omega(x), x) = 0.$$

This cannot happen unless $\|T_\omega(x^*) - x^*\| = 0$, a.e., because f is totally convex. Hence, x^* is an almost common fixed point of the family of operators T_ω, $\omega \in \Omega$.

(ii) The function $\omega \to D_f(z, T_\omega(x))$ is measurable because $D_f(z, \cdot)$ is continuous (since f is Fréchet differentiable). Thus, according to (46), the second inequality in (57) is satisfied. This implies the integrability of the function $\omega \to D_f(z, T_\omega(x))$. The first inequality in (57) results from the continuity and convexity of $D_f(z, \cdot)$ combined with Jensen's inequality. □

With these facts in mind we prove the following convergence result.

Theorem 5.3. *Suppose that the measurable family of operators T_ω, $\omega \in \Omega$ has an almost common fixed point. If the conditions (A)-(D) hold, then, for any initial point $x^0 \in C$, the sequence $\{x^k\}_{k\in\mathbb{N}}$ recursively generated according to (44) exists and it has the following properties:*

(i) The sequence $\{x^k\}_{k\in\mathbb{N}}$ has weak accumulation points,

$$\lim_{k\to\infty} \Upsilon_f(x^k) = 0$$

and, for almost all $\omega \in \Omega$, *we have*

$$\underline{\lim}_{k\to\infty} D_f(T_\omega(x^k), x^k) = 0; \tag{58}$$

(ii) If condition (E) *is also satisfied, then any weak accumulation point of* $\{x^k\}_{k\in\mathbb{N}}$ *is contained in* $Afix(T_\bullet)$*;*

(iii) If, in addition to the conditions $(A) - (E)$, *the function* $f^\circ : B \to B^*$ *is sequentially weakly-to-weak* continuous, then the sequence* $\{x^k\}_{k\in\mathbb{N}}$ *converges weakly to an almost common fixed point of the operators* T_ω, $\omega \in \Omega$.

Proof. Existence of the sequence $\{x^k\}_{k\in\mathbb{N}}$ results from Lemma 5.1. Let z be an element in $Afix(T_\bullet)$. Writing (57) for $x = x^k$ we deduce that the sequence $\{D_f(z, x^k)\}_{k\in\mathbb{N}}$ is nonincreasing and, hence, bounded by $\alpha = D_f(z, x^0)$. Therefore, the sequence $\{x^k\}_{k\in\mathbb{N}}$ is contained in the bounded set $R_\alpha^f(z)$ (see condition (D)). By consequence, the sequence $\{x^k\}_{k\in\mathbb{N}}$ has weakly convergent subsequences. Observe that, according to (57), for any nonnegative integer k, we have

$$0 \leq \Upsilon_f(x^k) \leq D_f(z, x^k) - D_f(z, x^{k+1}). \tag{59}$$

Since the sequence $\{D_f(z, x^k)\}_{k\in\mathbb{N}}$ is nonincreasing and bounded it is convergent and this, combined with (59), implies that $\lim_{k\to\infty} \Upsilon_f(x^k) = 0$. According to Fatou's lemma we have

$$0 \leq \int_\Omega \underline{\lim}_{k\to\infty} D_f(T_\omega(x^k), x^k) d\mu(\omega) \leq$$

$$\underline{\lim}_{k\to\infty} \Upsilon_f(x^k) = \lim_{k\to\infty} \Upsilon_f(x^k) = 0$$

and this proves (i). For proving (ii), let $\{x^{j_k}\}_{k\in\mathbb{N}}$ be any weakly convergent subsequence of $\{x^k\}_{k\in\mathbb{N}}$ and denote by x^* its weak limit. Let $\omega \in \Omega$ be such that (58) holds. According to the definition of the local modulus of convexity we have

$$\begin{aligned} \nu_f(x^*, \|T_\omega(x^*) - x^*\|) &\leq D_f(T_\omega(x^*), x^*) \leq \\ &\underline{\lim}_{k\to\infty} D_f(T_\omega(x^{j_k}), x^{j_k}) = 0, \end{aligned} \tag{60}$$

where the second inequality results from the condition (E). Consequently, we have

$$\nu_f(x^*, \|T_\omega(x^*) - x^*\|) = 0$$

and this cannot hold unless $T_\omega(x^*) = x^*$ because f is totally convex. Since $\omega \in \Omega$ is any of the elements for which (58) is satisfied, it follows that $x^* \in Afix(T_\bullet)$. This proves (ii). Now, we prove (iii). To this end, suppose that the sequence $\{x^k\}_{k\in\mathbb{N}}$ has two accumulation points

x' and x''. Since both x' and x'' are in $Afix(T_\bullet)$, it results that the (57) holds with z replaced by any of them. This means that the sequences $\{D_f(x',x^k)\}_{k\in\mathbb{N}}$ and $\{D_f(x'',x^k)\}_{k\in\mathbb{N}}$ are nonincreasing and bounded and, thus, convergent. Therefore, the following limits exist and we have

(61)
$$\lim_{k\to\infty}\left[D_f(x',x^k) - D_f(x'',x^k)\right] = f(x') - f(x'') - \lim_{k\to\infty} f^\circ(x^k, x' - x'').$$

Let $\{x^{i_k}\}_{k\in\mathbb{N}}$ and $\{x^{j_k}\}_{k\in\mathbb{N}}$ be subsequences of the sequence $\{x^k\}_{k\in\mathbb{N}}$ which converge weakly to x' and to x'', respectively. According to (61), we have

$$0 \le D_f(x',x'') = f(x') - f(x'') - \lim_{k\to\infty} f^\circ(x^{j_k}, x' - x'') =$$

$$f(x') - f(x'') - \lim_{k\to\infty} f^\circ(x^{i_k}, x' - x'') = -D_f(x'',x') \le 0.$$

This shows that $D_f(x'',x') = 0$ which, in turns, implies that

$$\nu_f(x', \|x'' - x'\|) = 0.$$

Since f is totally convex, that cannot hold unless $x' = x''$. By consequence, the sequence $\{x^k\}_{k\in\mathbb{N}}$ has an unique weak accumulation point which necessarily belongs to $Afix(T_\bullet)$, i.e. $\{x^k\}_{k\in\mathbb{N}}$ converges weakly to an almost common fixed point of the given family of operators. □

Theorem 4.1 has significant consequences. One of them is the next.

Corollary 5.4. *Suppose that B is a Banach space of finite dimension and that, for each $\omega \in \Omega$, the operator T_ω is continuous. If D_f is convex on $C \times C$ and if the conditions (A)-(D) hold, then the sequence $\{x^k\}_{k\in\mathbb{N}}$ generated according to (44) exists and converges to a point in $Afix(T_\bullet)$, provided that $Afix(T_\bullet) \neq \emptyset$.*

Proof. Condition (E) is satisfied because D_f is convex and continuous and each T_ω is continuous. Hence, application of Theorem 4.1 yields the conclusion. □

A particular instance in which Corollary 5.4 applies is that when B is a finite dimensional Hilbert space, $C = B$, $f(x) = \|x\|^2$ and the operators T_ω are *firmly nonexpansive*, i.e., for any $x, y \in C$,

$$\|T_\omega(x) - T_\omega(y)\|^2 \le \|x - y\|^2 - \|T_\omega(x) - T_\omega(y) - (x - y)\|^2. \tag{62}$$

Observe that, in this situation, $D_f(x,y) = \|x - y\|^2$ and, therefore, condition (62) implies (46). In this case, convergence of the sequence $\{x^k\}_{k\in\mathbb{N}}$ generated according to (44) to a fixed point of $\mathbf{T}$ was proven before in [10]. It should be noted that the application of Corollary

5.4 is not restricted to the case of nonexpansive operators T_ω. For example, if $B = \mathbb{R}^n$, $C \subset \mathbb{R}^n_{++}$ and f is the negentropy defined by (12), then, for a given measurable point-to-set mapping $Q : \Omega \to B$ with nonempty closed convex values, the *entropic projections*

$$T_\omega(x) = \arg\min \{D_f(y, x) : y \in Q_\omega\}, \tag{63}$$

are well-defined (see [7]) and they are not, necessarily, firmly nonexpansive. However, one can infer from the study of these operators done in [4], that Corollary 5.4 applies in this case too and guarantees existence and convergence of the sequence $\{x^k\}_{k\in\mathbb{N}}$ to a point in $Afix(T_\bullet)$, whenever such a point exists.

Applicability of Theorem 5.3 is not restricted to Banach spaces of finite dimension. A question of interest in practical problems is the following (see [3] and [6]). Let $Q : \Omega \to B$ be a measurable point-to-set mapping with closed convex nonempty values. Find a point $z \in B$ which belongs to almost all sets Q_ω, that is, such that

$$\mu(\{\omega \in \Omega : z \in Q_\omega\}) = 1,$$

provided that such a point exists. A way of answering this question is to extend the idea underlying the application of Corollary 5.4 to entropic projections. Precisely, choose the totally convex function $f : B \to (-\infty, +\infty]$ with closed domain $\mathcal{D}$ in such a manner that f is continuous on $\mathcal{D}$, Fréchet differentiable on $\mathcal{D}^\circ$ and

$$C := cl(conv \bigcup_{\omega\in\Omega} Q_\omega) \subset \mathcal{D}^\circ. \tag{64}$$

Define the operators $T_\omega : C \to C$ according to (63). These operators are well-defined (cf. [6, Lemma 2.2]) and are called *Bregman projections onto the sets Q_ω with respect to f*. In these circumstances, the conditions (A) and (C) are automatically satisfied (see [6, Lemma 5.4 and Lemma 5.5]). Therefore, Theorem 5.3 leads to the following improvement of Theorem 5.7(C) in [6].

Corollary 5.5. *If, in addition to the requirements above, the conditions (B), (D) and (E) also hold, then the sequence $\{x^k\}_{k\in\mathbb{N}}$ generated according to* (44), *with the Bregman projection operators as T_ω, has weak accumulation points and each weak accumulation point of this sequence belongs to almost all sets Q_ω. Moreover, if $f^\circ : B \to B^*$ is sequentially weakly-to-weak* continuous, then $\{x^k\}_{k\in\mathbb{N}}$ converges weakly.*

If B is a Hilbert space and $f(x) = \|x\|^2$, then the requirements of Corollary 5.5 are satisfied no matter how the measurable point-to-set mapping Q is chosen. In this case, the Bregman projection operators T_ω are exactly the metric projections onto the sets Q_ω and Corollary 5.5 reduces to Theorem 1 in [3] and shows that the sequence $\{x^k\}_{k\in\mathbb{N}}$ converges weakly to a point belonging to almost all sets Q_ω.

If the Banach space B is not a Hilbert space and/or if f is not the square of the norm, then the applicability of Corollary 5.5 depends on the realizability of the conditions (B), (D) and (E). According to [6, Lemma 5.5], if D_f is convex on $C \times C$, then (E) holds in this context. This fact is useful in the particular case when $B = \mathcal{L}^p([a,b])$ or $B = \ell^p$, $f(x) = \|x\|^p$ and $p \in (1,2)$ because, then, the conditions (B) and (D) are satisfied as follows from Theorem 4.1 and Corollary 4.9, respectively, and D_f is convex on the nonnegative orthant (cf. [6, Section 6.3]). It allows application of Corollary 5.5 in order to guarantee convergence of procedures of solving integral equations and best approximation problems as those described in [6, Sections 6.5 and 6.6].

In Hilbert spaces one can extrapolate the basic features of Corollary 5.5 to a more general class of operators than the metric projections. In fact, observe that, if B is a Hilbert space and if $f(x) = \|x\|^2$, then $D_f(x,y) = \|x-y\|^2$ and $\nu_f(x,t) = t^2$. Therefore, if all operators T_ω are firmly nonexpansive (see (62)), then the conditions (A)-(D) are automatically satisfied and we can apply Theorem 5.3 in order to deduce the following result which improves in some respects upon the main conclusion of [10].

Corollary 5.6. *If T_ω, $\omega \in \Omega$, is a measurable family of firmly nonexpansive operators defined on a closed convex subset C of a Hilbert space B, and if there exists an almost common fixed point of this family of operators, then, for any initial point $x^0 \in C$, the sequence $\{x^k\}_{k\in\mathbb{N}}$ generated according to (44) exists and converges weakly to a point in $Afix(T_\bullet)$.*

Proof. As noted above, under the actual hypothesis, the conditions (A)-(D) hold for $f(x) = \|x\|^2$. Thus, Theorem 5.3 guarantees that the sequence $\{x^k\}_{k\in\mathbb{N}}$ exists, has weak accumulation points,

$$\lim_{k\to\infty} \left\|x^k - x^{k+1}\right\| = 0, \tag{65}$$

and (58) holds for almost all $\omega \in \Omega$. Let x^* be a weak accumulation point of $\{x^k\}_{k\in\mathbb{N}}$ and let $\{x^{j_k}\}_{k\in\mathbb{N}}$ be a subsequence of $\{x^k\}_{k\in\mathbb{N}}$ which converges weakly to x^*. Let $\omega \in \Omega$ be an element such that (58) holds. Then, we can find a subsequence of $\{x^{j_k}\}_{k\in\mathbb{N}}$, which we denote by

$\{x^{h_k}\}_{k\in\mathbb{N}}$, such that

$$\lim_{k\to\infty} \left\| x^{h_k} - T_\omega(x^{h_k}) \right\| = 0. \tag{66}$$

For this subsequence, the nonexpansivity of T_ω yields

$$\left\| T_\omega(x^*) - x^{h_k} \right\| \leq \left\| x^{h_k} - T_\omega(x^{h_k}) \right\| + \left\| T_\omega(x^{h_k}) - T_\omega(x^*) \right\|$$

$$\leq \left\| x^{h_k} - T_\omega(x^{h_k}) \right\| + \left\| x^{h_k} - x^* \right\|$$

and this implies

$$\overline{\lim}_{k\to\infty} \left\| T_\omega(x^*) - x^{h_k} \right\| \leq \overline{\lim}_{k\to\infty} \left\| x^{h_k} - x^* \right\|$$

because of (66). Recall (cf. [11, Theorem 4.1]) that x^* is the unique asymptotic center of the subsequence $\{x^{h_k}\}_{k\in\mathbb{N}}$. On the other hand, the last inequality shows that $T_\omega(x^*)$ is an asymptotic center of this sequence. Therefore, $T_\omega(x^*) = x^*$. Since this happens for all $\omega \in \Omega$ for which (58) holds, it follows that $x^* \in Afix(T_\bullet)$. Hence, any weak accumulation point of the sequence $\{x^k\}_{k\in\mathbb{N}}$ is contained in $Afix(T_\bullet)$. Now one can reproduce without modification the argument in the proof of Theorem 5.3(*iii*) in order to show that $\{x^k\}_{k\in\mathbb{N}}$ has an unique weak accumulation point and this completes the proof. □

Acknowledgments

The authors are grateful to Professors Yair Censor and Simeon Reich and to the referees for comments which led to improvements of an earlier version of this work. Also, the authors wish to thank Dr. David Blanc for helping them prepare and revise the material.

References

[1] Alber, Ya., Metric and generalized projection operators in Banach spaces, in: M.E. Drakhlin and E. Litsyn (Eds.), "Functional Differential Equations", pp.1-21, *The Research Institute of the College of Judea and Samaria Publications*, Kedumim-Ariel, Israel, 1993.

[2] Alber, Y. and Butnariu, D., Convergence of Bregman-projection methods for solving convex feasibility problems in reflexive Banach spaces, *J. Optim. Theory Appl.*, to appear.

[3] Butnariu, D., The expected-projection methods: Its behavior and applications to linear operator equations and convex optimization, *J. Applied Analysis*, Vol. 1, No. 1, 1995, 95-108.

[4] Butnariu, D., Censor, Y. and Reich, S., Iterative averaging of entropic projections for solving stochastic convex feasibility problems, *J. Comput. Appl. Math.*, to appear.

[5] Butnariu, D. and Flåm, S.D., Strong convergence of expected projection methods in Hilbert spaces, *Numer. Funct. Anal. Optim.*, 16, 1995, 601-636.

[6] Butnariu, D., Iusem, A.N. and Burachik, R.S., Iterative methods of solving stochastic convex feasibility problems and applications, *preprint*.

[7] Censor, Y. and Lent, A., An iterative row action method for interval convex programming, *J. Optim. Theory Appl.*, 34(3), 1981, 321-353.

[8] Clarke, F.H., Optimization and Nonsmooth Analysis, *John Wiley and Sons*, New York, 1983.

[9] Diestel, J., Geometry of Banach Spaces - Selected Topics, *Springer-Verlag*, Berlin, 1975.

[10] Flåm, S.D., Successive averages of firmly nonexpansive mappings, *Math. Oper. Res.*, Vol. 20, No.2, 1995, 497-512.

[11] Goebel, K. and Reich, S., Uniform Convexity, Hyperbolic Geometry, and Nonexpansive Mappings, *Marcel Dekker*, New York, 1984.

[12] Phelps, R.R., Convex Functions, Monotone Operators and Differentiability, 2-nd Edition, *Springer Verlag*, Berlin, 1993.

[13] Reich, S., A weak convergence theorem for the alternating method with Bregman distances, in: A.G. Kartsatos (Ed.), "Theory and Applications of Nonlinear Operators of Accretive and Monotone Type", pp. 313-318, *Marcel Dekker*, New York, 1996.

[14] Vladimirov, A.A., Nesterov, Y.E. and Chekanov, Y.N., Uniformly convex functionals (Russian), *Vestnik Moskovskaya Universiteta, Series Mathematika i Kybernetika*, 3, 1978, 12-23.

DEPARTMENT OF MATHEMATICS AND COMPUTER SCIENCE, UNIVERSITY OF HAIFA, 31905 HAIFA, ISRAEL

INSTITUTO DE MATEMÁTICA PURA E APLICADA, ESTRADA DONA CASTORINA 110, JARDIM BOTÂNICO, RIO DE JANEIRO, R. J., CEP 22460-320, BRAZIL

Contemporary Mathematics
Volume **204**, 1997

FIXED POINT THEORY VIA NONSMOOTH ANALYSIS

F. H. CLARKE
CENTRE DE RECHERCHES MATHÉMATIQUES
UNIVERSITÉ DE MONTRÉAL
MONTRÉAL, QUÉBEC H3C 3J7, CANADA

YU. S. LEDYAEV
STEKLOV INSTITUTE OF MATHEMATICS
MOSCOW 117966, RUSSIA

R. J. STERN
DEPARTMENT OF MATHEMATICS AND STATISTICS
CONCORDIA UNIVERSITY
MONTREAL, QUEBEC H4B 1R6, CANADA

September 16, 1996

ABSTRACT. Results on the existence of zeros and fixed points of multifunctions in nonconvex sets are surveyed. Applications include results on the existence of equilibria in nonconvex sets which are weakly invariant with respect to a differential inclusion, as well as extensions of the classical fixed point theorems of Brouwer and Browder. The methods employed utilize tools of nonsmooth analysis.

1. INTRODUCTION

The classical Brouwer fixed point theorem asserts that if S is a compact homeomorphically convex subset of $\Re^n$ such that $g(S) \subseteq S$, where $g : \Re^n \to \Re^n$ is continuous, then S possesses a fixed point of g. That is, there exists $x^* \in S$ such that $g(x^*) = x^*$. A well-known dynamical application of the Brouwer theorem (and one which we shall subsequently call upon) pertains to the Cauchy problem

$$\dot{x}(t) = f(x(t)), \quad x(0) = x_0, \tag{1}$$

where f is a Lipschitz function; then for any initial data x_0, a unique solution $x(t) = x(t; x_0)$ is locally defined. We shall say that a set S is *invariant* with respect to (1) provided that for every $y \in S$ there is a globally defined (and of course, unique) solution $x(\cdot; y)$ such that $x(t; y) \in S$ for all $t \geq 0$. The result we have in mind is the following one:

Theorem 1.1. *Suppose that $S \subseteq \Re^n$ is a compact homeomorphically convex set which is invariant with respect to* (1), *where $f : S \to \Re^n$ is Lipschitz. Then S contains a point $\hat{x}$ such that $f(\hat{x}) = 0$.*

---

[0]MATHEMATICS SUBJECT CLASSIFICATION: 47H10, 26B05

[0]KEY WORDS: Multifunction, differential inclusion, zeros, fixed points, equilibria, weak invariance, tangency, nonsmooth analysis.

[0]The first and third authors were supported by the Natural Sciences Engineering Research Council of Canada and le fonds FCAR du Québec.

A brief sketch of the proof goes like this: For each positive T, one has $g_T(S) \subseteq S$, where

$$g_T(y) = x(T; y).$$

The fixed point $x_T \in S$ of g_T provided by the Brouwer fixed point theorem obviously corresponds to an orbit of period T (contained in S) of the differential equation. A limiting argument (this is where the work is) as $T \downarrow 0$ implies the existence in S of a point $\hat{x}$ such that $0 \in f(\hat{x})$; see e.g. Bhatia and Szego [4]. The point $\hat{x}$ is a *zero* of f, and in the dynamical context of the result, is also referred to as an *equilibrium* of the underlying differential equation, since the only solution of (1) emanating from $\hat{x}$ is the constant trajectory $x(t; \hat{x}) = \hat{x}$ for all $t \geq 0$.

The main thrust of the present article is towards providing an answer to the following

Question 1: What becomes of Theorem 1.1 in the multivalued setting?

By this we mean that the differential equation $\dot{x}(t) = f(x(t))$ is to be replaced by the *differential inclusion*

$$\dot{x}(t) \in F(x(t)). \tag{2}$$

Here $F : \Re^n \hookrightarrow \Re^n$ is a multifunction (i.e. a multivalued mapping), and a solution or *trajectory* $x(\cdot)$ on an interval $[0, T]$ is understood to be an absolutely continuous function satisfying (2) almost everywhere in $[0, T]$. One can view the above differential inclusion as a "generalized control system", since when $F(x) = f(x, U)$ (with U a given subset of some finite dimensional space), (2) takes the explicitly parametrized form

$$\dot{x}(t) = f(x(t), u(t)), \quad u(t) \in U. \tag{3}$$

It is well known that under mild assumptions, the systems (2) and (3) have the same trajectories. In the present article, however, we will stick to the more general system (2).

The notion of invariance that we have in mind is the following one: A set $S \subseteq \Re^n$ is said to be *weakly invariant* with respect to the differential inclusion (2) provided that for every initial point $x_0 \in S$, there exists a trajectory such that $x(0) = x_0$ and $x(t) \in S \;\; \forall\, t \geq 0$. We shall adopt abbreviated terminology and simply say that the pair (S, F) is weakly invariant. A corresponding notion of a point x being a zero of F (or equilibrium of (2)) is simply $0 \in F(x)$.

Our basic hypotheses on F are summarized in the following:

(H): The multifunction F satisfies

(a) $F(x)$ is a nonempty compact convex subset of $\Re^n \;\; \forall\, x \in \Re^n$.

(b) F is upper semicontinuous at every $x \in \Re^n$.

In regard to (b), upper semicontinuity of F at x means that for all $\varepsilon > 0$ there exists $\delta > 0$ such that

$$\|x - x'\| < \delta \Longrightarrow F(x') \subseteq F(x) + \varepsilon B,$$

where B denotes the open unit ball. When F is single-valued, the upper-semicontinuity property reduces to continuity.

The following result, due to Browder [5], provides for the existence of zeros of multifunctions in compact convex sets in terms of the "tangency" or "inwardness" condition

$$F(x) \cap T_S(x) \neq \phi \quad \forall\, x \in S, \tag{4}$$

in which $T_S(x)$ denotes the *tangent cone* to the closed convex set S at $x \in S$, familiar in convex analysis. This is given by

$$T_S(x) := \mathrm{cl}\{\alpha(y-x) : y \in S,\ \alpha \geq 0\}.$$

Theorem 1.2. *Let $S \subseteq \Re^n$ be compact and convex, and let the multifunction F satisfy* (H). *Assume that* (4) *holds. Then there exists $\hat{x} \in S$ such that $0 \in F(\hat{x})$.*

Although Browder's theorem holds true in a general Hausdorff locally convex linear topological space, we shall only require the $\Re^n$ version stated. For extensions of Browder's result and general investigations as to how inwardness implies the existence of zeros in convex sets, the reader is referred to Halpern [15], Reich [20], [21], [22], Aubin [1] and the references cited therein.

Browder's theorem easily yields the famous fixed point theorem of Kakutani, which one can phrase as follows: If G is an upper semicontinuous multifunction with closed convex values such that $G(x) \cap S \neq \phi$ for every x in the compact convex set S, then S contains a fixed point $\hat{x}$ of G (where this means that $\hat{x} \in G(\hat{x})$). The proof involves nothing more than defining $F(x) := G(x) \cap S - x$ and observing that the tangency condition (4) holds. Of course, Kakutani's theorem immediately implies that of Brouwer, upon specializing to the single-valued case. We may view Browder's theorem as a dynamic version of Kakutani's theorem, because of the following:

Theorem 1.3. *Let S and F be as in* Theorem 1.2. *Then condition* (4) *is necessary and sufficient for (S, F) to be weakly invariant.*

Theorem 1.3 is in fact a special case of Theorem 2.1 below, wherein S is permitted to be nonconvex, and a more general notion of tangency is employed. We now have at our fingertips a fairly satisfying answer to Question 1 posed above, which first appeared in Heymann and Stern [16]:

Theorem 1.4. *If F satisfies hypothesis* (H), *S is compact and convex, and the pair (S, F) is weakly invariant, then there exists $\hat{x} \in S$ such that $0 \in F(\hat{x})$.*

The discussion to this point summarizes the state of affairs on the issue of existence of zeros of a multifunction F in a compact convex set $S \subseteq \Re^n$, into the early 1990's. There remained, however, one important issue very worthy of investigation, and which provided the main motivation for the work to be described herein: While Brouwer's theorem and Theorem 1.1 require only *homeomorphic* convexity of S, their multivalued generalizations, namely Browder's (or Kakutani's) theorem and Theorem 1.4 require *true* convexity of S. We therefore are compelled to pose

Question 2: Can one relax convexity of S in Theorems 1.2 and 1.4?

Recent results providing alternate affirmative answers to this question will be surveyed in this article. The tools of nonsmooth analysis are central to our methods. The results and some extensions will be described in the sections following the next one, which provides a brief overview of requisite background material.

2. Preliminaries

2.1. Nonsmooth analysis: basic definitions and facts. For this subsection, general references are Clarke [6], [7], and Clarke, Ledyaev, Stern and Wolenski [8], [10].

Let S denote a closed subset of $\Re^n$, equipped with the standard inner product $\langle \cdot, \cdot \rangle$ along with the Euclidean norm $\| \cdot \|$. The distance function to S is denoted d_S. Given $u \in \Re^n$, the set of closest points to u in S is nonempty and is denoted

$$\mathrm{proj}_S(u) := \{x \in S : \|u - x\| = d_S(u)\}.$$

The multifunction proj_S is called the *metric projection* onto S. If $u \notin S$ and $x \in \mathrm{proj}_S(u)$, then the vector $u - x$ is referred to as a *perpendicular* to S at x. The set of all nonnegative multiples of such perpendiculars is denoted $N_S^P(x)$, and is called the *proximal normal cone* (or P-*normal cone*) to S at x. If $x \in \mathrm{int}(S)$ or if no perpendiculars to S exist at x, then by convention we set $N_S^P(x) = \{0\}$. One can show that the proximal normal cone is nonzero on a dense subset of $\mathrm{bdry}(S)$.

For $r \geq 0$, denote the "r-outer approximation" of S by

$$S^r := S + r\bar{B} = \{y \in \Re^n : d_S(y) \leq r\},$$

where $\bar{B}$ is the closed unit ball. Suppose that $d_S(y) = r > 0$, and let $0 \neq \zeta \in N_{S^r}^P(y)$. It can be proven that then $\mathrm{proj}_S(y)$ is a singleton, say x, and that

$$N_{S^r}^P(y) = \{\alpha(y - x) : \alpha \geq 0\} \subseteq N_S^P(x). \tag{5}$$

The *limiting normal cone* (or L-*normal cone*) to S at $x \in S$ is defined to be the cone

$$N_S^L(x) := \{\zeta : \zeta_i \to \zeta, \; \zeta_i \in N_S^P(x_i), \; x_i \to x\}.$$

This is always a nonzero for $x \in \mathrm{bdry}(S)$, while $N_S^L(x) = \{0\}$ if x is an interior point. The *Clarke tangent cone* (or C-*tangent cone*) to S at $x \in S$ is defined via polarity to be the closed convex cone

$$T_S^C(x) := [N_S^L(x)]^* = \left\{v \in \Re^n : \langle v, \zeta \rangle \leq 0 \;\; \forall\, \zeta \in N_S^L(x)\right\}.$$

We shall say that S is *epi-Lipschitz* at $x \in S$ provided that $\mathrm{int}[T_S^C(x)] \neq \phi$. This condition always holds at interior points, but may not hold on the boundary of S (i.e. there may be cusps). A set is said to be epi-Lipschitz provided that it is epi-Lipschitz at each of its points. Any convex set with nonempty interior is epi-Lipschitz. We refer the reader to Rockafellar [23] for a characterization (not required in the present article) of the epi-Lipschitz property in terms of S being locally isomorphic to the epigraph of a Lipschitz function—which of course justifies the terminology.

We shall require another kind of continuity for set–valued mappings. Let H be a multifunction on S, with values that are subsets of $\Re^n$. We say that H is *lower semicontinuous* at $x_0 \in S$ if the following holds: Let $y \in H(x)$. Then for any given $\varepsilon > 0$, there exists $\delta > 0$ such that

$$x \in \{x_0 + \delta B\} \cap S \Longrightarrow y \in H(x) + \varepsilon B.$$

We will require the fact that if S is epi-Lipschitz, then the C-tangent cone T_S^C is a lower semicontinuous multifunction on S.

The *D-tangent cone* to S at $x \in S$, referred to also as the Bouligand tangent cone in some references (e.g. Aubin [1], Aubin and Frankowska [3]), is defined to be the closed (but possibly nonconvex) cone

$$T_S^D(x) := \text{cone}\left\{\lim \frac{s_i - x}{\|s_i - x\|} : s_i \in S,\ s_i \to x\right\}.$$

One always has $T_S^C(x) \subseteq T_S^D(x)$. Should equality occur, then we say that S is *regular* at x, and if the property holds at every $x \in S$, then S is said to be a regular set. If the set S is convex, then it can be shown to be regular, and in fact, $T_S^D(x) = T_S^C(x) = T_S(x)$ for every $x \in S$.

2.2. Weak invariance.

We will make reference to the following result of [8]:

Theorem 2.1. *Let $S \subseteq \Re^n$ be compact, and let F satisfy* (H). *Then the following are equivalent:*

(a) *(S, F) is weakly invariant.*

(b) *One has*

$$F(x) \cap T_S^D(x) \neq \phi \ \ \forall\, x \in S. \tag{6}$$

(c) *For every $x \in S$ one has*

$$\min_{v \in F(x)} \langle \zeta, v\rangle \leq 0 \ \ \forall\, \zeta \in N_S^P(x). \tag{7}$$

The equivalence of (a) and (b) is attributed to Haddad [13], and is a generalization of a result of Nagumo [18] pertaining to differential equations without uniqueness; see [8] for further historic and bibliographic comments. The equivalence of (a) and (b) to the proximal criterion provided by (c), on the other hand, appeared in [8], and was proven by the technique of "proximal aiming", a technique which has antecedents in dynamic game strategy design techniques of Krasovskii and Subottin [17]; see also Veliov [24]. Condition (c) has the advantage of needing to be verified only at points where the P-normal cone is nonzero, and is important in methods of feedback control construction; again, see [8].

2.3. Proximal smoothness.

Some required material from Clarke, Stern and Wolenski [9] will now be reviewed. Although the results of [9] were formulated in Hilbert space, we here only require the finite dimensional versions. Like the epi-Lipschitz property, *proximal smoothness* generalizes convexity; proximal smoothness of S connotes bdry$[S^r]$ being a C^1-manifold for all small positive r. Characterizations of proximal smoothness were provided by Clarke, Stern and Wolenski in [9]; we will summarize those that are pertinent in the results to follow.

If one has

$$S \cap \operatorname{int}\left\{x + r\left(\frac{\zeta}{\|\zeta\|} + \bar{B}\right)\right\} = \phi \ \ \forall\, y \in S,$$

ζ is said to be *realized by an r-ball.* This means that the open ball of radius r centered at $x + r\frac{\zeta}{\|\zeta\|}$ has empty intersection with S. We denote $\Omega := \operatorname{bdry}(B)$, and

$$U(\gamma) := \{x \in \Re^n : 0 < d_S(x) < \gamma\}.$$

Theorem 2.2. *For a closed set $S \subseteq \Re^n$ and $\gamma > 0$, the following statements are equivalent:*

(a) *For every $x \in \operatorname{bdry}(S)$, every proximal normal $\zeta \in N_S^P(x)$ is realized by a γ-ball.*

(b) *The distance function d_S is continuously differentiable on $U(\gamma)$.*

(c) $\operatorname{proj}_S(x)$ *is a singleton for every $x \in U(\gamma)$.*

(d) *For each $r \in (0, \gamma)$ and any $y \in \Re^n$ such that $d_S(y) = r$, one has $N^P_{S^r}(y) \neq \{0\}$.*

Furthermore, if any of these conditions hold, then for each $r \in [0, \gamma)$ one has $N^P_{S^r}(y) = N^L_{S^r}(y)$, and if $d_S(y) = r \in (0, \gamma)$, then

$$N^P_{S^r}(y) \cap \Omega = \{\nabla d_S(y)\} \subseteq N^P_S(z),$$

where z is the unique closest point in S to y.

Should there exist γ as in the theorem, then S is said to be *proximally smooth*; we also call S "proximally smooth of radius γ" or "γ-proximally smooth". In this case, for each $r \in [0, \gamma)$, the outer approximation S^r is $(\gamma - r)$-proximally smooth. Also, a closed set $S \subseteq \Re^n$ is convex if and only if it is γ-proximally smooth for all positive γ.

Let us remark upon another interesting fact proven in [9], but one which we do not require presently: The following can be added to the list of equivalent conditions in Theorem 2.2:

(e) *The proximal subdifferential $\partial_P d_S(y)$ is nonempty on $U(\gamma)$.*

In this case one has $\partial_P d_S(y) = \nabla d_S(y)$ on $U(\gamma)$.

2.4. Multifunctions. We say that the multifunction G is *Lipschitz* on a domain Γ provided that there exists $K > 0$ such that

$$G(x') \subseteq G(x) + K\|x' - x\| \ \ \forall\, x, x' \in \Gamma.$$

The following result on upper approximation of multifunctions will be needed. (Our reference is Deimling [12], Lemma 2.2.)

Proposition 2.3. *Let S be a compact subset of $\Re^n$, and assume that the multifunction F satisfies* (H). *Let a sequence $\gamma_j \downarrow 0$ be given. Then there exists a sequence of multifunctions $\{F_j\}_{j=1}^{\infty}$, each Lipschitz on S and having compact convex images, such that for each $x \in S$ one has*

$$F(x) \subseteq F_{j+1}(x) \subseteq F_j(x) \subseteq F(\{x + \gamma_j B\} \cap S), \; j = 1, 2, \ldots . \tag{8}$$

The following result of [9] on the extension of Lipschitz multifunctions is also needed:

Proposition 2.4. *Suppose that $S \subseteq \Re^n$ is a compact γ-proximally smooth set, and let $F : S \hookrightarrow \Re^n$ be a Lipschitz multifunction having compact convex images. Then for any $\tilde{\gamma} \in (0, \gamma)$, there exists a Lipschitz multifunction $\tilde{F}$ with compact convex images and such that*

$$F(x) = \tilde{F}(x) \;\; \forall\, x \in S^{\tilde{\gamma}}.$$

Proposition 2.4 follows from the fact (proven in [9]) that under the stated hypotheses, the metric projection proj_S, which is single-valued on the tube $U(\gamma)$ (by Theorem 2.2), is in fact Lipschitz on $U(\gamma)$. One then considers $\tilde{F}(x) = F(\text{proj}_S(x))$ to prove the assertion.

3. The case of Lipschitzian sets

This section is based upon Clarke, Ledyaev and Stern [11]. We shall require the familiar fact that corresponding to any finite covering $\bigcup_{i=1}^{k} C_i$ of a compact set $S \subseteq \Re^n$, with each C_i open and bounded, there exists a *Lipschitz partition of unity* $\{p_1, p_2, \ldots, p_k\}$. (See e.g. Aubin and Cellina [2].) This means that each p_i is a Lipschitz function and the following hold:

(i) $0 \leq p_i(x) \leq 1 \;\; \forall\, x \in S, \; i = 1, 2, \ldots k.$

(ii) $x \notin C_i \Longrightarrow p_i(x) = 0, \; i = 1, 2, \ldots, k.$

(iii) $\sum_{i=1}^{k} p_i(x) = 1 \;\; \forall\, x \in S.$

The next result generalizes Browder's theorem to the nonconvex setting. We remark that, bearing in mind the properties of the Clarke tangent cone, the manner in which a partition of unity is applied is similar to the technique employed in Haddad and Lasry [14], where the existence of periodic trajectories in compact convex sets was treated.

Theorem 3.1. *Let hypothesis* (H) *hold, and assume that S is compact, homeomorphically convex, and epi-Lipschitz. Suppose that*

$$F(x) \cap T_S^C(x) \neq \phi \;\; \forall\, x \in S. \tag{9}$$

Then F has a zero in S.

Proof: In view of Proposition 2.3, it is not hard to confirm that F can without loss of generality be assumed to be Lipschitz on S. Also note that due to the epi-Lipschitz assumption on S, we have

$$\{F(x)+\varepsilon B\}\cap \mathrm{int}[T_S^C(x)]\neq \phi \quad \forall\, x\in S,$$

for any $\varepsilon>0$. The lower semicontinuity of the multifunction T_S^C then implies that for each $y\in F(x)\cap\{\mathrm{int}(T_S^C(x))\}$, there exists $\delta(x,y)>0$ such that

$$x'\in x+\delta(x,y)B := B(x,\delta(x,y)) \Longrightarrow y\in\{F(x')+\varepsilon B\}\cap\{\mathrm{int}(T_S^C(x'))\}.$$

Therefore the family of balls $B(x,\delta(x,y))$ forms an open covering of S. Let us denote a finite subcover by $\bigcup_{i=1}^k B(x_i,\delta(x_i,y_i))$, and let $\{p_1,p_2,\dots,p_k\}$ be a corresponding Lipschitz partition of unity. We now define Lipschitz function

$$f_\varepsilon(x) := \sum_{i=1}^k p_i(x)y_i.$$

In view of the convexity of the C-tangent cone and the sets $F(x)$, it is a straightforward matter to check that the selection f_ε satisfies

$$f_\varepsilon(x)\in\{F(x)+\varepsilon B\}\cap\{\mathrm{int}[T_S^C(x)]\} \quad \forall\, x\in S. \tag{10}$$

Let us now consider the ordinary differential equation

$$\dot{x}(t)=f_\varepsilon(x(t)). \tag{11}$$

In view of (10) and the general containment $T_S^C\subseteq T_S^D$, Theorem 2.1 implies that S is invariant with respect to (11), and Theorem 1.1 then yields the existence of $x_\varepsilon\in S$ such that $f_\varepsilon(x_\varepsilon)=0$. Since

$$0\in F(x_\varepsilon)+\varepsilon B.$$

In order to complete the proof, consider any sequence $\varepsilon_i\downarrow 0$. Then since S is compact, for a subsequence (not relabeled) one has $x_{\varepsilon_i}\to\hat{x}\in S$, and $\hat{x}$ is clearly a zero of F. $\square$.

Having generalized Browder's theorem, we can refer to Theorem 2.1 in order to immediately obtain the following generalization of Theorem 1.4:

Theorem 3.2. *Let F satisfy (H), and let $S\subseteq\Re^n$ be compact, homeomorphically convex, epi-Lipschitz and regular. Then weak invariance of (S,F) implies the existence of a point $\hat{x}\in S$ such that $0\in F(\hat{x})$.*

Remarks 3.3.

(a) In Theorem 3.1, the C-tangency condition (9) is not replaceable by the D-tangency condition (6). Essentially, the reason for this is that the D-tangent cone can be "too big". An example which illustrates this phenomenon is the following:

Let $n=3$, and let

$$S = cl[\{\sqrt{2}B\}\backslash K],$$

where K is the cone

$$K = \{x = (x_1, x_2, x_3) : \sqrt{x_1^2 + x_2^2} \leq x_3\}.$$

One defines an upper semicontinuous multivalued vector field F on S as follows: Introduce the sets

$$W = \{(x_1, x_2, 1) : x_1^2 + x_2^2 = 1\},$$
$$Q = \{(x_1, x_2, 1) : x_1^2 + x_2^2 \leq 2\},$$

and let

$$F(x) = \begin{cases} Q & \text{if } x \in S \backslash W \\ co\{Q, (-x_2, x_1, 0)\} & \text{if } x \in W, \end{cases}$$

where *co* denotes convex hull. The set S is compact, homeomorphically convex, and epi-Lipschitz. Furthermore, the D-tangency condition (6) holds at every $x \in S$; equivalently, one can just as well verify that the proximal normality condition (7) holds. (In verifying the latter, note that there is nothing to check at the origin, since no nonzero proximal normals exist there.) On the other hand, the C-tangency (9) fails only at the origin; this is the sole point where S is not regular. Yet S does not contain a zero. In view of Proposition 2.3, there in fact exists a *Lipschitz* multifunction which exhibits the same pathology as the F constructed above.

(b) The example above illustrates another important fact: The C-tangency condition (9), while sufficient for weak invariance of (S, F) (by Theorem 2.1), is not necessary. Of course the tangency criteria (6) and (9) are equivalent when S is a regular set.

(c) In Theorem 3.1, the assumption that the images $F(x)$ are compact can be relaxed to closedness (with convexity maintained). In particular, under this milder hypothesis, the proof of the theorem given in [11] utilizes a version of Proposition 2.3 in which the approximating family is lower semicontinuous. Theorems 2.1 and 3.2, however, do require the compactness assumption.

(d) The Lipschitz function f_ε constructed in the proof of Theorem 3.1 is connected to the idea of *feedback control* in the control system case, where $F(x) = f(x, U)$. Since f_ε is a selection in $f_\varepsilon(x, U) := f(x, U) + \varepsilon B$, there exists a "feedback law" $\mu : S \to U$ such that $f_\varepsilon(x, \mu(x)) = f_\varepsilon(x)$. Hence when this feedback is implemented, the unique solution of the associated ordinary differential equation $\dot{x}(t) = f_\varepsilon(x(t), \mu(x(t)))$ with $x(0) \in S$ exists on $[0, \infty)$ and satisfies $x(t) \in S \;\; \forall\, t \geq 0$.

In [11], it was pointed out that in Theorem 3.1, when F is a single valued (and consequently continuous) function f, the tangency condition (9) takes the following form, in terms of the function $g(x) := f(x) - x$:

$$\lim_{\substack{x' \to x \\ x' \in S \\ \lambda \downarrow 0}} \frac{d_S[\lambda g(x') + (1 - \lambda)x]}{\lambda} = 0. \tag{12}$$

Therefore one has the following version of the Brouwer fixed point theorem:

Corollary 3.4. *If $S \subseteq \Re^n$ is a compact epi-Lipschitz homemorphically convex set and $g : S \to \Re^n$ is continuous and satisfies* (12), *then g has a fixed point in S.*

We also have a result in terms of the following "complementary weak invariance" type of property: A set is said to be *weakly avoidable* if $\hat{S} := \Re^n \backslash \text{int}(S)$ is weakly invariant.

Theorem 3.5. *Let F satisfy (H), and assume that S is compact, epi-Lipschitz, weakly avoidable, and that $\hat{S}$ is regular. Then S contains a zero of F.*

This conclusion can be drawn from Theorem 3.2 upon invoking the fact that the epi-Lipschitz assumption implies $T_S^C(x) = -T_{\hat{S}}^C(x)$ for every $x \in \text{bdry}(S) = \text{bdry}(\hat{S})$.

Distinct from the epi-Lipschitz property, another type of Lipschitz condition that one can impose upon S is the requirement that it be *bi-Lipschitz.* This connotes S being homeomorphic to a compact convex set, such that both the homeomorphism and its inverse are Lipschitz mappings. (The graph of a Lipschitz function is an example of a bi-Lipschitz set which is not epi-Lipschitz.) In [11] the following result was proven:

Theorem 3.6. *Let F satisfy* (H) *and let S be a bi-Lipschitz subset of $\Re^n$ such that* (9) *holds. Then F has a zero in S.*

The proof of Theorem 3.6 is quite different from that of Theorem 3.1, and furthermore, it extends to general Banach spaces. The method involves properties of the Clarke generalized derivative of d_S, as exposited in [6].

4. Remarks on the non-Lipschitzian case

The next proposition provides a mechanism for obtaining analogs of the results described in the preceding section, when Lipschitz hypotheses on S do not hold. It follows from results in Plaskacz [19], but we will provide an independent proof more in line with our methods:

Proposition 4.1. *Let $S \subseteq \Re^n$ be compact, and let the multifunction F satisfy* (H). *Assume that S satisfies the following conditions:*

(A1): *S is proximally smooth.*
(A2): *S^r is homeomorphically convex for all small positive r.*

Then the tangency condition (9) *implies the existence of a point $\hat{x} \in S$ such that* $0 \in F(\hat{x})$.

Note: Since proximal smoothness implies regularity (see [9]), the C-tangency condition (9) in the statement is equivalent to the D-tangency condition (6).

Proof of the proposition: We first consider

Case 1: *F is Lipschitz on S.*

In view of hypothesis (A1) and Proposition 2.4, there exists $r_0 > 0$ such that S is r_0-proximally smooth and such that F has a Lipschitz extension $\tilde{F}$, with compact convex images, defined on S^{r_0}. Consider $\rho \in (0, r_0)$, and let $y \in \text{bdry}[(S^\rho]$; that is, $d_S(y) = \rho$. Let z be the unique projection of y onto S. By Theorem 2.2, one has

$$N^P_{S^\rho}(y) \cap \Omega = \{\nabla d_S(y)\} \subseteq N^P_S(z).$$

Then (9) implies the existence of $v \in \tilde{F}(z) = F(z)$ such that

$$\langle \nabla d_S(y), v \rangle \leq 0. \tag{13}$$

Denote a Lipschitz constant for $\tilde{F}$ on $S + r_0\bar{B}$ by K. Then there exists $w \in \tilde{F}(y)$ such that $\|w - v\| \leq K\|y - z\| = K\rho$, and since $\|\nabla d_S(y)\| = 1$, (13) implies

$$\langle \nabla d_S(y), w \rangle \leq K\rho. \tag{14}$$

It follows that

$$\left.\begin{array}{c} 0 < r < r_0 \\ y \in S^r \backslash S \end{array}\right\} \Longrightarrow \min_{u \in \tilde{F}(y)} \langle \nabla d_S(y), u \rangle \leq Kr, \tag{15}$$

and therefore

$$\left.\begin{array}{c} 0 < r < r_0 \\ y \in S^r \backslash S \end{array}\right\} \Longrightarrow \min_{u \in \tilde{F}(y) + 3Kr\bar{B}} \langle \nabla d_S(y), u \rangle \leq -2Kr. \tag{16}$$

Let $r \in (0, r_0)$, denote the minimizing set of u in (16) by $M(y)$, and consider the multifunction $G : S^r \hookrightarrow \Re^n$ given by

$$G(y) := \begin{cases} M(y) & \text{if } y \in S^r \backslash S \\ \tilde{F}(y) + 3Kr\bar{B} & \text{if } y \in S \end{cases}$$

It is not difficult to verify that G is upper semicontinuous on S^r and that for each $y \in S^r$, the image $G(y)$ is a nonempty compact convex subset of $\tilde{F}(y) + 3Kr\bar{B}$; one requires continuity of ∇d_S on $S^r \backslash S$.)

Due to Theorem 2.2, for every $y \in \text{bdry}(S^r)$, one has $v \in T^C_{S^r}(y)$ if and only if $\langle \nabla d_S(y), v \rangle \leq 0$. Hence Theorem 3.1 implies that there exists $x_r \in S^r$ such that $0 \in \tilde{F}(x_r) + 3Kr\bar{B}$. Letting $r \downarrow 0$ and invoking the compactness of S, we deduce the existence of $\hat{x} \in S$ such that $0 \in \tilde{F}(\hat{x}) = F(\hat{x})$.

<u>Case 2</u>: *General case.*

In view of Proposition 2.3, for any given sequence $\gamma_j \downarrow 0$ there exists a sequence of Lipschitz multifunctions $F_j : S \hookrightarrow \Re^n$ with compact convex images and such that (8) holds. Since each F_j satisfies the tangency condition (9), Case 1 implies that for each j there exists $x_j \in S$ such that

$$0 \in F_j(x_j) \subseteq F\left(\{x_j + \gamma_j B\} \cap S\right).$$

Invoking the compactness of S, we choose a subsequence (not relabeled) such that $x_j \to \hat{x} \in S$. Upper semicontinuity of F then implies that for any given $\varepsilon > 0$, one has

$$0 \in F_j(x_j) \subseteq F(\hat{x}) + \varepsilon B$$

for all j sufficiently large. Since ε is arbitrary and $F(\hat{x})$ is closed, we conclude that $\hat{x}$ is a zero of F. $\square$

Remark 4.2. Proposition 4.1 requires homeomorphic convexity of nearby outer approximations of S, but not of S itself. As an illustration, observe that the proposition is applicable to the compact set

$$S = \left\{(x,y) \in \Re^2 : \sqrt{|y|} \leq x \leq 1\right\} \cup \{(x,0) : -1 \leq x \leq 0\}.$$

Note that S is neither epi- nor bi-Lipschitz, but (A1) and (A2) hold; also note that S is not homeomorphically convex, whereas its outer approximations are.

A simple application of the preceding corollary is the star-shaped case. A set $S \subseteq \Re^n$ is said to be *star-shaped* provided that there exists a point $\hat{s} \in S$ with the property that for any $s \in S$ and $0 < \lambda \leq 1$,

$$\lambda\hat{s} + (1-\lambda)s \in \operatorname{relint}(S),$$

where the relative interior is taken with respect to span(S). Clearly convexity implies the star-shaped property, and any compact star-shaped set is homeomorphic to $\bar{B}$. One can easily show that S^r is star-shaped for all $r > 0$. Hence we have the following result, which generalizes the case of compact convex S:

Corollary 4.3. *Let $S \subset \Re^n$ be a compact, star-shaped, proximally smooth set, and suppose that the multifunction F satisfies* (H). *Then the tangency condition* (9) *implies the existence of $\hat{x} \in S$ such that $0 \in F(\hat{x})$.*

Remark 4.4. The following example shows that the preceding corollary is false without the assumed proximal smoothness. Let

$$S = \left\{\sqrt{2}\bar{B}\right\} \cap \left\{x \in \Re^3 : x_3 \leq (x_1^2 + x_2^2)^{\frac{1}{4}}\right\},$$

and take F as in Remark 3.3(a). Observe that S is compact and star-shaped, with proximal smoothness failing only at the origin. Checking that the tangency condition (9) holds is a straightforward matter. However, S contains no zeros of F.

References

[1] J. P. Aubin. *Viability Theory.* Birkhauser, Boston, 1991.

[2] J. P. Aubin and A. Cellina. *Differential Inclusions.* Springer–Verlag, Berlin, 1984.

[3] J. P. Aubin and H. Frankowska. *Set-valued Analysis.* Birkhauser, Boston, 1990.

[4] N. P. Bhatia and G. P. Szego. *Stability Theory of Dynamical Systems.* Springer–Verlag, Berlin, 1970.

[5] F. Browder. The fixed point theory of multivalued mappings in topological vector spaces. *Math. Ann.*, 177:283–301, 1968.

[6] F. H. Clarke. *Optimization and Nonsmooth Analysis.* Wiley-Interscience, New York, 1983. Republished as Vol. 5 of Classics in Applied Mathematics, S.I.A.M., Philadelphia, 1990.

[7] F. H. Clarke. *Methods of Dynamic and Nonsmooth Optimization*, volume 57 of *CBMS-NSF Regional Conference Series in Applied Mathematics.* S.I.A.M., Philadelphia, 1989.

[8] F. H. Clarke, Yu. S. Ledyaev, R. J. Stern, and P. R. Wolenski. Qualitative properties of trajectories of control systems: a survey. *J. Dyn. Control Sys.*, 1:1–48, 1995.

[9] F. H. Clarke, R. J. Stern, and P. R. Wolenski. Proximal smoothness and the lower-C^2 property. *J. Convex Anal.*, 2:117–145, 1995.

[10] F.H. Clarke, Yu. S. Ledyaev, R.J. Stern, and P.R. Wolenski. *Introduction to Nonsmooth Analysis.* (In preparation).

[11] F.H. Clarke, R.J. Stern, and Yu. S. Ledyaev. Fixed points and equilibria in nonconvex sets. *Nonlinear Analysis*, 25:145–161, 1995.

[12] K. Deimling. *Multivalued Differential Equations.* de Gruyter, Berlin, 1992.

[13] G. Haddad. Monotone trajectories of differential inclusions and functional differential inclusions with memory. *Israel J. Math.*, 39:83–100, 1981.

[14] G. Haddad and J. M. Lasry. Periodic solutions of functional differential inclusions and fixed points of σ-selectionable correspondences. *J. Math. Anal. Appl.*, 96:295–312, 1983.

[15] B. Halpern. Fixed point theorems for set-valued maps in infinite-dimensional spaces. *Math. Ann.*, 189:87–98, 1970.

[16] M. Heymann and R. J. Stern. Omega rest points in autonomous control systems. *J. Diff. Eq.*, 20:389–398, 1976.

[17] N. N. Krasovskii and A. I. Subottin. *Game-Theoretical Control Problems.* Springer-Verlag, New York, 1988.

[18] M. Nagumo. Uber die Lage der Integralkurven gewohnlicher Differnentialgleichungen. *Proc. Phys. Math. Soc. Japan*, 24:551–559, 1942.

[19] S. Plaskacz. On the solution sets for differential inclusions. *Boll. U.M.I.*, 7:387–394, 1992.

[20] S. Reich. Fixed points in locally convex spaces. *Math. Z.*, 125:17–31, 1972.

[21] S. Reich. Approximate selections, best approximations, fixed points, and invariant sets. *J. Math. Anal. Appl.*, 62:104–113, 1978.

[22] S. Reich. Fixed point theorems for set-valued mappings. *J. Math. Anal. Appl.*, 69:353–358, 1979.

[23] R. T. Rockafellar. Clarke's tangent cones and boundaries of closed sets in R^n. *Nonlinear Analysis*, 3:145–154, 1979.

[24] V. M. Veliov. Sufficient conditions for viability under imperfect measurement. *Set Valued Analysis*, 1:305–317, 1993.

Contemporary Mathematics
Volume **204**, 1997

RANGES OF GENERALIZED PSEUDO-MONOTONE PERTURBATIONS OF MAXIMAL MONOTONE OPERATORS IN REFLEXIVE BANACH SPACES

Zhengyuan Guan and Athanassios G. Kartsatos

ABSTRACT. Let X be a real Banach space with dual X^* and normalized duality mapping J. Various results are given for ranges of sums $A+B$, where $A : X \supset D(A) \to 2^{X^*}$ is usually maximal monotone and $B : X \to 2^{X^*}$ is at least generalized pseudo-monotone. These results are extensions and/or improvements of recent relevant results by the authors as well as Browder, Browder and Hess, et al. The basic lemma in this theory uses the method of generalized Galerkin approximants. Another range of sums theorem is also given with $B : D(A) \to X^*$ bounded and such that $B(\lambda A + J)^{-1}$ is compact, for some $\lambda > 0$. This result generalizes a recent result by Hirano and Kalinde and uses homotopies with the Leray-Schauder degree theory.

1. Introduction and Preliminaries

In what follows, X is a real Banach space with dual space X^* and normalized duality mapping J. We denote the norms of X and X^* by $\|\cdot\|$. For $x \in X$ and $x^* \in X^*$, we use the symbol (x^*, x) or the symbol (x, x^*) to denote the value of x^* at x. Let Y be another real Banach space. For a multi-valued mapping $T : X \to 2^Y$, we set $D(T) = \{x \in X \; ; \; Tx \neq \emptyset\}$ and $R(T) = \cup\{Tx \; ; \; x \in D(T)\}$. When we write $: T : X \to 2^Y$ we mean that the operator T has domain $D(T) = X$. Unless otherwise stated, or implied, the term "continuous" means strongly continuous. An operator $T : X \supset D(T) \to Y$, is "demicontinuous" if it is continuous from the strong topology of X to the weak topology of Y. $T : X \to 2^Y$ is "finitely-continuous" if T is upper semicontinuous from each finite-dimensional subspace F of X to the weak topology on Y, i.e., for any given element $x_0 \in F$ and a weak neighborhood V of Tx_0 in Y there exists a neighborhood U of x_0 in F such that $Tx \subset V$ for all $x \in U$. An operator $T : X \to 2^{X^*}$ is said to be bounded if it maps bounded sets onto bounded sets. It is said to be "quasi-bounded" if for each $M > 0$ there exists $K(M) > 0$ such that for $x \in X$ and $y \in Tx$ with $\|x\| \le M$ and $(y, x) \le M\|x\|$,

1991 *Mathematics Subject Classification.* Primary 47H17; Secondary 47B07, 47H05, 47H06, 47H10.

Key words and phrases. Maximal monotone operator, generalized pseudo-monotone operator, range of sums, Leray-Schauder degree theory.

we have $\|y\| \leq K(M)$. It is said to be "strongly quasi-bounded" if for each $M > 0$ there exists $K(M) > 0$ such that for $x \in X$ and $y \in Tx$ with $\|x\| \leq M$ and $(y, x) \leq M$, we have $\|y\| \leq K(M)$.

A mapping $T : X \supset D(T) \to 2^{X^*}$ is said to be "monotone" if for every $x,\ y \in D(T)$, $u \in Tx$ and $v \in Ty$ we have

$$(u - v, x - y) \geq 0.$$

A monotone mapping $T : X \supset D(T) \to 2^{X^*}$ is said to be "maximal monotone", if $R(T + \lambda J) = X^*$ for every $\lambda > 0$. This is equivalent to saying that: T is monotone and

$$(u - u_0, x - x_0) \geq 0,$$

for every $x \in D(T)$ and $u \in Tx$, implies $x_0 \in D(T)$ and $u_0 \in Tx_0$.

A mapping $T : X \supset D(T) \to 2^{X^*}$ is said to be "pseudo-monotone" if

a) Tx is bounded, closed and convex for all $x \in D(T)$;

b) T is finitely continuous;

c) for every pair of sequences $\{x_n\}$, $\{y_n\}$ such that $\{x_n\} \subset D(T)$, $y_n \in Tx_n$, $x_n \rightharpoonup x_0 \in D(T)$ and

$$\limsup_{n \to \infty} (y_n, x_n - x_0) \leq 0$$

we have the following property: for every $x \in X$ there exists $y(x) \in Tx_0$ such that

$$(y(x), x_0 - x) \leq \liminf_{n \to \infty} (y_n, x_n - x).$$

A mapping $T : X \supset D(T) \to 2^{X^*}$ is said to be "generalized pseudo-monotone" if a) of above is satisfied and for every pair of sequences $\{x_n\}$, $\{y_n\}$ such that $\{x_n\} \subset D(T)$, $y_n \in Tx_n$, $x_n \rightharpoonup x_0 \in D(T)$, $y_n \rightharpoonup y_0 \in X^*$ and

$$\limsup_{n \to \infty} (y_n, x_n - x_0) \leq 0,$$

we have $y_0 \in Tx_0$ and $(y_n, x_n) \to (y_0, x_0)$.

$T : X \supset D(T) \to X^*$ is said to be of "type (S_+)" if for every sequence $\{x_n\} \subset D(T)$ with $x_n \rightharpoonup x_0 \in X$ and $\limsup_{n \to \infty} (Tx_n, x_n - x_0) \leq 0$, we have $x_n \to x_0$. It is said to be "quasi-monotone" if for every sequence $\{x_n\} \subset D(T)$, with $x_n \rightharpoonup x_0 \in X$, we have

$$\limsup_{n \to \infty} (Tx_n, x_n - x_0) \geq 0.$$

For the basic properties of mappings of monotone type and the duality mapping J, we refer to Barbu [1], Browder [5], Cioranescu [7], Pascali and Sburlan [14] and Zeidler [16]. For operators in Hilbert spaces, we cite the book of Brézis [2].

The main purpose of this paper is to present several range results for operators of the type $A + B$, where $A : X \supset D(A) \to 2^{X^*}$ is maximal monotone and $B : X \to 2^{X^*}$ is, mainly, generalized pseudo-monotone. Equivalently, we consider the solvability of the inclusion

$$Ax + Bx \ni y, \tag{$*$}$$

for y in certain specific subsets of X^*. Let $A_\lambda = (A^{-1} + \lambda J^{-1})^{-1}$, for $\lambda > 0$. Since the operator A is not necessarily defined on all of X, we consider first the inclusion

$$A_\lambda x + Bx + (1/n)Jx \ni y.$$

The operator A_λ has considerably better properties than those of A itself. We use the method of generalized Galerkin approximations (proof of Lemma 2.1) to show the existence of solutions of this inclusion. The solvability of the original problem $(*)$ above follows then by further approximations. Most of the results in this paper are obtained via versions of this method.

To a large extent, the results in this paper are improvements of the results in the Section 2 of Guan and Kartsatos [8]. However, our methodology in this paper is generally different from that of [8].

Finally, we give a theorem (Theorem 3.3) for ranges of sums which uses the fact that $B : D(A) \to X^*$ is just a bounded operator such that $B(\lambda A+J)^{-1}$ is compact, for some $\lambda > 0$. The novelty in this result lies in the fact that the operator B is not assumed to be continuous. Such results originated in the paper of Hirano and Kalinde [10] who studied zeros and ranges involving m-accretive operators A. The "zeros" parts of [10] were improved and extended by Kartsatos in [11] and [12].

The following simple lemma is a direct consequence of the Banach-Steinhaus theorem and can be found in Browder [5, Lemma 1].

LEMMA 1.1. *Let X be a Banach space, $\{x_n\}$ a sequence in X, and $\{\alpha_n\}$ a sequence of positive constants with $\alpha_n \to 0$ as $n \to \infty$. Fix $r > 0$ and assume that for every $h \in X^*$ with $\|h\| \leq r$ there exists a constant C_h such that $(h, x_n) \leq \alpha_n\|x_n\| + C_h$, for all n. Then the sequence $\{x_n\}$ is bounded.*

2. Main Results

We assume everywhere that X, X^* are reflexive and locally uniformly convex. Then the duality mapping J is single-valued, bicontinuous and of type (S_+).

We denote by Γ the set of all functions $\beta : R^+ \to R^+$ such that $\beta(r) \to 0$ as $r \to \infty$. We need the following basic lemma.

LEMMA 2.1. *Let X be a real reflexive Banach space, $A : X \to 2^{X^*}$ maximal monotone, and $B : X \to 2^{X^*}$ with*

$$(v, x - x_0) \geq -\beta(\|x\|)\|x\|^2$$

for some $\beta \in \Gamma$, $x_0 \in X$, all $x \in X$ with $\|x\|$ sufficiently large, and all $v \in Bx$. Furthermore, assume that B satisfies one of the following conditions.

1) B is quasi-bounded, finitely continuous and generalized pseudo-monotone.

2) B is pseudo-monotone.

Then $R(A + B + \lambda J) = X^*$ for all $\lambda > 0$.

PROOF. 1) We prove the result for $x_0 = 0$. If $x_0 \neq 0$, we consider instead the operators $\widetilde{A}(x) \equiv A(x+x_0)$, $\widetilde{B}(x) \equiv B(x+x_0)$ and, if necessary, $\widetilde{J}(x) \equiv J(x+x_0)$. Let Λ be the family of all finite-dimensional subspace F of X, partially ordered by inclusion. For $F \in \Lambda$, let $j_F : F \to X$ be the inclusion mapping of F into X, $j_{F^*} : X^* \to F^*$ the dual projection mapping of X^* onto F^*. Let

$$S_F \equiv j_{F^*}(A + B + \lambda J)j_F.$$

Since A is maximal monotone, it is well known that, for every $x \in X$, Ax is a bounded, closed and convex subset of X^*, and A is upper semicontinuous from the strong topology on X to the weak topology on X^*. It is easy to see that, for any $x \in F$, $S_F(x)$ is a bounded, closed and convex subset of F^*, and S_F is upper semicontinuous from F to 2^{F^*}. For every $x \in F$, $y \in S_F(x)$,

$$\begin{aligned}(y,x) &= (u_x + v_x + \lambda Jx, x)\\ &\geq -\|u_0\|\|x\| - \beta(\|x\|)\|x\|^2 + \lambda\|x\|^2,\\ &\to \infty\end{aligned}$$

as $\|x\| \to \infty$, where $u_x \in Ax$, $v_x \in Bx$ and $u_0 \in A(0)$. By Proposition 10 of Browder and Hess [6], $R(S_F) = F^*$.

To show that $R(A+B+\lambda J) = X^*$, we only have to show that $0 \in R(A+B+\lambda J)$. This is because, for every $y \in X$, the operator $A - y$ is also maximal monotone. Given $F \in \Lambda$, there exists $x_F \in F$ such that

$$j_{F^*}(u_F + v_F + \lambda Jx_F) = 0,$$

for some $u_F \in Ax_F$, $v_F \in Bx_F$. Thus,

$$\begin{aligned}0 &= (j_{F^*}(u_F + v_F + \lambda Jx_F), x_F)\\ &= (u_F + v_F + \lambda Jx_F, x_F)\\ &\geq -\|u_0\|\|x_F\| - \beta(\|x_F\|)\|x_F\|^2 + \lambda\|x_F\|^2,\end{aligned}$$

which implies that $\{x_F\}$ is uniformly bounded in F.

For any $F \in \Lambda$, let

$$V_F = \cup_{F' \in \Lambda, F \subset F'}\{x_{F'}\},$$

where $x_{F'}$ is a solution of the inclusion $S_{F'}(x) \ni 0$. The set V_F is bounded in F and $\{wcl(V_F)\}$ has the finite intersection property, where "wcl" denotes weak closure. Thus, $\cap_{F \in \Lambda}\{wcl(V_F)\} \neq \emptyset$.

Let $x_0 \in \cap_{F \in \Lambda}\{wcl(V_F)\}$. Choose $F \in \Lambda$ such that $x_0 \in F$. Then there exists a sequence $\{x_n\} \subset V_F$ and a sequence $\{F_n\} \subset \Lambda$ such that $F \subset F_n \in \Lambda$, $x_n \in F_n$, and $x_n \rightharpoonup x_0$. We have

$$j_{F_n^*}(u_n + v_n + \lambda Jx_n) = 0,$$

for some $u_n \in Ax_n$, and $v_n \in Bx_n$. Thus,

$$(u_n + v_n + \lambda Jx_n, x_n - x_0) = 0$$

and

$$(v_n, x_n) = -(u_n, x_n) - \lambda(Jx_n, x_n) \leq -(u_0, x_n) \leq \|u_0\|\|x_n\|,$$

for some $u_0 \in A(0)$. Since $x_n \rightharpoonup x_0$, $\|x_n\| \leq M$ for some $M > 0$. By the quasi-boundedness of B, $\{v_n\}$ is bounded. By passing to a subsequence, if necessary, we may assume that $v_n \rightharpoonup v_0$ for some $v_0 \in X^*$. Also,

$$\begin{aligned}(v_n, x_n - x_0) &= -(u_n, x_n - x_0) - \lambda(Jx_n, x_n - x_0)\\ &\leq -(u_0, x_n - x_0) - \lambda(Jx_0, x_n - x_0)\\ &\to 0\end{aligned}$$

as $n \to \infty$. By the fact that B is generalized pseudo-monotone, $v_0 \in Bx_0$ and $(v_n, x_n) \to (v_0, x_0)$. Moreover,

$$\begin{aligned}\lambda(Jx_n, x_n - x_0) &= -(u_n, x_n - x_0) - (v_n, x_n - x_0)\\ &\leq -(u_0, x_n - x_0) - (v_n, x_n) + (v_n, x_0)\\ &\to 0,\end{aligned}$$

as $n \to \infty$. Since J is type (S_+), we have $x_n \to x_0$ and $Jx_n \to Jx_0$. By the local boundedness of A at point x_0, $\{u_n\}$ is bounded in X^*. We may assume that $u_n \rightharpoonup u_0$ for some $u_0 \in X^*$. Since A is maximal monotone, $u_0 \in Ax_0$.

Now, we claim that $0 \in Ax_0 + Bx_0 + \lambda Jx_0$. Suppose this is not true. Then 0 is separated from nonempty, closed and convex set $Ax_0 + Bx_0 + \lambda Jx_0$ (since $u_0 + v_0 + \lambda Jx_0 \in Ax_0 + Bx_0 + \lambda Jx_0$), i.e., there exists $y_0 \in X$ such that

$$\inf_{z \in Ax_0 + Bx_0 + \lambda Jx_0} (z, y_0) > 0.$$

Choose $F \in \Lambda$ such that $x_0 \in F$ and $y_0 \in F$. Let $\{x_n\} \subset V_F$, $F \subset F_n \in \Lambda$, $x_n \in F_n$ and $x_n \rightharpoonup x_0$. As above, there exist a subsequence of $\{x_n\}$, which we call $\{x_n\}$ again, and $u_n \in Ax_n$, $v_n \in Bx_n$ such that $u_n + v_n + \lambda Jx_n \rightharpoonup u_0 + v_0 + \lambda Jx_0 \in Ax_0 + Bx_0 + \lambda Jx_0$. Since $y_0 \in F \subset F_n$, we have

$$(u_n + v_n + \lambda Jx_n, y_0) = 0.$$

Letting $n \to \infty$, we have

$$(u_0 + v_0 + \lambda Jx_0, y_0) = 0,$$

which is a contradiction.

2) The proof of this part follows directly from Theorem 3 of Browder and Hess [6]. ∎

Using Lemma 1.1 and Lemma 2.1, we have the following basic result for the range of the sum $A + B$.

THEOREM 2.1. *Let $A : X \supset D(A) \to 2^{X^*}$ be maximal monotone with $0 \in D(A)$, and let $B : X \to 2^{X^*}$ be quasi-bounded, finitely continuous and generalized pseudo-monotone. Let S be a subset of X^* such that: for every $s \in S$, there exist $x_s \in X$ and $\beta = \beta_s \in \Gamma$ such that*

$$(*) \qquad (u + v - s, x - x_s) \geq -\beta(\|x\|)\|x\|,$$

for all $x \in D(A)$ with $\|x\|$ sufficiently large, and all $u \in Ax$, $v \in Bx$. Moreover,

$$(**) \qquad (v, x - x_0) \geq -\beta_1(\|x\|)\|x\|^2,$$

for some $x_0 \in D(A)$, $\beta_1 \in \Gamma$, all x with $\|x\|$ sufficiently large, and all $v \in Bx$. Then $S \subset \overline{R(A + B)}$ and $\mathrm{int} S \subset \mathrm{int} R(A + B)$.

PROOF. We may assume $0 \in A(0)$. If this is not true, then we consider instead the operators $\widetilde{A}(x) \equiv A(x) - u_0$, $\widetilde{B}(x) \equiv B(x) + u_0$ and the sum $\widetilde{A} + \widetilde{B}$, where u_0 is a fixed point in $A(0)$. Fix $s \in S$ and consider the inclusion

$$Ax + Bx + \frac{1}{n}Jx \ni s.$$

For $\lambda > 0$, let $A_\lambda \equiv (A^{-1} + \lambda J^{-1})^{-1}$. Then $A_\lambda : X \to X^*$ is single-valued, bounded, maximal monotone and demicontinuous with $D(A_\lambda) = X$ and $0 = A_\lambda(0)$ (See Pascali and Sburlan [14]). By Lemma 2.1, there exists x_λ, such that

$$A_\lambda x_\lambda + Bx_\lambda + \frac{1}{n}Jx_\lambda \ni s,$$

or

$$A_\lambda x_\lambda + v_\lambda + \frac{1}{n}Jx_\lambda = s$$

for some $v_\lambda \in Bx_\lambda$.

We claim that there exists $\lambda_0 > 0$ such that $\{x_\lambda\}$ is bounded for $\lambda \in (0, \lambda_0)$. In fact,

$$\begin{aligned} 0 &= (A_\lambda x_\lambda + v_\lambda + \frac{1}{n}Jx_\lambda - s, x_\lambda - x_0) \\ &= (A_\lambda x_\lambda - A_\lambda x_0, x_\lambda - x_0) + (A_\lambda x_0, x_\lambda - x_0) + (v_\lambda, x_\lambda - x_0) \\ &\quad + (\frac{1}{n}Jx_\lambda - s, x_\lambda - x_0) \\ &\geq -\|A_\lambda x_0\|\|x_\lambda - x_0\| - \beta_1(\|x_\lambda\|)\|x_\lambda\|^2 + \frac{1}{n}\|x_\lambda\|^2 - \frac{1}{n}\|x_\lambda\|\|x_0\| \\ &\quad - \|s\|\|x_\lambda - x_0\|. \end{aligned}$$

Since $x_0 \in D(A)$, $\|A_\lambda x_0\| \leq \|u_0\|$ for any $u_0 \in Ax_0$ (see Pascali and Sburlan [14]) and $\beta(r) \to 0$ as $r \to \infty$, we have the boundedness of $\{x_\lambda\}$. Furthermore, since

$$(v_\lambda, x_\lambda) = -(A_\lambda x_\lambda, x_\lambda) - (\frac{1}{n}Jx_\lambda, x_\lambda) + (s, x_\lambda) \leq \|s\|\|x_\lambda\|,$$

by the quasi-boundedness of B, we have the boundedness of $\{v_\lambda\}$. Consequently, $\{A_\lambda x_\lambda\}$ is bounded as well. By Proposition 13 of Browder and Hess [6], we have $s \in R(A + B + \frac{1}{n}J)$.

Let x_n be a solution of $Ax + Bx + (1/n)Jx \ni s$ for each $n \in Z^+$. We have

$$(1/n)Jx_n = -u_n - v_n + s,$$

for some $u_n \in Ax_n$, $v_n \in Bx_n$. Assuming that $\|x_n\|$ is sufficiently large, we obtain

$$(\frac{1}{n}Jx_n, x_n - x_s) = -(u_n + v_n - s, x_n - x_s) \leq \beta(\|x_n\|)\|x_n\|,$$

which implies

$$\frac{1}{n}\|x_n\|^2 \leq \frac{1}{n}\|x_s\|\|x_n\| + \beta(\|x_n\|)\|x_n\|.$$

which says that $(1/n)Jx_n \to 0$. It follows that $s \in \overline{R(A + B)}$.

Now, we are going to show $intS \subset intR(A+B)$. To this end, fix $s \in intS$. Then there exists $r > 0$ such that for any $h \in \overline{B_r(0)}$ we have $s + h \in S$. Let x_n denote a solution of $Ax + Bx + (1/n)Jx \ni s$. We want to show that the sequence $\{x_n\}$ is bounded. Assume that this is not true. Then we may also assume, without loss of generality, that $\|x_n\| \to \infty$ as $n \to \infty$. We have

$$u_n + v_n + \frac{1}{n} Jx_n - (s+h) = -h, \text{ for some } u_n \in Ax_n, v_n \in Bx_n$$

and

$$(h, x_n - x_{s+h}) = -(u_n + v_n - (s+h), x_n - x_{s+h}) - \frac{1}{n}(Jx_n, x_n - x_{s+h}).$$

Hence

$$\begin{aligned}(h, x_n) =&(h, x_{s+h}) - \frac{1}{n}(Jx_n, x_n - x_{s+h}) \\ &- (u_n + v_n - (s+h), x_n - x_{s+h}) \\ &\le (h, x_{s+h}) + \beta(\|x_n\|)\|x_n\| \\ &- \frac{1}{n}(\|x_n\|^2 - \|x_n\|\|x_{s+h}\|) \\ &\le (h, x_{s+h}) + \beta(\|x_n\|)\|x_n\| \\ &- \frac{1}{n}(\|x_n\|^2 - \frac{1}{2}(\|x_n\|^2 + \|x_{s+h}\|^2)) \\ &\le (h, x_{s+h}) + \beta(\|x_n\|)\|x_n\| + \frac{1}{2n}\|x_{s+h}\|^2,\end{aligned}$$

for all large n, which, by Lemma 1.1, is a contradiction to the unboundedness of $\{x_n\}$. Thus, $\{x_n\}$ is bounded. Since X is reflexive, we may assume that $x_n \rightharpoonup \widetilde{x}_0 \in X$. We have

$$(v_n, x_n) = -(u_n, x_n) - \frac{1}{n}(Jx_n, x_n) + (s, x_n) \le \|s\|\|x_n\|,$$

which, by the quasi-boundedness of B, implies the boundedness of $\{v_n\}$. We assume that $v_n \rightharpoonup v_0 \in X^*$.

Since $v_n = -u_n - (1/n)Jx_n + s$, we have, for $x \in D(A),\ y \in Ax$,

$$(v_n, x_n - \widetilde{x}_0) = (v_n, x_n - x) - (v_n, \widetilde{x}_0 - x)$$

and

$$\begin{aligned}(v_n, x_n - x) &\le (v_n + u_n - y, x_n - x) \\ &= (s - \frac{1}{n}Jx_n - y, x_n - x) \\ &\to (s - y, \widetilde{x}_0 - x)\end{aligned}$$

and $(v_n, \widetilde{x}_0 - x) \to (v_0, \widetilde{x}_0 - x)$, where we have used the monotonicity of the operator A. Thus,

$$\begin{aligned}\limsup_{n\to\infty}(v_n, x_n - \widetilde{x}_0) &\le (s - y, \widetilde{x}_0 - x) - (v_0, \widetilde{x}_0 - x) \\ &= (s - v_0 - y, \widetilde{x}_0 - x),\end{aligned}$$

for every $x \in D(A),\ y \in Ax$. We also have $\inf\{(s - v_0 - y, \widetilde{x}_0 - x) : x \in D(A),\ y \in Ax\} \le 0$. If this is not true, then $(s - v_0 - y, \widetilde{x}_0 - x) \ge c > 0$ for every $x \in D(A),\ y \in Ax$. Since A is maximal monotone, we have $\widetilde{x}_0 \in D(A)$ and $s - v_0 \in A\widetilde{x}_0$. Let $x = \widetilde{x}_0,\ y = s - v_0$. Then $(s - v_0 - y, \widetilde{x}_0 - x) = 0$, which is a contradiction to the assumed inequality. It follows that $\limsup_{n\to\infty}(v_n, x_n - \widetilde{x}_0) \le 0$. Since B is generalized pseudo-monotone, we have $v_0 \in B\widetilde{x}_0$ and $(v_n, x_n) \to (v_0, \widetilde{x}_0)$.

Now, for any $x \in D(A)$, $y \in Ax$,

$$\begin{aligned}(s - v_0 - y, \widetilde{x}_0 - x) &= (s - y, \widetilde{x}_0 - x) - (v_0, \widetilde{x}_0 - x) \\ &= (s - y, \widetilde{x}_0 - x) - \lim_{n\to\infty} (v_n, x_n - x) \\ &= \lim_{n\to\infty} (s - y - v_n, x_n - x) \\ &= \lim_{n\to\infty} (u_n + \frac{1}{n} J x_n - y, x_n - x) \\ &= \lim_{n\to\infty} (u_n - y, x_n - x) + \lim_{n\to\infty} (\frac{1}{n} J x_n, x_n - x) \\ &\geq 0.\end{aligned}$$

By the fact that A is a maximal monotone, we have $\widetilde{x}_0 \in D(A)$ and $s - v_0 \in A\widetilde{x}_0$. Thus, $s \in A\widetilde{x}_0 + B\widetilde{x}_0$, which says that $\text{int}S \subset R(A+B)$. We conclude that $\text{int}S \subset \text{int}R(A+B)$. ∎

In Theorem 2.1, the condition $0 \in D(A)$ could be replaced by the following weaker condition: for x_0 in $(**)$, there exists $x' \in D(A)$ such that $x_0 - x' \in D(A)$.

Letting $S = X^*$ in theorem 2.1, we have the following result.

THEOREM 2.2. *Let $A : X \supset D(A) \to 2^{X^*}$ be maximal monotone with $0 \in D(A)$, and $B : X \to 2^{X^*}$ quasi-bounded, finitely continuous and generalized pseudo-monotone. Assume that for every $s \in X^*$ there exists $\beta = \beta_s \in \Gamma$ such that:*

$$(v - s, x - x_0) \geq -\beta(\|x\|)\|x\|,$$

for some fixed $x_0 \in D(A)$, all $x \in X$ with $\|x\|$ sufficiently large, and all $v \in Bx$. Then $R(A+B) = X^*$.

PROOF.. We let $S = X^*$ in Theorem 2.1. Then for every $s \in X^*$,

$$\begin{aligned}(u - v - s, x - x_0) &= (u - u_0, x - x_0) + (v - (s - u_0), x - x_0) \\ &\geq -\beta(\|x\|)\|x\|,\end{aligned}$$

for all $x \in D(A)$ with sufficiently large norm, $u \in Ax$, $v \in Bx$. Here $u_0 \in Ax_0$. We have $(*)$ of Theorem 2.1, and the condition $(**)$ is obviously satisfied. Thus, Theorem 2.2 follows from Theorem 2.1. ∎

Theorem 2.2 may be viewed as a complement of Browder's basic result on pseudomonotone perturbations of maximal monotone operators (see Browder [4], page 92, or Pascali and Sburlan [14]). Browder's result assumes B to be bounded and pseudomonotone. We only need B to be quasi-bounded and generalized pseudomonotone. Also, Browder assumed that B is coercive with respect to $s \in X^*$, i.e., $(v - s, x) > 0$ for all $x \in X^*$ with $\|x\|$ sufficiently large, and all $v \in Bx$. We only need $(v - s, x) \geq -\beta(\|x\|)\|x\|$. Not only $-\beta(\|x\|)\|x\|$ is negative, but also $-\beta(\|x\|)\|x\|$ could tend to $-\infty$ as $\|x\| \to \infty$. Because of this weak coercivity assumption, Theorem 2.2 does not follow from any related results in Browder and Hess [6].

Letting $S = R(A) + R(B)$ in Theorem 2.1, we obtain the following result.

THEOREM 2.3. *Let $A : X \supset D(A) \to 2^{X^*}$ be maximal monotone with $0 \in D(A)$, and $B : X \to 2^{X^*}$ quasi-bounded, finitely continuous and generalized pseudo-monotone. Assume that for every $(x_1, x_2) \in D(A) \times X$ there exists $\beta \in \Gamma$ such that*

$$(v - v_2, x - x_1) \geq -\beta(\|x\|)\|x\|,$$

for all $x \in X$ with $\|x\|$ sufficiently large, $v \in Bx$ and $v_2 \in Bx_2$. Then $\overline{R(A+B)} = \overline{R(A) + R(B)}$ and $\mathrm{int}R(A+B) = \mathrm{int}(R(A) + R(B))$.

PROOF. As in Theorem 2.1, let $S = R(A) + R(B)$. Given $s \in S$, we have $s = u_s + v_s$, for some $u_s \in Ax_s$, and $v_s \in By_s$ with some $x_s \in D(A)$ and $y_s \in X$. Given $x \in D(A)$ with $\|x\|$ sufficiently large, $u \in Ax$ and $v \in Bx$, we have

$$\begin{aligned}(u - v - s, x - x_s) &= (u + v - u_s - v_s, x - x_s) \\ &= (u - u_s, x - x_s) + (v - v_s, x - x_s) \\ &\geq 0 - \beta(\|x\|)\|x\|.\end{aligned}$$

For some $x_1 \in D(A)$, $v_1 \in Bx_1$,

$$\begin{aligned}(v, x - x_1) &= (v - v_1, x - x_1) + (v_1, x - x_1) \\ &\geq -\beta(\|x\|)\|x\| - \|v_1\|\|x - x_1\| \\ &\geq -\beta_1(\|x\|)\|x\|^2,\end{aligned}$$

for all $x \in X$ with $\|x\|$ sufficiently large, and some $\beta_1 \in \Gamma$. By Theorem 2.1, we have $\overline{R(A) + R(B)} \subset \overline{R(A+B)}$ and $\mathrm{int}(R(A) + R(B)) \subset \mathrm{int}R(A+B)$. Trivially, $R(A+B) \subset R(A) + R(B)$. The proof is complete. ∎

Theorem 2.3 is an extension of Theorem 2 of Browder [5]. In Browder's result B is bounded and pseudo-monotone.

THEOREM 2.4. *Let $A : X \supset D(A) \to 2^{X^*}$ be maximal monotone with $0 \in D(A)$ and $B : X \to 2^{X^*}$ quasi-bounded, finitely continuous and generalized pseudo-monotone. Assume that for every $x_0 \in D(A)$ there exists $\beta = \beta_{x_0} \in \Gamma$ such that*

$$(***) \qquad (v, x - x_0) \geq -\beta(\|x\|)\|x\|,$$

for all $x \in X$ with $\|x\|$ sufficiently large, and $v \in Bx$. Then $\overline{R(A)} \subset \overline{R(A+B)}$ and $\mathrm{int}R(A) \subset \mathrm{int}R(A+B)$.

PROOF. Let $S = R(A)$ in Theorem 2.1. Then, for $s \in R(A)$, there exists $x_s \in D(A)$ such that $s \in Ax_s$. Thus, for every $x \in D(A)$ with sufficiently large $\|x\|$ and every $u \in Ax$ and $v \in Bx$, we have

$$\begin{aligned}(u + v - s, x - x_s) &= (y - s, x - x_s) + (Bx, x - x_s) \\ &\geq -\beta(\|x\|)\|x\|.\end{aligned}$$

Condition $(**)$ of Theorem 2.1 follows from $(***)$. The conclusion follows from Theorem 2.1. ∎

The assumption that $0 \in D(A)$ was made in Theorem 2.4 for convenience. It can be removed by considering new operators $\widetilde{A}$, $\widetilde{B}$ with translated domains. Theorem 2.1 to 2.4 are extensions of various results of Guan and Kartsatos [8]. In [8], B is assumed to be bounded and pseudomonotone. Just as in [8], it is quite clear that a greater variety of results can be derived from Theorem 2.1.

In the following result we assume that B is pseudo-monotone.

THEOREM 2.5. *Let* $A : X \supset D(A) \to 2^{X^*}$ *be maximal monotone and bounded, with* $0 \in D(A)$, *and* $B : X \to 2^{X^*}$ *pseudo-monotone. Let* S *be a subset of* X^* *such that: for every* $s \in S$, *there exist* $x_s \in X$ *and* $\beta = \beta_s \in \Gamma$ *such that*

$$(*) \qquad (u + v - s, x - x_s) \geq -\beta(\|x\|)\|x\|,$$

for all $x \in D(A)$ *with* $\|x\|$ *sufficiently large and all* $u \in Ax$, $v \in Bx$. *Moreover, assume that*

$$(**) \qquad (v, x - x_0) \geq -\beta_1(\|x\|)\|x\|^2,$$

for some $x_0 \in D(A)$, $\beta_1 \in \Gamma$ *and all* x *with* $\|x\|$ *sufficiently large and* $v \in Bx$. *Then* $S \subset \overline{R(A+B)}$ *and* $\mathrm{int}S \subset \mathrm{int}R(A+B)$.

PROOF. As in the proof of Theorem 2.1, we may assume $0 \in A(0)$. Let $s \in S$ and we consider the inclusion

$$Ax + Bx + \frac{1}{n}Jx \ni s.$$

Let $\lambda > 0$, $A_\lambda = (A^{-1} + \lambda J^{-1})^{-1}$. Lemma 2.1, b), insures the solvability of

$$A_\lambda x + Bx + \frac{1}{n}Jx \ni s.$$

Let x_λ denote a solution of the above inclusion. Then

$$A_\lambda x_\lambda + v_\lambda + \frac{1}{n}Jx_\lambda = s,$$

for some $v_\lambda \in Bx_\lambda$. As in the proof of Theorem 2.1, there exists $\lambda_0 > 0$ such that $\{x_\lambda\}$ is bounded for $\lambda \in (0, \lambda_0)$. Let $M > 0$ be such that $\|x_\lambda\| \leq M$, $\lambda \in (0, \lambda_0)$. We define $R_\lambda : X \to D(A)$ by

$$J(x - R_\lambda x) = \lambda A_\lambda x$$

and notice that we have $A_\lambda x \in AR_\lambda x$ (see Pascal and Sburlan [14]). For every $x_0 \in D(A)$, $u_0 \in A(x_0)$ and $x \in X$ we have

$$\begin{aligned}(u_0, R_\lambda x - x_0) &\leq (A_\lambda x, R_\lambda x - x_0) \\ &= -\frac{1}{\lambda}(J(R_\lambda x - x), R_\lambda x - x_0) \\ &= -\frac{1}{\lambda}(J(R_\lambda x - x), R_\lambda x - x) - \frac{1}{\lambda}(J(R_\lambda x - x), x - x_0)\end{aligned}$$

by the monotonicity of A. This is equivalent to

$$\|R_\lambda x - x\|^2 \leq -\lambda(u_0, R_\lambda x - x_0) - (J(R_\lambda x - x), x - x_0),$$

which implies the uniformly boundedness of $\{R_\lambda x - x\}$ for $\lambda \in (0, \lambda_0)$ and x in any bounded subset of X. Thus, for $\lambda \in (0, \lambda_0)$ and $x \in X$ with $\|x\| < M$, there exists $K_M > 0$ such that $\|R_\lambda x - x\| \leq K_M$. This says that

$$\|R_\lambda x_\lambda\| \leq \|R_\lambda x_\lambda - x_\lambda\| + \|x_\lambda\| \leq K_M + M.$$

Since $A_\lambda x \in AR_\lambda x$ and A is bounded, $\{A_\lambda x_\lambda\}$ is also bounded. Consequently, by

$$A_\lambda x_\lambda + v_\lambda + \frac{1}{n} J x_\lambda = s,$$

we have the boundedness of $\{v_\lambda\}$. It is not difficult to see that $B + \frac{1}{n}J$ is pseudo-monotone. As such, it is also generalized pseudo-monotone. By Proposition 13 of Browder and Hess [6], we have $s \in R(A + B + \frac{1}{n}J)$.

The rest of proof will be same as in the proof of Theorem 2.1. ∎

Theorem 2.5 show that in results like Theorem 2.1 of Guan and Kartsatos [8], and other theorems which follow from that, the boundedness assumption of B may be replaced by the boundedness assumption of A. In this vein, we get Theorem 2.6 below, which parallels the result of Browder in [5, p.92].

THEOREM 2.6. *Let $A : X \supset D(A) \to 2^{X^*}$ be maximal monotone and bounded with $0 \in D(A)$, and $B : X \to 2^{X^*}$ pseudo-monotone. Assume that for every $s \in X^*$, there exists $\beta = \beta_s \in \Gamma$ such that*

$$(v - s, x - x_0) \geq -\beta(\|x\|)\|x\|,$$

for some fixed $x_0 \in D(A)$, all $x \in X$ with $\|x\|$ sufficiently large and all $v \in Bx$. Then $R(A + B) = X^*$. ∎

If the inequality in Theorem 2.6 is assumed to hold for all x, and not just for all x with sufficiently large $\|x\|$, then we can use a simple argument by Browder and Hess [8] in which the boundedness of A is replaced by its strong quasi-boundedness. This is the content of the next result.

THEOREM 2.7. *Let $A : X \supset D(A) \to 2^{X^*}$ be maximal monotone and strongly quasi-bounded, with $0 \in D(A)$, and $B : X \to 2^{X^*}$ pseudo-monotone. Assume that for every $s \in X^*$, there exists a bounded $\beta = \beta_s \in \Gamma$ such that*

$$(v - s, x) \geq -\beta(\|x\|)\|x\|,$$

for all $x \in X$ and all $v \in Bx$. Then $R(A + B) = X^*$.

PROOF. As in the proof of Theorem 2.5, we have the boundedness of $\{x_\lambda\}$ and $\{R_\lambda x_\lambda\}$. Since

$$J(x_\lambda - R_\lambda x_\lambda) = \lambda A_\lambda x_\lambda,$$

we have

$$(\lambda A_\lambda x_\lambda, x_\lambda - R_\lambda x_\lambda) \geq 0$$

and

$$\begin{aligned} (A_\lambda x_\lambda, R_\lambda x_\lambda) &\leq (A_\lambda x_\lambda, x_\lambda) \\ &\leq -(v_\lambda - s, x_\lambda) - \frac{1}{n}(J x_\lambda, x_\lambda) \\ &\leq \beta(\|x_\lambda\|)\|x_\lambda\| \\ &\leq M, \end{aligned}$$

for some $M > 0$, where we have used the fact that $\{x_\lambda\}$ and β are bounded. Since $A_\lambda x_\lambda \in AR_\lambda x_\lambda$ and $\{R_\lambda x_\lambda\}$ is bounded, the strong quasi-boundedness of A implies that $\{A_\lambda x_\lambda\}$ is bounded. The rest of the proof follows from the Proof of Theorem 2.5. It is therefore omitted. ∎

Form the Theorem 2.5, we also have

THEOREM 2.8. *Let $B : X \to 2^{X^*}$ be pseudo-monotone. Assume that for every $s \in X^*$ there exist $x_s \in X$ and $\beta = \beta_s \in \Gamma$ such that*

$$(y - s, x - x_s) \geq -\beta(\|x\|)\|x\|,$$

for all $x \in X$ with $\|x\|$ sufficiently large and all $y \in Bx$. Then $R(B) = X^*$.

PROOF. Let $S = X^*$ and $Ax = 0$ in Theorem 2.5. ■

Theorem 2.8 is an improvement of Theorem 3 of Browder and Hess [6]. In [5], B is coercive, i.e., $(u, x) \geq q(\|x\|)\|x\|$ for all $x \in X$ and $u \in Bx$ with $q(r) \to \infty$ as $r \to \infty$.

THEOREM 2.9. *Let $A : X \to X^*$ be single-valued, demicontinuous and quasi-monotone. Assume, further, that $A(\overline{B(0,r)})$ is closed, for any ball $B(0,r) \subset X$, and that A satisfies one of the following conditions:*

1) *for every $s \in X^*$ there exists $x_s \in X$ and $\beta = \beta_s \in \Gamma$ such that*

$$(Ax - s, x - x_s) \geq -\beta(\|x\|)\|x\|,$$

for all $x \in X$ with $\|x\|$ sufficiently large;

2) *for every $s \in X^*$ there exist $x_s \in X$, $K(s) > 0$ such that*

$$(Ax - s, x - x_s) \geq -K(s)\|x\|,$$

for all $x \in X$ with $\|x\|$ sufficiently large, and A^{-1} is bounded.

Then $R(A) = X^*$.

PROOF. Under the assumptions of the theorem, $A + \frac{1}{n}J$ is of type (S_+) (see, for example, Pascali and Sburlan [14]). It is easy to see that this operator is also pseudo-monotone. For any $s \in X^*$, by Theorem 2.8, there exists $\{x_n\} \subset X$ such that

$$Ax_n + \frac{1}{n}Jx_n = s.$$

We want to show $\{x_n\}$ is bounded.

If 1) holds, then, as in the proof of Theorem 2.1,

$$Ax_n + \frac{1}{n}Jx_n - (s + h) = -h,$$

for any $h \in X$, and

$$(h, x_n - x_{s+h}) = -(Ax_n - (s + h), x_n - x_{s+h}) - \frac{1}{n}(Jx_n, x_n - x_{s+h}).$$

Hence

$$\begin{aligned}
(h, x_n) =& (h, x_{s+h}) - \frac{1}{n}(Jx_n, x_n - x_{s+h}) - (Ax_n - (s + h), x_n - x_{s+h}) \\
&\leq (h, x_{s+h}) + \beta(\|x_n\|)\|x_n\| \\
&- \frac{1}{n}(\|x_n\|^2 - \|x_n\|\|x_{s+h}\|) \\
&\leq (h, x_{s+h}) + \beta(\|x_n\|)\|x_n\| \\
&- \frac{1}{n}(\|x_n\|^2 - \frac{1}{2}(\|x_n\|^2 + \|x_{s+h}\|^2)) \\
&\leq (h, x_{s+h}) + \beta(\|x_n\|)\|x_n\| + \frac{1}{2n}\|x_{s+h}\|^2,
\end{aligned}$$

By Lemma 1.1, we have the boundedness of $\{x_n\}$.

If 2) holds, then

$$\begin{aligned}(\frac{1}{n}Jx_n, x_n - x_s) &= -(Ax_n - s, x_n - x_s)\\ &\leq K(s)\|x_n\|.\end{aligned}$$

Hence,

$$\frac{1}{n}\|x_n\|^2 \leq \frac{1}{n}\|x_n\|\|x_s\| + K(s)\|x_n\|,$$

$$(\frac{1}{n}\|x_n\|)^2 \leq \frac{1}{n}\|x_s\|\frac{1}{n}\|x_n\| + \frac{1}{n}K(s)\|x_n\| \leq M_1\frac{1}{n}\|x_n\| + M_2,$$

for some $M_1 > 0$ and $M_2 > 0$. This implies that $\{\frac{1}{n}x_n\}$ is bounded. The boundedness of A^{-1} and

$$Ax_n = s - \frac{1}{n}x_n,$$

imply the boundedness of $\{x_n\}$. With the boundedness of $\{x_n\}$, say $\|x_n\| < M$ for some $M > 0$, we have $\frac{1}{n}Jx_n \to 0$ and $Ax_n \to s$. Since $A(\overline{B(0,M)})$ is closed, there exists $x_0 \in \overline{B(0,M)}$ such that $Ax_0 = s$. ∎

Theorem 2.9 complements Theorem 5.1.2 of Pascali and Sburlan [14, p. 227] which assumes that A^{-1} is locally bounded and $(Ax,x) \geq 0$ for large $\|x\|$. In Theorem 2.9, A^{-1} is bounded, but the inequality in 2) is much weaker. In fact, if we take $x_s = 0$, then $(Ax, x) \geq -K\|x\|$, for a constant $K > 0$, is sufficient for the inequality to hold.

If $A : X \to X^*$ is pseudo-monotone, then A is quasi-monotone and $A(\overline{B(0,r)})$ is closed for any $B(0,r) \subset X$. In Theorem 2.8, if we replace $\beta(\|x\|)$ by a constant K_s' and assume, in addition, that B^{-1} is bounded, then we also have $R(B) = X^*$.

3. Perturbations of Maximal Monotone Operators Involving Compactness

The following theorem was given by Hirano and Kalinde in [10].

THEOREM 3.1. *Let $A : X \supset D(A) \to 2^X$ be m-accretive with $(A+I)^{-1}$ compact. Let $B : D(A) \to X$ be bounded and such that the operator $B(I+\lambda A)^{-1} : X \to X$ is condensing for some $\lambda > 0$. Let $p \in X$ and assume that there exist positive constants b, r and $z \in D(A)$ such that $\|z\| < b$ and*

$$(u + Bx - p, j) \geq 0,$$

for every $x \in D(A)$ with $\|x\| \geq b$, $u \in Ax$ and $j \in J(x-z)$. Then $p \in R(A+B)$. ∎

Kartsatos improved this theorem in [12] by assuming that the above boundary condition holds only on the boundary of a certain bounded, open set, and that B is only bounded on a certain subset of X. This result of Kartsatos has also been extended to the case of maximal monotone operators A in Theorem 7 of [11]. We quote the latter below because we need to apply it to the range of sums result (Theorem 3.3 below).

THEOREM 3.2. *Let $A : X \supset D(A) \to 2^{X^*}$ be maximal monotone and $B : D(A) \to X^*$. Let $(A+J)^{-1}$ be compact. Let $G \subset X$ be open, bounded and such that, for some $z \in D(A) \cap G$ and some $v^* \in Az$,*

$$(*) \qquad Ax - v^* \not\ni 0 \text{ and } (u^* + Bx, x - z) > 0, \ (x, u^*) \in (D(A) \cap \partial G) \times Ax.$$

Assume, further, that the operator $B(\lambda A + J)^{-1}$ is compact, where λ is a fixed positive constant, and the set $B(D(A) \cap \overline{G})$ is bounded. Then $p \in (A+B)(D(A) \cap \overline{G})$. ∎

We are now ready for a result on ranges of sums that does not assume that C is a continuous mapping. We do assume instead that the operator $A+B$ is closed.

THEOREM 3.3. *Let $A : X \supset D(A) \to 2^{X^*}$ be maximal monotone with $(A+J)^{-1}$ compact. Let $B : D(A) \to X^*$ be bounded and such that $B(\lambda A + J)^{-1}$ is compact, where λ is a fixed positive constant. Let $S \subset X^*$ be such that: for every $s \in S$ there exist $z \in D(A)$, $\beta = \beta_s \in \Gamma$ such that*

$$(1) \qquad (u^* + Bx - s, x - z) \geq -\beta(\|x\|)\|x\|,$$

for every $x \in D(A)$ with $\|x\|$ sufficiently large and every $u^* \in Ax$. Then $S \subset \overline{R(A+B)}$. If, moreover, $A+B$ is closed, then $\text{int}S \subset R(A+B)$.

PROOF. We consider the operators $\widetilde{A}x \equiv Ax + (1/n)Jx$ and $\widetilde{B}x \equiv Bx - (1/n)Jz$. It is easy to see that $\widetilde{A}$ is a maximal monotone operator. From the paper of the authors [8, discussion preceding Theorem 3.1] we also know that the resolvent operator

$$(\widetilde{A} + J)^{-1} = \left(A + \frac{n+1}{n}J\right)^{-1}$$

is compact. The point x_0 in that reference can be taken to be zero. To see that $\widetilde{B}(\lambda\widetilde{A} + J)^{-1}$ is also compact, we observe first that from Lemma 3 of Kartsatos [11] we have

$$(2) \qquad y \equiv J_\lambda x = (I + \lambda J^{-1}A)^{-1}x = (\lambda A + J)^{-1}(Jy + J(x-y))$$

and

$$(3) \qquad x \equiv (\lambda A + J)^{-1}x^* = (I + \lambda J^{-1}A)^{-1}(x + J^{-1}(x^* - Jx)).$$

In addition, from the resolvent identity for maximal monotone operators we have

$$(4) \qquad J_\mu x = J_\lambda\left(\frac{\lambda}{\mu}x + \frac{\mu - \lambda}{\mu}J_\mu x\right).$$

Consequently, we compute

$$
\begin{aligned}
x \equiv (\lambda \widetilde{A} + J)^{-1} x^* &= \left(\lambda A + \frac{\lambda + n}{n} J\right)^{-1} x^* \\
&= \left(\frac{\lambda n}{\lambda + n} A + J\right)^{-1} \left(\frac{n}{\lambda + n} x^*\right) \\
&= \left(I + \frac{\lambda n}{\lambda + n} J^{-1} A\right)^{-1} \left(x + J^{-1}\left(\frac{n}{\lambda + n} x^* - Jx\right)\right) \qquad (5)\\
&= J_\lambda \left(\frac{\lambda + n}{n} u - \frac{\lambda}{n} J_{\frac{\lambda n}{\lambda + n}} u\right) \\
&= \left(I + \lambda J^{-1} A\right)^{-1} v \\
&= (\lambda A + J)^{-1}(Jx + J(v - x)),
\end{aligned}
$$

where

$$u \equiv x + J^{-1}\left(\frac{n}{\lambda + n} x^* - Jx\right)$$

and

$$v \equiv \left(\frac{\lambda + n}{n} u - \frac{\lambda}{n} J_{\frac{\lambda n}{\lambda + n}} u\right).$$

All identities (2)-(4) were used in (5). It is easy to see from (5) that the operator

$$\widetilde{B}(\lambda \widetilde{A} + J)^{-1} = B(\lambda \widetilde{A} + J)^{-1} - (1/n) Jz$$

is compact.

We are now going to establish a boundary condition like $(*)$ of Theorem 3.2, but for the operators $\widetilde{A}$, $\widetilde{B}$. In fact, we observe that

$$
\begin{aligned}
(\widetilde{u}^* + \widetilde{B}x - s, x - z) &= (u^* + \widetilde{B}x - s + (1/n)(Jx - Jz), x - z) \\
&\geq (1/n)(Jx - Jz, x - z) - \beta(\|x\|)\|x\| \qquad (6)\\
&\geq (1/n)(\|x\| - \|z\|)^2 - \beta(\|x\|)\|x\|,
\end{aligned}
$$

for all $x \in D(A)$ with $\|x\|$ sufficiently large and all $\widetilde{u}^* \in \widetilde{A}x$, where $\widetilde{u}^* = u^* + (1/n)Jx$ with $u^* \in Ax$. It is easy to see that (6) implies

$$(\widetilde{u}^* + \widetilde{B}x - s, x - z) > 0, \qquad (7)$$

for all $x \in D(A)$ with sufficiently large norm and all $\widetilde{u}^* \in \widetilde{A}x$. Thus we may assume that (7) holds for all $x \in D(A)$ with $\|x\| \geq Q$, where Q is a positive constant such that $Q > \|z\|$. Then (7) holds for all $x \in \partial B_Q(0)$. We also observe that $u^* \in \widetilde{A}x$, $v^* \in \widetilde{A}y$ and $u^* = v^*$ imply that $x = y$. This follows from the strict monotonicity of the duality mapping. Thus, if $v^* \in \widetilde{A}z$, then it is impossible to have $v^* \in Ax$, for some $x \in \partial B_Q(0)$, because z lies in the ball $B_Q(0)$. It follows that all the assumptions of Theorem 3.2 are satisfied with $G = B_Q(0)$. Consequently, the inclusion

$$Ax + Bx + (1/n)(Jx - Jz) \ni s \qquad (8)$$

has a solution $x_n \in D(A)$. It should be noted that the point $z \in D(A)$ does not depend on the integer n.

Let the sequence $\{x_n\}$ solve (8). It is obvious that if $\{x_n\}$ has a bounded subsequence, then $s \in \overline{R(A+B)}$. Otherwise, we have that $\|x_n\| \to \infty$ as $n \to \infty$. This and

$$Ax_n + Bx_n + (1/n)(Jx_n - Jz) \ni s \tag{9}$$

imply

$$(1/n)(\|x_n\| - \|z\|)^2 \leq \beta(\|x_n\|)\|x_n\|,$$

for all large n. This in turn implies that $(1/n)\|x_n\| \to 0$ as $n \to \infty$ and finishes the proof of the first part of our conclusion.

To show the second part of our conclusion, we assume that $s \in \text{int}S$ and let $\{x_n\}$ solve (9). If $\{x_k\}$ is a bounded subsequence of $\{x_n\}$, then

$$x_k = (A+J)^{-1}(s + [(k-1)/k]Jx_k + (1/k)Jz - Bx_k)$$

implies that $\{x_k\}$ lies in a compact set. We may thus assume that $x_k \to x_0$ as $k \to \infty$. Then since $v_k + Bx_k \to s$, for some $v_k \in Tx_k$, and $A+B$ is closed, we have $x_0 \in D(A)$ and $Ax_0 + Bx_0 \ni s$. Now, let us assume that $\|x_n\| \to \infty$ as $n \to \infty$. Since $s \in \text{int}S$, there exists $r > 0$ such that $B_r(s) \subset S$. Let $h \in B_r(0) \subset X^*$. Then $s + h \in S$ and from

$$Ax_n + Bx_n + (1/n)(Jx_n - Jz) - (s+h) \ni h$$

and (1) we obtain

$$(x_n - z, h) \leq \beta(\|x_n\|)\|x_n\|,$$

where $\beta = \beta_{s+h} \in \Gamma$ and n is sufficiently large. Thus, Lemma 1.1 implies that $\{x_n - z\}$, hence $\{x_n\}$, is a bounded sequence. This contradiction completes the proof of the theorem. ∎

References

1. V. Barbu, *Nonlinear Semigroups and Differential Equations in Banach Spaces*, Noordhoff Int. Publ., Leyden (The Netherlands), 1975.
2. H. Brézis, *Opérateurs Maximaux Monotones et Semi-groupes de Contractions dans les Espaces de Hilbert*, Math. Studies 5, North Holland, Amsterdam, 1973.
3. H. Brézis and A. Haraux, *Image d'une somme d'opérateurs monotones et applications*, Israel J. Math. **23** (1976,), 165 - 186.
4. F. E. Browder, *Nonlinear operators and nonlinear equations of evolution in Banach spaces*, Proc. Sympos. Pure Math. **18 (2)** (1976).
5. F. E. Browder, *On a principle of H. Brézis and its applications*, J. Func. Anal. **25** (1977), 356-365.
6. F. E. Browder and P. Hess, *Nonlinear Mappings of Monotone Type in Banach Spaces*, J. Func. Anal. **11** (1972), 251-294.
7. I. Cioranescu, *Geometry of Banach Spaces, Duality Mappings and Nonlinear Problems*, Kluwer Acad. Publ., Boston, 1990.
8. Z. Guan and A. G. Kartsatos, *Ranges of Perturbed Maximal Monotone and M-Accretive Operators in Banach Spaces*, Trans. Amer. Math. Soc. **347** (1995), 2403-2435.
9. Z. Guan and A. G. Kartsatos, *Solvability of nonlinear equations with coercivity generated by compact perturbations of m-accretive operators in Banach spaces*, Houston J. Math. **21** (1995), 149-188.
10. N. Hirano and A. K. Kalinde, *On perturbations of m-accretive operators in Banach spaces*, Proc. Amer. Math. Soc. (to appear).
11. A. G. Kartsatos, *New results in the perturbation theory of maximal monotone and m-accretive operators in Banach spaces*, Trans. Amer. Math. Soc. **348** (1996), 1663-1707.
12. A. G. Kartsatos, *On the perturbation theory of m-accretive operators in Banach spaces*, Proc. Amer. Math. Soc. **124** (1996), 1811-1820.

13. V. Lakshmikantham and S. Leela, *Nonlinear Differential Equations in Abstract Spaces*, Pergamon Press, Oxford, 1981.
14. D. Pascali and Sburlan, *Nonlinear Mappings of Monotone Type*, Sijthoff and Noordhoof, Bucharest, 1978.
15. S. Reich, *The range of sums of accretive and monotone operators*, J. Math. Anal. Appl. **68** (1979), 310 - 317.
16. E. Zeidler, *Nonlinear Functional Analysis and its Applications*, **II/B**, Springer-Verlag, New York, 1990.

DEPARTMENT OF MATHEMATICS, UNIVERSITY OF WISCONSIN-EAU CLAIRE, EAU CLAIRE, WISCONSIN 54702
E-mail address: guanz@uwec.edu

DEPARTMENT OF MATHEMATICS, UNIVERSITY OF SOUTH FLORIDA, TAMPA, FLORIDA 33620-5700
E-mail address: hermes@gauss.math.usf.edu

Contemporary Mathematics
Volume **204**, 1997

Determining Projections and Functionals for Weak Solutions of the Navier-Stokes Equations

M. J. Holst and E. S. Titi

ABSTRACT. In this paper we prove that an operator which projects weak solutions of the two- or three-dimensional Navier-Stokes equations onto a finite-dimensional space is determining if it annihilates the difference of two "nearby" weak solutions asymptotically, and if it satisfies a single appoximation inequality. We then apply this result to show that the long-time behavior of weak solutions to the Navier-Stokes equations, in both two- and three-dimensions, is determined by the long-time behavior of a finite set of bounded linear functionals. These functionals are constructed by local surface averages of solutions over certain simplex volume elements, and are therefore well-defined for weak solutions. Moreover, these functionals define a projection operator which satisfies the necessary approximation inequality for our theory. We use the general theory to establish lower bounds on the simplex diameters in both two- and three-dimensions. Furthermore, in the three dimensional case we make a connection between their diameters and the Kolmogoroff dissipation small scale in turbulent flows.

1. Introduction

Consider a viscous incompressible fluid in $\Omega \subset \mathbb{R}^d$, where Ω is an open bounded domain with Lipschitz continuous boundary, and where $d = 2$ or $d = 3$. Given the kinematic viscosity $\nu > 0$, and the vector volume force function $f(x,t)$ for each $x \in \Omega$ and $t \in (0,\infty)$, the governing Navier-Stokes equations for the fluid velocity vector $u = u(x,t)$ and the scalar pressure field $p = p(x,t)$ are:

$$\frac{\partial u}{\partial t} - \nu\Delta u + (u\cdot\nabla)u + \nabla p = f \quad \text{in} \quad \Omega \times (0,\infty), \tag{1.1}$$

$$\nabla \cdot u = 0 \quad \text{in} \quad \Omega \times (0,\infty). \tag{1.2}$$

Also provided are initial conditions $u(0) = u_0$, as well as appropriate boundary conditions on $\partial\Omega \times (0,\infty)$.

1991 *Mathematics Subject Classification.* Primary 35B40, 35Q30; Secondary 65D05, 65M60.

The first author was supported in part by the NSF under Cooperative Agreement No. CCR-9120008. The work of the second author was supported in part by NSF Grant No. DMS-93-08774 and by the University of California-Irvine Graduate Council Research Fund. The second author would like to thank the CNLS and the IGPP at the Los Alamos National Laboratory for their kind hospitality while this work was completed.

The notion of *determining modes* for the Navier-Stokes equations was first introduced in [**FP67**] as an attempt to identify and estimate the number of degrees of freedom in turbulent flows (cf. [**CFMT85**] for a thorough discussion of the role of determining sets in turbulence theory). This concept later led to the notion of *Inertial Manifolds* [**FST88**]. An estimate of the number of determining modes was given in [**FMTT83**] and later improved in [**JTi93**]. The notion of *determining nodes*, and other more general determining concepts, were introduced in [**FT83**]. In [**FT84**] the notion of determining nodes was discussed in detail, and estimates for their number were reported in [**JTi92b**], and later improved in [**JTi93**]. In [**FTi91**] (see also [**JTi92a**]) the concept of *determining volume elements* was presented, and a connection was established between this concept and Inertial Manifolds. A generalized and unified theory of all of the above was recently presented in [**CJTi95, CJTi97**].

Bounds on the number of determining modes, nodes, and volumes are usually phrased in terms of a generalized *Grashof number*, which is defined for the two-dimensional Navier-Stokes equations as:

$$Gr = \frac{\rho^2 F}{\nu^2} = \frac{F}{\lambda_1 \nu^2},$$

where λ_1 is the smallest eigenvalue of the Stokes operator and $\rho = \sqrt{\lambda_1}$ is the related (best) Poincaré constant. Here, $F = \limsup_{t\to\infty}(\int_\Omega |f(x,t)|^2)^{1/2}$ if $f \in L^2(\Omega)$ for almost every t, or $F = \limsup_{t\to\infty} \sqrt{\lambda_1}\|f\|_{H^{-1}(\Omega)}$ if $f \in H^{-1}(\Omega)$ for almost every t.

The best known estimate for the determining set size for the two-dimensional Navier-Stokes equations with periodic boundary conditions and H^2-regular solutions is of order Gr [**JTi93**]. In obtaining their estimate, the authors relied on the fact that the domain had no physical boundaries to shed vorticity, which made available some convenient properties of H^2-regular solutions. However, in the two-dimensional case with no-slip boundary conditions, to our knowledge the best estimate on the cardinal of any determining set (modes, nodes, or volumes) that can be obtained is of order Gr^2, even for H^2-regular solutions.

Due to the Sobolev Imbedding Theorem $H^2 \hookrightarrow C^0$ (which holds in dimensions 1, 2, and 3), or rather due to the failure of the imbedding $H^1 \hookrightarrow C^0$ in dimensions 2 and 3, determining node analysis is necessarily restricted to H^2-regular solutions to make sense of point-wise values. However, when talking about determining modes or volume elements, it is sufficient for functions to be H^1-regular, so that these concepts also make sense for weaker solutions. To construct a general analysis framework for the case of weak H^1 solutions, we can begin by defining notions of *determining projections* and *determining functionals* for weak solutions. (The standard spaces H, V, and V' are reviewed fully in §2.)

DEFINITION 1.1. *Let $f(t), g(t) \in V'$ be any two forcing functions satisfying*

$$\lim_{t\to\infty} \|f(t) - g(t)\|_{V'} = 0, \tag{1.3}$$

and let $u, v \in V$ be corresponding weak solutions to (1.1)–(1.2). The projection operator $R_N : V \mapsto V_N \subset L^2(\Omega)$, $N = \dim(V_N) < \infty$, is called a determining projection *for weak solutions of the d-dimensional Navier-Stokes equations if*

$$\lim_{t\to\infty} \|R_N(u(t) - v(t))\|_{L^2(\Omega)} = 0, \tag{1.4}$$

implies that

$$\lim_{t\to\infty} \|u(t) - v(t)\|_H = 0. \tag{1.5}$$

Given a basis $\{\phi_i\}_{i=1}^N$ for the finite-dimensional space V_N, and a set of bounded linear functionals $\{l_i\}_{i=1}^N$ from V', we can construct a projection operator as:

$$R_N u = \sum_{i=1}^N l_i(u)\phi_i. \tag{1.6}$$

The assumption (1.4) is then implied by:

$$\lim_{t\to\infty} |l_i(u(t) - v(t))| = 0, \qquad i = 1, \ldots, N \tag{1.7}$$

so that we can ask equivalently whether the set $\{l_i\}_{i=1}^N$ forms a set of *determining functionals* (see [**CJTi95, CJTi97**]). The analysis of whether R_N or $\{l_i\}_{i=1}^N$ are determining can be reduced to an analysis of the approximation properties of R_N. Note that in this construction, the basis $\{\phi_i\}_{i=1}^N$ need not span a subspace of the solution space V, so that the functions ϕ_i need not be divergence-free for example. Note that Definition 1.1 encompasses each of the notions of determining modes, nodes, and volumes by making particular choices for the sets of functions $\{\phi_i\}_{i=1}^N$ and $\{l_i\}_{i=1}^N$ (see [**JTi92a, JTi92b**]).

In this paper, we will employ Definition 1.1 to extend the results of [**CJTi95, CJTi97**] to the more general setting of H^1-regular solutions. In particular, we will show that if a projection operator $R_N : V \mapsto V_N \subset L^2(\Omega)$, $N = \dim(V_N) < \infty$, satisfies an approximation inequality for $\gamma > 0$ of the form,

$$\|u - R_N u\|_{L^2(\Omega)} \leq C_1 N^{-\gamma} \|u\|_{H^1(\Omega)}, \tag{1.8}$$

then the operator R_N is a determining projection in the sense of Definition 1.1, provided N is large enough. We will also derive explicit bounds on the dimension N which guarantees that R_N is determining. While we gain generality in our approach here, we also lose something in the balance: the bounds obtained here are generally of order Gr^2, whereas the bounds in [**CJTi95, CJTi97**] (requiring H^2-regularity) are of order Gr.

Outline of the paper. Preliminary material is presented in §2, including some inequalities for bounding the nonlinear term appearing in weak formulations of the Navier-Stokes equations. In §3, a finite element interpolant due to Scott and Zhang is presented, which (unlike nodal interpolation) is well-defined for H^1-functions. It is shown that the interpolant satisfies the approximation assumption (1.8) for H^1-functions on arbitrary polyhedral domains in both two and three dimensions; most of the details are relegated to the Appendix. In §4, we consider the two-dimensional Navier-Stokes equations, and derive bounds on the dimension N of the space V_N, employing only the approximation assumption (1.8). As an application of this general result, we employ some standard assumptions about simplex triangulations of the domain (discussed in §3) and derive lower bounds on the simplex diameters, sufficient to ensure that the SZ-interpolant is a determining projection (equivalently, that the simplex surface integrals forming SZ-interpolant coefficients are a determining set of linear functionals). We extend these results to three dimensions in §5, by requiring (following [**CDTi95**]) that weak solutions satisfy an additional technical assumption (due to the lack of appropriate global *a priori* estimates), which is related to the natural notion of mean dissipation rate of energy.

2. Preliminary Material

We briefly review some background material following the notation of [**CF88, Lio69, Tem77, Tem83**]. Let $\Omega \subset \mathbb{R}^d$ denote an open bounded set. The imbedding results we will need are known to hold for example if the domain Ω has a locally Lipschitz boundary, denoted as $\Omega \in \mathcal{C}^{0,1}$ (cf. [**Ada78**]). For example, open bounded convex sets $\Omega \subset \mathbb{R}^d$ satisfy $\Omega \in \mathcal{C}^{0,1}$ (Corollary 1.2.2.3 in [**Gri85**]), so that convex polyhedral domains (which we consider here) are in $\mathcal{C}^{0,1}$.

Let $H^k(\Omega)$ denote the usual Sobolev spaces $W^{k,2}(\Omega)$. Employing multi-index notation, the distributional partial derivative of order $|\alpha|$ is denoted D^α, so that the (integer-order) norms and semi-norms in $H^k(\Omega)$ may be denoted

$$\|u\|^2_{H^k(\Omega)} = \sum_{j=0}^{k} |\Omega|^{\frac{j-k}{d}} |u|^2_{H^j(\Omega)}, \qquad |u|^2_{H^j(\Omega)} = \sum_{|\alpha|=j} \|D^\alpha u\|_{L^2(\Omega)}, \quad 0 \le j \le k,$$

where $|\Omega|$ represents the measure of Ω. Fractional order Sobolev spaces and norms may be defined for example through Fourier transform and extension theorems, or through interpolation. A fundamentally important subspace is the $k = 1$ case of

$$H_0^k(\Omega) = \text{closure of } C_0^\infty(\Omega) \text{ in } H^k(\Omega),$$

in which the Poincaré Inequality reduces to: If Ω is bounded, then

$$\|u\|_{L^2(\Omega)} \le \rho(\Omega) |u|_{H^1(\Omega)}, \qquad \forall u \in H_0^1(\Omega). \tag{2.1}$$

The spaces above extend naturally (cf. [**Tem77**]) to product spaces of vector functions $u = (u_1, u_2, \dots, u_d)$, which are denoted with the same letters but in bold-face; for example, $\mathbf{H}_0^k(\Omega) = \left(H_0^k(\Omega)\right)^d$. The inner-products and norms in these product spaces are extended in the natural Euclidean way; the convention here will be to subscript these extended vector norms the same as the scalar case.

Define now the space $\mathcal{V}$ of divergence free $\mathbf{C}^\infty$ vector functions with compact support as

$$\mathcal{V} = \{\phi \in \mathbf{C}_0^\infty(\Omega) \mid \nabla \cdot \phi = 0\}.$$

The following two subspaces of $\mathbf{L}^2(\Omega)$ and $\mathbf{H}_0^1(\Omega)$ are fundamental to the study of the Navier-Stokes equations.

$$H = \text{closure of } \mathcal{V} \text{ in } \mathbf{L}^2(\Omega), \qquad V = \text{closure of } \mathcal{V} \text{ in } \mathbf{H}_0^1(\Omega).$$

To simplify the notation, it is common (cf. [**CF88, Tem77**]) to use the following notation for inner-products and norms in H and V:

$$(u, v) = (u, v)_H, \quad |u| = \|u\|_H, \quad ((u, v)) = (u, v)_V, \quad \|u\| = \|u\|_V. \tag{2.2}$$

The Navier-stokes equations (1.1)–(1.2) are equivalent to the functional differential equation:

$$\frac{du}{dt} + \nu A u + B(u, u) = f, \qquad u(0) = u_0. \tag{2.3}$$

The Stokes operator A and bilinear form B are defined as

$$Au = -P\Delta u, \qquad B(u, v) = P[(u \cdot \nabla) v],$$

where the operator P is the Leray orthogonal projector, $P : H_0^1 \mapsto V$ and $P : L^2 \mapsto H$, respectively.

Weak formulations, which we consider shortly, will use the bilinear Dirichlet form $((\cdot,\cdot))$ and trilinear form $b(\cdot,\cdot,\cdot)$ as:

$$((u,v)) = (\nabla u, \nabla v), \qquad b(u,v,w) = (B(u,v),w) = (P((u\cdot\nabla)v),w).$$

(Note that thanks to the Poincaré inequality (2.1), the form $((\cdot,\cdot))$ is actually an inner-product on V, and the induced norm $\|\cdot\| = ((\cdot,\cdot))^{1/2}$ is in fact a norm on V, equivalent to the H^1-norm.)

A priori bounds can be derived for the form $b(\cdot,\cdot,\cdot)$ (cf. [**CF88, Lad69, Tem77**]). In particular, if $\Omega \subset \mathbb{R}^d$, then the trilinear form $b(u,v,w)$ is bounded on $V\times V\times V$ as follows:

$$(2.4)\ d=2: \quad |b(u,v,w)| \le 2^{1/2}\|u\|^{1/2}_{L^2(\Omega)}|u|^{1/2}_{H^1(\Omega)}|v|_{H^1(\Omega)}\|w\|^{1/2}_{L^2(\Omega)}|w|^{1/2}_{H^1(\Omega)},$$

$$(2.5)\ d=3: \quad |b(u,v,w)| \le 2\|u\|^{1/4}_{L^2(\Omega)}|u|^{3/4}_{H^1(\Omega)}|v|_{H^1(\Omega)}\|w\|^{1/4}_{L^2(\Omega)}|w|^{3/4}_{H^1(\Omega)}.$$

Moreover, from Hölder inequalities we have for $d=2$ or $d=3$:

$$(2.6) \qquad |b(v,u,v)| \le \|\nabla u\|_{L^\infty(\Omega)}\|v\|^2_{L^2(\Omega)}.$$

3. Polynomial interpolation in $\mathbf{H}^1_0(\Omega)$

An example of a projection operator which satisfies the approximation assumption (1.8) is that used for defining determining volumes [**JTi92a**]; we examine now powerful alternative operator. Let $\Omega \subset \mathbb{R}^d$ be a d-dimensional polygon, exactly triangulated by (for example) Delaunay triangulation [**Ede87**], with quasi-uniform, shape-regular simplices, the vertices of which will form a set of N generalized interpolation points in our analysis. Note that for quasi-uniform, shape-regular triangulations in $\mathbb{R}^d$ (see [**Cia78**] for detailed discussions), it holds that

$$(3.1) \qquad C_0|\Omega|h^{-d} \le N \le C_0'|\Omega|h^{-d},$$

where h is the maximum of the diameters of the simplices, and where C_0 and C_0' are universal constants, independent of both N and h. The parameter h will be referred to as the characteristic parameter, or characteristic length scale, of such a quasi-uniform shape-regular mesh.

It should be noted that given some initial triangulation satisfying (3.1), repeated bi-section [**Bän91**] or octa-section [**Zha88**] (quadra-section in 2D) of each simplex can be performed in such a way as to guarantee non-degeneracy asymptotically, in that the quasi-uniformity and shape-regularity are preserved. Therefore, inequality (3.1) can be made to hold, for the same universal constants, for finer and finer meshes in a nested sequence of simplex triangulations.

To properly define a continuous piecewise-linear nodal interpolant of a function $u\in H^1(\Omega)$ based on the nodes of a triangulation of Ω, the particular function u must be bounded point-wise. This will be true if the function u is continuous in Ω, hence uniformly continuous on $\bar{\Omega}$. One of the Sobolev imbedding results (cf. [**Ada78**]) states that if $\Omega\subset\mathbb{R}^d$ satisfies $\Omega\in\mathcal{C}^{0,1}$, then for nonnegative real numbers k and s it holds that $H^k(\Omega)\hookrightarrow C^s(\bar{\Omega})$, $k > s+\frac{d}{2}$. This implies that for $d=1$, the interpolant can be correctly defined, since $H^1(\Omega)$ is continuously imbedded in $C^0(\bar{\Omega})$. However, in higher dimensions, $H^{1+\alpha}(\Omega)\hookrightarrow C^0(\bar{\Omega})$ only if $\alpha>0$ when $d=2$, or if $\alpha>1/2$ when $d=3$. While it may be possible to use the nodal interpolant and a regularity assumption such as $u\in H^{1+\alpha}(\Omega)$ for appropriate $\alpha>0$, an alternative approach is taken here.

The generalized interpolant due to Scott and Zhang [**SZ90**] can be defined for H^1-functions in both two and three spatial dimensions. The SZ-interpolant I_h is constructed from a combination linear interpolation and local averaging on faces and edges of simplices, and has optimal approximation properties even in the case of H^1-functions.

LEMMA 3.1. *For the SZ-interpolant of* $u \in \mathbf{H}_0^{1+\alpha}(\Omega)$, $\alpha \geq 0$, *it holds that*

$$\|u - I_h u\|_{L^2(\Omega)} \leq C_1 h^{1+\alpha} |u|_{H^{1+\alpha}(\Omega)}.$$

PROOF. See the appendix for a condensed proof following [**BS94, SZ90**]. □

Note that both the usual nodal interpolant and the SZ-interpolant I_h can be written as a linear combination of linear functionals:

$$I_h u(x) = \sum_{i=1}^{N} \phi_i(x) l_i(u).$$

In either case, the set of functions $\{\phi_i\}_{i=1}^N$ is the usual continuous piecewise-polynomial nodal finite element basis defined over the simplicial mesh, satisfying the *Lagrange* property at the vertices of the mesh:

$$\phi_i(x_j) = \delta_{ij}.$$

The difference between the two interpolants is simply the choice of the linear functionals: in the case of the nodal interpolant, the functionals are delta functions centered at the vertices of the mesh; in the case of the SZ-interpolant, they are defined in terms of a bi-orthogonal dual basis (see the Appendix).

4. The Two-dimensional Navier-Stokes Equations

A general weak formulation of the Navier-Stokes equations (1.1)–(1.2) can be written as (cf. [**CF88, Tem77**]):

DEFINITION 4.1. *Given* $f \in L^2([0,T]; V')$, *a weak solution of the Navier-Stokes equations satisfies* $u \in L^2([0,T]; V) \cap C_w([0,T]; H)$, $du/dt \in L^1_{\text{loc}}((0,T]; V')$, *and*

$$(4.1) \quad < \frac{du}{dt}, v > + \nu((u,v)) + b(u,u,v) = < f, v >, \quad \forall v \in V, \quad \text{for almost every } t,$$

$$(4.2) \qquad u(0) = u_0.$$

Here, the space $C_w([0,T]; H)$ is the subspace of $L^\infty([0,T]; H)$ of weakly continuous functions, and $< \cdot, \cdot >$ denotes the duality pairing between V and V', where H is the Riesz-identified pivot space in the Gelfand triple $V \subset H = H' \subset V'$. Note that since the Stokes operator can be uniquely extended to $A : V \mapsto V'$, and since it can be shown that $B : V \times V \mapsto V'$ (cf. [**CF88, Tem83**] for both results), the functional form (2.3) still makes sense for weak solutions, and the total operator represents a mapping $V \mapsto V'$.

In the two-dimensional case, for a forcing function $f \in L^\infty([0,T]; V')$, there exists a unique weak solution $u \in L^2([0,T]; V) \cap C_w([0,T]; H)$ (cf. [**CF88, Tem83**]). Consider now two forcing functions $f, g \in L^2([0,\infty]; V')$ and corresponding weak solutions u and v to (2.3) in either the two- or three-dimensional case. Subtracting the equations (2.3) for u and v yields an equation for the difference function $w = u - v$, namely

$$\frac{dw}{dt} + \nu A w + B(u,u) - B(v,v) = f - g. \tag{4.3}$$

Since the residual of equation (4.3) lies in the dual space V', for almost every t, we can consider the dual pairing of each side (4.3) with a function in V, and in particular with $w \in V$, which yields

$$< \frac{dw}{dt}, w > + \nu \|w\|^2 + b(u,u,w) - b(v,v,w) =< f-g, w > \quad \text{for almost every } t.$$

It can be shown (cf. [**Tem77**], Chapter 3, Lemma 1.2) that

$$\frac{1}{2}\frac{d}{dt}|w|^2 =< \frac{dw}{dt}, w >$$

in the distribution sense. It can also be shown [**CF88, Tem77**] that $b(u,v,w) = -b(u,w,v)$, $\forall u,v,w \in V$, so that $b(w,u,w) = b(u,u,w) - b(v,v,w)$. Therefore, the function $w = u - v$ must satisfy

$$\frac{1}{2}\frac{d}{dt}|w|^2 + \nu\|w\|^2 + b(w,u,w) =< f-g, w > . \tag{4.4}$$

The following generalized Gronwall inequality will be a key tool in the analysis to follow (see [**FMTT83**] and [**JTi92a**]).

LEMMA 4.1. *Let $T > 0$ be fixed, and let $\alpha(t)$ and $\beta(t)$ be locally integrable and real-valued on $(0,\infty)$, satisfying:*

$$\liminf_{t\to\infty} \frac{1}{T}\int_t^{t+T} \alpha(\tau)d\tau = m > 0, \qquad \limsup_{t\to\infty} \frac{1}{T}\int_t^{t+T} \alpha^-(\tau)d\tau = M < \infty,$$

$$\lim_{t\to\infty} \frac{1}{T}\int_t^{t+T} \beta^+(\tau)d\tau = 0,$$

where $\alpha^- = \max\{-\alpha, 0\}$ and $\beta^+ = \max\{\beta, 0\}$. If $y(t)$ is an absolutely continuous non-negative function on $(0,\infty)$, and $y(t)$ satisfies the following differential inequality:

$$y'(t) + \alpha(t)y(t) \le \beta(t), \quad \text{a.e. on } (0,\infty),$$

then $\lim_{t\to\infty} y(t) = 0$.

The main two-dimensional results are now given; we assume that $\Omega \subset \mathbb{R}^2$ is an open bounded domain with Lipschitz continuous boundary.

THEOREM 4.1. *Let $f(t), g(t) \in V'$ be any two forcing functions satisfying*

$$\lim_{t\to\infty} \|f(t) - g(t)\|_{V'} = 0,$$

and let $u, v \in V$ be the corresponding weak solutions to (1.1)–(1.2) for $d = 2$. If there exists a projection operator $R_N : V \mapsto V_N \subset L^2(\Omega)$, $N = \dim(V_N)$, satisfying

$$\lim_{t\to\infty} \|R_N(u(t) - v(t))\|_{L^2(\Omega)} = 0,$$

and satisfying for $\gamma > 0$ the approximation inequality

$$\|u - R_N u\|_{L^2(\Omega)} \le C_1 N^{-\gamma} \|u\|_{H^1(\Omega)},$$

then

$$\lim_{t\to\infty} |u(t) - v(t)| = 0$$

holds if N is such that

$$\infty > N > C\left(\frac{1}{\nu^2}\limsup_{t\to\infty}\|f(t)\|_{V'}\right)^{\frac{1}{\gamma}},$$

where C is a constant independent of ν and f.

PROOF. Using the notation (2.2), we begin with equation (4.4), employing the inequality (2.4) along with Cauchy-Schwarz and Young's inequalities to yield

$$\begin{aligned}\frac{1}{2}\frac{d}{dt}|w|^2+\nu\|w\|^2 &\le \|u\|\;|w|\;\|w\|+\|f-g\|_{V'}\|w\|\\ &\le \frac{1}{\nu}\|u\|^2|w|^2+\frac{1}{\nu}\|f-g\|_{V'}^2+\frac{\nu}{2}\|w\|^2.\end{aligned}$$

Equivalently, this is

$$\frac{d}{dt}|w|^2+\nu\|w\|^2-\frac{2}{\nu}\|u\|^2|w|^2\le\frac{2}{\nu}\|f-g\|_{V'}^2.$$

To bound the second term on the left from below, we employ the approximation assumption on R_N, or rather the following inequality which follows from it:

$$|w|^2\le 2N^{-2\gamma}C_1^2\|w\|^2+2\|R_Nw\|_{L^2(\Omega)}^2,$$

which yields

$$\frac{d}{dt}|w|^2+\left(\frac{\nu N^{2\gamma}}{2C_1^2}-\frac{2}{\nu}\|u\|^2\right)|w|^2\le\frac{2}{\nu}\|f-g\|_{V'}^2+\frac{\nu N^{2\gamma}}{C_1^2}\|R_Nw\|_{L^2(\Omega)}^2.$$

This is of the form

$$\frac{d}{dt}|w|^2+\alpha|w|^2\le\beta,$$

with obvious definition of α and β.

The generalized Gronwall Lemma 4.1 can now be applied. Recall that both $\|f-g\|_{V'}\to 0$ and $\|R_Nw\|_{L^2(\Omega)}\to 0$ as $t\to\infty$ by assumption. Since it is assumed that u and v, and hence w, are in V, so that all other terms appearing in α and β remain bounded, it must hold that

$$\lim_{t\to\infty}\frac{1}{T}\int_t^{t+T}\beta^+(\tau)d\tau=0,\qquad \limsup_{t\to\infty}\frac{1}{T}\int_t^{t+T}\alpha^-(\tau)d\tau<\infty.$$

It remains to verify that for some fixed $T>0$,

$$\limsup_{t\to\infty}\frac{1}{T}\int_t^{t+T}\alpha(\tau)d\tau>0.$$

This means we must verify the following inequality for some fixed $T>0$:

$$N^{2\gamma}>\frac{2C_1^2}{\nu}\left(\limsup_{t\to\infty}\frac{1}{T}\int_t^{t+T}\frac{2\|u\|^2}{\nu}d\tau\right)=\frac{4C_1^2}{\nu^2}\limsup_{t\to\infty}\frac{1}{T}\int_t^{t+T}\|u\|^2d\tau. \tag{4.5}$$

The following *a priori* bound on any weak solution can be shown to hold (this is a simple generalization to $f\in V'$ of the bound in [**CF88**] for $f\in H$):

$$\limsup_{t\to\infty}\frac{1}{T}\int_t^{t+T}\|u(\tau)\|^2d\tau\le\frac{2}{\nu^2}\limsup_{t\to\infty}\|f(t)\|_{V'}^2,$$

for $T = \rho^2/\nu > 0$, where ρ is the best constant from the Poincaré inequality (2.1). Therefore, if

$$N^{2\gamma} > 8C_1^2 \left(\frac{1}{\nu^2} \limsup_{t\to\infty} \|f(t)\|_{V'} \right)^2 \geq \frac{4C_1^2}{\nu^2} \left(\frac{2}{\nu^2} \limsup_{t\to\infty} \|f(t)\|_{V'}^2 \right), \tag{4.6}$$

implying that (4.5) holds, then by the Gronwall Lemma 4.1, it follows that

$$\lim_{t\to\infty} |w(t)| = \lim_{t\to\infty} |u(t) - v(t)| = 0.$$

□

Assume now that $\Omega \subset \mathbb{R}^2$ is also polyhedral, and can be exactly triangulated with a quasi-uniform, shape-regular set of simplices of maximal diameter $h = O(N^{-1/2})$, where N is the number of vertices in the triangulation (see §3). As an application of the general result above, we establish a lower bound on the simplex diameters of such a triangulation, which ensures that the SZ-interpolant is a determining projection (equivalently, that the simplex surface integrals forming SZ-interpolant coefficients are a determining set of linear functionals).

COROLLARY 4.1. *The SZ-interpolant is determining for the two-dimensional Navier-Stokes equations if the diameter h of the simplices is small enough so that*

$$\infty > h^{-2} > C \left(\frac{1}{\nu^2} \limsup_{t\to\infty} \|f(t)\|_{V'} \right)^2.$$

PROOF. Since $h = O(N^{-1/2})$ for quasi-uniform, shape-regular triangulations in two dimensions, taking $\alpha = 0$ in Lemma 3.1 yields

$$\|u - I_h u\|_{L^2(\Omega)} \leq C_1 h |u|_{H^1(\Omega)} \leq \tilde{C}_1 N^{-1/2} \|u\|_{H^1(\Omega)}.$$

Therefore, the SZ-interpolant I_h satisfies the approximation inequality (1.8) for $\gamma = 1/2$. The corollary then follows by application of Theorem 4.1. □

REMARK 4.1. If $f \in H$, then we have in fact a strong solution, i.e. $u \in H^2(\Omega)$, and the interpolation Lemma 3.1 may be applied with $\alpha = 1$. This falls into the theoretical framework of [**CJTi95, CJTi97**], and in the periodic case they have shown that $N \approx Gr$, whereas the above result for the no-slip case states that $N \approx Gr^2$. Whether the no-slip case may be improved to $N \approx Gr$ with additional regularity ($f \in H$) is unclear, due to the lack of an analogous identity to

$$(B(w, w), Aw) = 0,$$

which holds for the two-dimensional periodic case. In physical terms, in two dimensions this identity illustrates the lack of a boundary vorticity shedding source when the boundary is absent.

5. The Three-dimensional Navier-Stokes Equations

The lack of appropriate *a priori* estimates in the three-dimensional case requires a modification of the approach taken for the two-dimensional case in the previous section. However, the interpolation results we have employed are dimension-independent, and by following the analysis approach of [**CDTi95**] very closely, we can obtain similar results for the three-dimensional case. Again we require only that $f \in V'$, but we also assume the existence of a unique weak solution to the three-dimensional Navier-Stokes equations. An additional technical assumption is that some measure of the mean rate of energy dissipation be finite, namely:

$$\epsilon_\infty = \inf_{T>0} \limsup_{t\to\infty} \frac{\nu}{T} \int_t^{t+T} \|\nabla u\|_\infty d\tau < \infty.$$

This assumption implies that eventually the weak solution for the three-dimensional Navier-Stokes equations becomes unique, and also in the case $f \in H$ the weak solution eventually becomes strong. But this assumption does not imply anything about the transients, since the quantity is required to be finite only for large time. We assume again that $\Omega \subset \mathbb{R}^3$ is an open bounded domain with Lipschitz continuous boundary.

THEOREM 5.1. *Let $f(t), g(t) \in V'$ be any two forcing functions satisfying*

$$\lim_{t\to\infty} \|f(t) - g(t)\|_{V'} = 0,$$

and let $u, v \in V$ be the corresponding weak solutions to (1.1)–(1.2) for $d = 3$. If there exists a projection operator $R_N : V \mapsto V_N \subset L^2(\Omega)$, $N = \dim(V_N)$, satisfying

$$\lim_{t\to\infty} \|R_N(u(t) - v(t))\|_{L^2(\Omega)} = 0,$$

and satisfying for $\gamma > 0$ the approximation inequality

$$\|u - R_N u\|_{L^2(\Omega)} \leq C_1 N^{-\gamma} \|u\|_{H^1(\Omega)},$$

then

$$\lim_{t\to\infty} |u(t) - v(t)| = 0$$

holds if N is such that

$$\infty > N > C\left(\frac{1}{\nu}\inf_{T>0}\left\{\limsup_{t\to\infty} \frac{1}{T}\int_t^{t+T} \|\nabla u(s)\|_{L^\infty(\Omega)} ds\right\}\right)^{\frac{1}{2\gamma}},$$

where C is a constant independent of ν, f, and u.

PROOF. Beginning with equation (4.4), the inequality (2.6) is employed along with Cauchy-Schwarz and Young's inequalities to yield

$$\begin{aligned}\frac{1}{2}\frac{d}{dt}|w|^2 + \nu\|w\|^2 &\leq \|\nabla u\|_{L^\infty(\Omega)}|w|^2 + \|f-g\|_{V'}\|w\| \\ &\leq \|\nabla u\|_{L^\infty}|w|^2 + \frac{1}{2\nu}\|f-g\|_{V'}^2 + \frac{\nu}{2}\|w\|^2\end{aligned}$$

Equivalently,

$$\frac{d}{dt}|w|^2 + \nu\|w\|^2 - \|\nabla u\|_{L^\infty(\Omega)}|w|^2 \leq \frac{1}{\nu}\|f-g\|_{V'}^2.$$

To bound the second term on the left from below, we employ a consequence of the approximation assumption on R_N, namely the inequality

$$|w|^2 \leq 2N^{-2\gamma}C_1^2\|w\|^2 + 2\|R_N w\|_{L^2(\Omega)}^2,$$

which yields

$$\frac{d}{dt}|w|^2 + \left(\frac{\nu N^{2\gamma}}{2C_1^2} - \|\nabla u\|_{L^\infty}\right)|w|^2 \leq \frac{1}{\nu}\|f-g\|_{V'}^2 + \frac{\nu N^{2\gamma}}{C_1^2}\|R_N w\|_{L^2(\Omega)}^2.$$

This has the form

$$\frac{d}{dt}|w|^2 + \alpha|w|^2 \leq \beta,$$

with again obvious definition of α and β.

The analysis now proceeds exactly as in the proof of Theorem 4.1, so that all that remains is to check again that for some fixed $T > 0$,

$$\limsup_{t\to\infty} \frac{1}{T}\int_t^{t+T} \alpha(\tau)d\tau > 0.$$

Thus, we must prove our assumption on N guarantees for a fixed $T > 0$ that

$$N^{2\gamma} > \frac{2C_1^2}{\nu} \limsup_{t\to\infty} \frac{1}{T}\int_t^{t+T} \|\nabla u\|_{L^\infty(\Omega)}d\tau. \tag{5.1}$$

If we select $T_* > 0$ such that

$$2\inf_{T>0}\left(\limsup_{t\to\infty} \frac{1}{T}\int_t^{t+T} \|\nabla u(s)\|_{L^\infty(\Omega)}ds\right) \geq \limsup_{t\to\infty} \frac{1}{T_*}\int_t^{t+T_*} \|\nabla u(s)\|_{L^\infty(\Omega)}ds,$$

then our assumption gives

$$N^{2\gamma} > \frac{4C_1^2}{\nu}\inf_{T_*>0}\left(\limsup_{t\to\infty} \frac{1}{T_*}\int_t^{t+T_*} \|\nabla u(s)\|_{L^\infty(\Omega)}ds\right) \tag{5.2}$$

which implies (5.1). The theorem then follows by the Gronwall Lemma 4.1. □

Assume now that $\Omega \subset \mathbb{R}^3$ is also polyhedral, and can be exactly triangulated with a quasi-uniform, shape-regular set of simplices of maximal diameter $h = O(N^{-1/3})$, where N is the number of vertices in the triangulation. As an application of the general three-dimensional result above, we will establish a lower bound on the simplex diameters of such a triangulation, which ensures that the SZ-interpolant is a determining projection (and that the simplex surface integrals forming SZ-interpolant coefficients are a determining set of linear functionals).

COROLLARY 5.1. *The SZ-interpolant is determining for the three-dimensional Navier-Stokes equations if the diameter h of the simplices is small enough so that*

$$\infty > h^{-2} > C\left(\frac{1}{\nu}\inf_{T>0}\left\{\limsup_{t\to\infty} \frac{1}{T}\int_t^{t+T} \|\nabla u(s)\|_{L^\infty(\Omega)}ds\right\}\right).$$

PROOF. Since $h = O(N^{-1/3})$ for quasi-uniform, shape-regular triangulations in three dimensions, taking $\alpha = 0$ in Lemma 3.1 yields

$$\|u - I_h u\|_{L^2(\Omega)} \leq C_1 h|u|_{H^1(\Omega)} \leq \tilde{C}_1 N^{-1/3}\|u\|_{H^1(\Omega)}.$$

Therefore, the SZ-interpolant I_h satisfies the approximation inequality (1.8) for $\gamma = 1/3$. The corollary then follows by application of Theorem 5.1. □

Appendix: Approximability of the Scott-Zhang Interpolant

We will sketch the proof of the approximability result for the SZ-interpolant given as Lemma 3.1; we will follow quite closely the proof given in [**BS94, SZ90**]. As throughout this paper, we assume that $\Omega \in \mathcal{C}^{0,1}$, and that the given exact simplicial triangulation of Ω is both shape-regular and quasi-uniform.

The proof of Lemma 3.1 will follow easily from the following result (see the comments at the end of this appendix).

LEMMA 5.1. *For the SZ-interpolant of $u \in H_0^{1+\alpha}(\Omega)$, $\alpha \geq 0$, it holds that*

$$\|u - I_h u\|_{L^2(\Omega)} \leq C_1 h^{1+\alpha}|u|_{H^{1+\alpha}(\Omega)}.$$

To prove Lemma 5.1, we will begin by defining carefully the SZ-interpolant. Let $\mathcal{T}_h = \{\tau_i\}_{i=1}^L$ be the given quasi-uniform, shape-regular mesh of d-simplices which exactly triangulate the underlying domain Ω, and let $\Omega_h = \{x_i\}_{i=1}^N$ be the set of vertices of these d-simplices. Define

$$V_h = \text{span}\{\phi_i(x)\}_{i=1}^N \subset H^1(\Omega),$$

where $\{\phi_i(x)\}$ is the set of standard continuous piecewise linear (nodal) basis functions. The nodal basis satisfies the Lagrange relationship at the vertices (which are exactly the "nodes" in this setting):

$$\phi_i(x_j) = \delta_{ij}.$$

Now, for each vertex x_i, we select (arbitrarily) an associated $(d-1)$-simplex σ_i from the given simplicial mesh satisfying only:

$$(1)\ x_i \in \bar{\sigma}_i, \qquad \text{and} \qquad (2)\ \sigma_i \subset \partial\Omega \text{ if } x_i \in \partial\Omega.$$

In other words, for a given vertex x_i we pick an arbitrary $(d-1)$-simplex from edges or faces of the d-simplices which contain x_i as a vertex. In two-dimensions, we are picking the edge of one of the triangles that have x_i as a vertex; in three-dimensions, we are picking the face of one of the tetrahedra which have x_i as a vertex. The only restriction on this choice is near the boundary: if x_i is on the boundary, then the $(d-1)$-simplex we pick must be one of the edges or faces of the a simplex which lies exactly on the boundary (such a choice is always possible).

In each $(d-1)$-simplex σ_i, we number the generating vertex x_i first in the set of vertices of σ_i, denoted $\{x_{i,j}\}_{j=1}^d$. (I.e., we set $x_{i,1} = x_i$.) For each σ_i, we also have a $(d-1)$-dimensional nodal basis $\{\phi_{i,j}\}_{j=1}^d$, where again we set $\phi_{i,1} = \phi_i$. There exists an associated $L^2(\sigma_i)$-dual (bi-orthogonal) basis $\{\psi_{i,j}\}$ satisfying

$$\int_{\sigma_i} \psi_{i,j}(x)\phi_{i,k}(x)dx = \delta_{jk}, \qquad j,k = 1,\dots,d.$$

Again we take $\psi_{i,1} = \psi_i$, $\forall x_i \in \Omega_h$. Note that ψ_i and ϕ_j also satisfy a bi-orthogonal relationship, namely $\int_{\sigma_i} \psi_i\phi_j dx = 0$, $i \neq j$. We define now the SZ-interpolant as

$$I_h : H^1(\Omega) \mapsto V_h(\Omega), \qquad I_h u(x) = \sum_{i=1}^N \phi_i(x) l_i(u), \qquad l_i(u) = \int_{\sigma_i} \psi_i(\xi)u(\xi)d\xi.$$

Thanks to the Trace Theorem [**Ada78**], the interpolant $I_h u(x)$ is well-defined at nodal values even for $u \in H^1(\Omega)$, since $H^1(\Omega) \hookrightarrow L^2(\sigma_i)$. Almost by construction, one can show [**SZ90**] that

- $I_h : H^1(\Omega) \mapsto V_h(\Omega)$ is a projection
- $I_h : H_0^1(\Omega) \mapsto V_{0h}(\Omega)$

where V_{0h} is the subset of V_h having zero trace on the boundary of Ω. Thus, I_h preserves homogeneous Dirichlet boundary conditions. Using homogeneity arguments, the following stability result for the interpolant is established in [**SZ90**].

LEMMA 5.2. *For any $\tau \in \mathcal{T}_h$, if the support region of τ is defined as the set $S_\tau =$ interior $(\cup\{\bar{\tau}_i \mid \bar{\tau}_i \cap \bar{\tau} \neq \emptyset,\ \tau_i \in \mathcal{T}_h\})$, then it holds that*

$$\|I_h u\|_{H^m(\tau)} \leq C \sum_{k=0}^{l} h^{k-m} |u|_{H^k(S_\tau)}, \quad 0 \leq m \leq l, \quad l > 1/2.$$

PROOF. See the proof of Theorem 3.1 in [**SZ90**]. □

The proof of the Scott and Zhang [**SZ90**] approximation result is as follows.

PROOF. (Lemma 5.1) Since I_h is a projector from $H^1(\Omega)$ onto $V_h(\Omega)$, it follows that on each element, I_h is a projector from $H^1(\tau)$ onto $\mathcal{P}_1(\tau)$, the space of linear polynomials over τ. Thus, $I_h p = p$, $\forall p \in \mathcal{P}_1(\tau)$, and employing also the stability result in Lemma 5.2 we have that for $0 \le m \le k \le 2$,

$$\|u - I_h u\|_{H^m(\tau)} \le \|u - p\|_{H^m(\tau)} + \|I_h(p-u)\|_{H^m(\tau)} \le C \sum_{k=0}^{m} h^{k-m} \|u-p\|_{H^k(S_\tau)},$$

where S_τ is the element support region surrounding τ as defined in Lemma 5.2. Employing the modified Bramble-Hilbert lemma developed in [**DS80**] to estimate the terms of the sum gives

$$\inf_{p \in \mathcal{P}_1(\tau)} \|u-p\|_{H^m(S_\tau)} \le C h^{k-m} |u|_{H^k(S_\tau)}, \quad 0 \le m \le k \le 2,$$

where due to the assumptions about the domain and the mesh, the constant C depends only on the spatial dimension d. Together with the equation above this is

$$\|u - I_h u\|_{H^m(\tau)} \le C h^{k-m} |u|_{H^k(S_\tau)} \quad 0 \le m \le k \le 2.$$

Since the set

$$Q = \sup_{\tau \in \mathcal{T}_h} \{\text{card}\{\tau \in \mathcal{T}_h | \tau \cap S_\tau \ne \emptyset\}\}$$

is finite due to the quasi-uniformity and shape-regularity of the mesh, we have finally that for $0 \le m \le k \le 2$, it holds that

$$\|u - I_h u\|^2_{H^m(\Omega)} = \sum_{\tau \in \mathcal{T}_h} \|u - I_h u\|^2_{H^m(\tau)} \le C h^{2(k-m)} \|u\|^2_{H^k(\Omega)}.$$

The result for non-integer exponents k and m follows by the usual norm interpolation arguments between $L^2(\Omega)$ and $H^2(\Omega)$, which completes the proof. □

Lemma 5.1 can be easily extended to the vector case, which provides finally the proof of Lemma 3.1.

PROOF. (Lemma 5.1) For $u \in \mathbf{H}_0^{1+\alpha}(\Omega) = (H_0^{1+\alpha}(\Omega))^d$, we have that

$$\|u - I_h u\|^2_{L^2} = \sum_{i=1}^{d} \|u_i - I_h^{(i)} u_i\|_{L^2(\Omega)} \le C_1^2 h^{2(1+\alpha)} \sum_{i=1}^{d} |u_i|^2_{H^{1+\alpha}(\Omega)},$$

where $I_h^{(i)}$ denotes the scalar SZ-interpolant applied to u_i. Thus,

$$\|u - I_h u\|_{L^2(\Omega)} \le C_1 h^{1+\alpha} |u|_{H^{1+\alpha}(\Omega)}.$$

□

References

[Ada78] R. A. Adams, *Sobolev spaces*, Academic Press, San Diego, CA, 1978.

[Bän91] E. Bänsch, *Local mesh refinement in 2 and 3 dimensions*, Impact of Computing in Science and Engineering **3** (1991), 181–191.

[BS94] S. C. Brenner and L. R. Scott, *The mathematical theory of finite element methods*, Springer-Verlag, New York, NY, 1994.

[CDTi95] P. Constantin, C. R. Doering, and E. S. Titi, *Rigorous estimates of small scales in turbulent flows*, Journal of Mathematical Physics (1995), (Submitted).

[CF88] P. Constantin and C. Foias, *Navier-stokes equations*, University of Chicago Press, Chicago, IL, 1988.

[CFMT85] P. Constantin, C. Foias, O.P. Manley, and R. Temam, *Determining modes and fractal dimension of turbulent flows*, J. Fluid Mech. **150** (1985), 427–440.

[Cia78] P. G. Ciarlet, *The finite element method for elliptic problems*, North-Holland, New York, NY, 1978.

[CJTi95] B. Cockburn, D. A. Jones, and E. S. Titi, *Degrés de liberté déterminants pour équations nonlinéaires dissipatives*, C.R. Acad. Sci. Paris, Série I **321** (1995), 563–568.

[CJTi97] B. Cockburn, D. A. Jones, and E. S. Titi, *Estimating the number of asymptotic degrees of freedom for nonlinear dissipative systems*, Math. Comp. (1997), (To appear).

[DS80] T. Dupont and R. Scott, *Polynomial approximation of functions in sobolev spaces*, Math. Comp. **34** (1980), 441–463.

[Ede87] H. Edelsbrunner, *Algorithms in combinatorial geometry*, EATCS Monographs on Theoretical Computer Science, vol. 10, Springer-Verlag, Berlin, 1987.

[FMTT83] C. Foias, O. P. Manley, R. Temam, and Y. Treve, *Asymptotic analysis of the Navier-Stokes equations*, Physica D9 (1983), 157–188.

[FP67] C. Foias and G. Prodi, *Sur le comportement global des solutions non stationnaires des équations de Navier-Stokes en dimension two*, Rend. Sem. Mat. Univ. Padova **39** (1967), 1–34.

[FST88] C. Foias, G. Sell, and R. Temam, *Inertial manifolds for nonlinear evolutionary equations*, J. Diff. Eq. **73** (1988), 309–353.

[FT83] C. Foias and R. Temam, *Asymptotic numerical analysis for the Navier-Stokes equations*, Nonlinear Dynamics and Turbulence (1983).

[FT84] C. Foias and R. Temam, *Determining the solutions of the Navier-Stokes equations by a set of nodal values*, Math. Comp. **43** (1984), 177–183.

[FTi91] C. Foias and E. S. Titi, *Determining nodes, finite difference schemes and inertial manifolds*, Nonlinearity **4** (1991), 135–153.

[Gri85] P. Grisvard, *Elliptic problems in nonsmooth domains*, Pitman Publishing, Marshfield, MA, 1985.

[JTi92a] D. A. Jones and E. S. Titi, *Determing finite volume elements for the 2D Navier-Stokes equations*, Physica D60 (1992), 117–133.

[JTi92b] D. A. Jones and E. S. Titi, *On the number of determining nodes for the 2D Navier-Stokes equations*, J. Math. Anal. Appl. **168** (1992), 72–88.

[JTi93] D. A. Jones and E. S. Titi, *Upper bounds on the number of determining modes, nodes, and volume elements for the Navier-Stokes equations*, Indiana University Mathematics Journal **42** (1993), no. 3, 875–887.

[Lad69] O. A. Ladyženskaja, *The mathematical theory of viscous incompressible flow*, Gordon and Breach, New York, NY, 1969.

[Lio69] J. L. Lions, *Quelques méthodes de résolution de problèmes aux limites nonlinéaires*, Dunod, Paris, 1969.

[SZ90] L. R. Scott and S. Zhang, *Finite element interpolation of nonsmooth functions satisfying boundary conditions*, Math. Comp. **54** (1990), no. 190, 483–493.

[Tem77] R. Temam, *Navier-stokes equations: Theory and numerical analysis*, North-Holland, New York, NY, 1977.

[Tem83] R. Temam, *Navier-stokes equations and nonlinear functional analysis*, CBMS-NSF Regional Conference Series in Applied Mathematics, SIAM, Philadelphia, PA, 1983.

[Zha88] S. Zhang, *Multi-level iterative techniques*, Ph.D. thesis, Dept. of Mathematics, Pennsylvania State University, 1988.

Applied Mathematics 217-50, Caltech, Pasadena, CA 91125, USA.
E-mail address: `holst@ama.caltech.edu`

Department of Mathematics and Department of Mechanical and Aerospace Engineering, University of California, Irvine, CA 92697-3875, USA.
E-mail address: `etiti@math.uci.edu`

Contemporary Mathematics
Volume **204**, 1997

Parametric Morse lemmas for $C^{1,1}$-functions *

Alexander Ioffe and Efim Schwartzman
Department of mathematics
The Technion
Haifa 32000, Israel†

Abstract

We prove extensions (involving parametric and global versions) of the Morse lemma to $C^{1,1}$ functions on Hilbert spaces.

Introduction. In this note we prove several extensions of the Morse lemma to $C^{1,1}$-functions (i.e. C^1-functions with locally Lipschitz gradient mappings) on Hilbert spaces, including global and local parametric versions of the lemma. Of course, the transformations which produce the normal form of the function are no longer diffeomorphisms but Lipschitz homeomorphisms.

The standard assumption for "textbook" proofs of the Morse lemma is that the function belongs to the class C^k for $k \geq 3$ (e.g. Palais [13], Nirenberg [12], Schwartz [14], Aubin - Ekeland [2], Arnold - Varchenko - Gusein-Zade [1]) and the standard way of proving it is by using the famous Hadamard lemma. For the C^2-case proofs were given by Takens [15] (who actually proved analogues of the lemma for arbitrary k-jet, $k \geq 2$, of functions on R^n; the result of Takens also covers the $C^{1,1}$ case in R^n but gives only a homeomorphism, not a Lipschitz homeomorphism), Kuiper [10], Cambini [3], Mahwin-Willem [11], Hofer [7] and others - see Chang [4] for probably the latest and most general formulation.

Our proof follows the classical scheme of Thom's homotopic method (see e.g. [1]). It is based on a construction similar to that used by Hofer [7] but which is simpler because we do not need the Lyapunov-Schmidt-type reduction for the description of the construction. A more attentive look at the construction allows to conclude that it works under weaker differentiability

---

*1991 *Mathematics Subject Classification.* Primary 58E05, Secondary 49K.

†The first author was supported in this research by the US-Israel BSF grant 90-00455 and by the NSF grant DMS-9404128, and the second author by the Israel Ministry of Science and Technology grant 3501–1–91.

requirements ($C^{1,1}$ instead of C^2) and, with the help of simple a priori estimates, obtain non-local formulations which easily incorporate parametric dependences. All the results and proofs carry over to the C^2 case as well, with normalizing transformations automatically becoming diffeomorphisms, in this case (cf [7, 4]). It seems that also some C^2 results in this paper have not been mentioned earlier.

We use the following notation:

$(x \mid y)$ for the inner product;
B_r for the ball of radius r around the origin;
$\nabla f(x)$ for the gradient of f at x;
$\nabla^2 f(x)$ for the Hessian of f at x.

Basic Lemma. *Consider the function*

$$f(x) = (1/2)(Tx|x) + \phi(x),$$

where T is a bounded linear self-adjoint invertible operator and ϕ is a function of class $C^{1,1}$ defined on the ball B_{R_0} of radius R_0 around the origin. We assume that

$$\phi(0) = 0, \quad \nabla\phi(0) = 0,$$

and that the mapping $x \to \nabla\phi(x)$ satisfies on B_{R_0} the Lipschitz condition with constant $L < ||T^{-1}||^{-1}$. Then there are $a > 0$, $M > m > 0$, depending only on $\| T \|$, $\| T^{-1} \|$ and L, and a zero preserving Lipschitz homeomorphism $\Gamma(x)$ defined at least on the ball B_r of radius $r = aR_0$ and such that

$$f(\Gamma(x)) = (1/2)(Tx, x);$$
$$m \| x - x' \| \leq \| \Gamma(x) - \Gamma(x') \| \leq M \| x - x' \| .$$

To prove the Basic Lemma we need the following

Proposition. *Set*

$$y_t = y_t(x) = Tx + t\nabla\phi(x), t \in [0, 1].$$

and

$$R = cR_0/C; \quad c = (||T^{-1}||^{-1} - L); \; C = (||T|| + L).$$

Then the mapping $y_t|_{B_R}$ is homeomorphism and moreover, for any x and x' from B_R the following inequalities hold true

$$c||x - x'|| \leq ||y_t(x) - y_t(x)|| \leq C||x - x'||. \tag{1}$$

Proof. Consider the equation

$$y_t(x) = v. \tag{2}$$

The mapping

$$x \mapsto T^{-1}v - tT^{-1}\nabla\phi(x)$$

maps the ball B_{R_0} into itself whenever $v \in B_{cR_0}$ (with c as in the statement) and the standard contraction mapping principle applied to this mapping gives that (2) has a unique solution $x = y_t^{-1}(v)$ and moreover, y_t^{-1} being defined on B_{cR_0} is a Lipschitz mapping with the Lipschitz constant c^{-1}. Now the desired statement follows from the observation that for any $x \in B_R$, $y_t(x) \in B_{cR_0}$.

Proof of the Basic Lemma. We have

$$\int_0^1 ||\nabla\phi(tx) - \nabla\phi(tx')||dt \leq L||x - x'||. \tag{3}$$

It follows that

$$\begin{array}{rcl} |\phi(x) - \phi(x')| & = & |\int_0^1[(\nabla\phi(tx)|x) - (\nabla\phi(tx')|x')]dt| \\ & = & |\int_0^1[(\nabla\phi(tx)|x - x') - (\nabla\phi(tx') - \nabla\phi(tx)|x')]dt| \\ & \leq & L||x|| \cdot ||x - x'|| + L||x - x'|| \cdot ||x'|| \\ & = & L(||x|| + ||x'||)||x - x'||. \end{array} \tag{4}$$

Combining this with (1), we get (taking into account that $y_t(0) = 0$)

$$|\phi(x) - \phi(x')| \leq (L/c)(||y_t|| + ||y_t'||)||x - x'|| \leq (L/c^2)(||y_t|| + ||y_t'||)||y - y_t'|| \tag{5}$$

for x and x' in B_R. We also have from (4), (5)

$$|\phi(x)| \leq L||x||^2 \leq Lc^{-2}||y_t||^2 \tag{6}$$

for any $x \in B_R$. Set

$$v_t(x) = -\phi(x)||y_t(x)||^{-2}y_t(x), \tag{7}$$

if $x \neq 0$ and $v_t(0) = 0$.

We shall show that $v_t(\cdot)$ satisfies the Lipschitz condition with constant depending only on L and the norms of T and T^{-1}. Take $x, x' \in B_R$ and assume that $||y_t(x)|| \geq ||y_t(x')||$. Then (writing for brevity ϕ, ϕ' instead of $\phi(x), \phi(x')$ etc.) we have by (4), (1), (2) and (6),

$$\begin{array}{l} ||\phi||y_t'||^2y_t - \phi'||y_t||^2y_t'|| \\ \quad \leq ||y_t'||^2||y_t|||\phi - \phi'| + |\phi'|||y_t||^2||y_t - y_t'|| + |\phi'|||y_t||(||y_t||^2 - ||y_t'||^2) \\ \quad \leq (L/c)||y_t'||^2||y_t||(||y_t|| + ||y_t'||)||x - x'|| \\ \qquad\qquad + (L/c^2)||y_t'||^2(2||y_t||^2 + ||y_t'||||y_t||)C||x - x'|| \\ \quad \leq (L/c)(2 + 3C/c)||y_t'||^2||y_t||^2||x - x'|| = a_1||y_t'||^2||y_t||^2||x - x'||, \end{array}$$

which means that $v_t(\cdot)$ satisfies the Lipschitz condition with constant $a_1 = (L/c)(2+3C/c))$.

Let $\Gamma_t(x)$ be the flow defined by the equation

$$\dot{x} = v_t(x), \ x(0) = x. \tag{8}$$

If x is so small that $\Gamma_t(x) \in B_R$ for all $t \in [0,1]$, then it follows that

$$||\Gamma_t(x)|| \leq ||x||e^{a_1 t}, \ t \in [0,1].$$

Conversely, for the same reason, $||\Gamma_t(x)|| \leq R$ for all t if $||x|| \leq Re^{-a_1} = r$.

For $||x|| < r$ consider the function

$$g_t(x) = (1/2)(T\Gamma_t(x)|\Gamma_t(x)) + t\phi(\Gamma_t(x)).$$

We have $g_0(x) = (1/2)(Tx|x)$ and by the definition of v_t in (7)

$$(\partial/\partial t)g_t(x) = (T\Gamma_t(x)|(\partial/\partial t)\Gamma_t(x)) + \phi(\Gamma_t(x)) + t(\nabla\phi(\Gamma_t(x))|(\partial/\partial t)\Gamma_t(x))$$

$$= (y_t(\Gamma_t(x)|v_t(\Gamma_t(x))) + \phi(\Gamma_t(x)) = 0.$$

Therefore for $\Gamma(x) = \Gamma_1(x)$ we have

$$(1/2)(Tx|x) \ = g_0(x) = g_1(x) = (1/2)(T\Gamma(x)|\Gamma(x)) + \phi(\Gamma(x)) = f(\Gamma(x)).$$

It remains to note that, being a flow generated by (8), $\Gamma(\cdot)$, as every $\Gamma_t(\cdot)$, is a Lipschitz homeomorphism with Lipschitz constants (m and M) defined by the Lipschitz constant of $v_t(\cdot)$. This completes the proof of the lemma.

The proof with little change (actually simplification) extends to the case when f is a C^2-function. In that case taking T to be the second derivative of f at zero we have a $\phi(\cdot)$ of class C^2 with $\nabla^2\phi(0) = 0$. Then taking an $L < ||T^{-1}||^{-1}$ (of course under the assumption that T is invertible and $\nabla\phi(0) = 0$) we find that $||\nabla^2\phi(x)|| \leq L$ if $||x||$ is small enough, so the assumptions of the Basic Lemma will be fulfilled.

But in this case the mapping $y_t(\cdot)$ will be continuously differentiable, hence a diffeomorphism, and simple arguments show that $v_t(\cdot)$ is continuously differentiable as well. (Indeed, by (1) $v_t(\cdot)$ is continuously differentiable at any point $x \neq 0$. On the other hand, using the freedom of choosing L (with the corresponding decrease in R_0), we can apply the above argument with arbitrarily small L. Then (6) shows that $\varphi(x) = o(||y_t||^2)$ from which differentiability of $v_t(\cdot)$ at zero, with the derivative equal to zero, and continuity of the derivative at zero follow immediately.) So (8) in this case is a differential equation with a right-hand side of class C^1 which means that $\Gamma_t(\cdot)$ is a diffeomorphism for any t. Thus, as a corollary we get the Morse lemma for C^2-functions.

Corollary 1. (the C^2-Morse lemma). *Let f be a C^2-function defined in a neighborhood of zero. Assume that zero is a nondegenerate critical point of f, that is to say, that $\nabla f(0) = 0$ and $T = \nabla^2 f(0)$ is an invertible operator. Then there is a zero preserving diffeomorphism $\Gamma(x)$ defined in a neighborhood of zero such that $f(\Gamma(x)) = (1/2)(Tx|x) + f(0)$.*

We note, however, that our proof cannot guarantee $(k-1)$-smoothness of $\Gamma(x)$ if $f \in C^k, k > 2$, that can be obtained by the classical proof involving the Hadamard lemma.

The two principal results of the paper, the global and the local parametric lemmas are also easy corollaries of the basic lemma.

Theorem 1 (global parametric Morse lemma). *Let P be a topological space. Consider the function*

$$f(x,p) = (1/2)(T(p)x|x) + \eta(x,p)$$

on $H \times P$, and assume that the following conditions are satisfied:

(a_1) *$T(p)$ is a continuous family of self-adjoint operators in H such that the norms of $T(p)$ and $T^{-1}(p)$ are uniformly bounded;*

(a_2) *$\eta(x,p)$ and its gradient $\nabla\eta$ with respect to x are defined and continuous on $N \times P$;*

(a_3) *there is a positive constant $L < inf_p ||T(p)^{-1}||^{-1} = K$ such that*

$$||\nabla\eta\ (x,p) - \nabla\eta\ (x',p)|| \leq L||x - x'||,$$

for any $x, x' \in H, p \in P$.

Then there is a continuous family of Lipschitz homeomorphism $\Gamma_p(\cdot) : H \mapsto H$ such that

$$f(\Gamma_p(x),p) = (1/2)(T(p)x|x) + g(p), \tag{9}$$

$$m||x - x'|| \leq ||\Gamma_p(x) - \Gamma_p(x')|| \leq M||x - x'||, \tag{10}$$

for any $x, x' \in H, p \in P$ with $M > m > 0$ depending only on L and the bounds for the norms of $T(p)$ and $T^{-1}(p)$.

Proof. By (a_3) the mapping $x \to T(p)x + \nabla\eta\ (x,p)$ is, for any p, a Lipschitz homeomorphism with the Lipschitz constant of the inverse mapping not exceeding $(K-L)^{-1}$ (apply the Proposition with $R_0 = \infty$), so there is a uniquely defined continuous mapping $x(p)$ such that $T(p)x(p) + \nabla\eta\ (x(p),p) \equiv 0$. Set

$$\bar{f}(x,p) = f(x + x(p),p) - f(x(p),p) = (1/2)(T(p)x|x) + \bar{\eta}(x,p), \tag{11}$$

where

$$\bar{\eta}(x,p) = (T(p)x(p)|x) + \eta(x + x(p), p) - \eta(x(p), p).$$

Then $\bar{\eta}$ obviously satisfies (a_2) and (a_3) and, in addition

$$\bar{\eta}(0,p) \equiv 0, \quad \nabla\bar{\eta} \equiv 0. \tag{12}$$

This means that for every fixed p we can apply the Basic Lemma with $T = T(p)$, $\phi(x) = \bar{\eta}(x,p)$ and $R_0 = \infty$ which implies that there is a zero preserving Lipschitz homeomorphism $\bar{\Gamma}(x)$ from H into itself satisfying

$$\bar{f}(\bar{\Gamma}_p(x), p) = (1/2)(T(p)x|x) + \bar{f}(0,p).$$

We note further that in the case when $\bar{\Gamma}_p$ is as in the proof of the Basic Lemma, that is to say, if the flow is generated by (8) with $v_t(x) = v_t(x,p)$ defined by (7) with $\phi(x) = \bar{\eta}(x,p)$, then v_t is continuous and therefore $\bar{\Gamma}_p(p)$ is jointly continuous in (x,p).

We complete the proof of the theorem, by setting

$$\Gamma_p(x) = \bar{\Gamma}_p(x) + x(p), \quad g(p) = f(x(p), p).$$

Theorem 2 (local parametric Morse lemma). *Let p and f be as in Theorem 1. Assume that (a_1) holds but $(a_2), (a_3)$ are valid only for $(x,p) \in B_R \times P$ for some $R > 0$. Assume also that there is a $\bar{p} \in P$ such that $\eta(0,\bar{p}) = 0, \nabla\eta\ (0,\bar{p}) = 0$. Then there are $r > 0$, a neighborhood U of $\bar{p}$ in P and a continuous family Γ_p of Lipschitz homeomorphism from B_r into B_R defined for $p \in U$ and such that*

$$f(\Gamma_p(x)) = (1/2)(T(p)x|x) + g(p),$$

for any $x \in B_r, p \in U$ and (10) holds (for $x, x' \in B_r, p \in U$) with $M > m > 0$ depending only on L and the bounds for norms of $T(p)$ and $T^{-1}(p)$.

Proof. Define $x(p)$ as in the proof of Theorem 1. It is clear that $x(\bar{p}) = 0$, so we can choose a neighborhood U of $\bar{p}$ such that $x(p) \in B_{R/2}$ for $p \in U$. For such p the function $\bar{f}(\cdot, p)$ defined by (11) is defined on $B_{R/2}$ and (12) is valid. So it remains (as in the proof of Theorem 1) to apply the Basic Lemma.

Applying as before Corollary 1, we can deduce from the Basic Lemma the following C^2 version of Theorems 1 and 2.

Corollary 2. *If under the assumptions of either Theorem 1 or Theorem 2, $f(\cdot,p)$ belongs to class C^2 and $\nabla^2 f(x,p)$ is continuous jointly in (x,p), then the conclusions of the theorems can be strengthened by the statement that each $\Gamma_p(\cdot)$ is a diffeomorphism.*

Theorems 1 and 2 can be further specified if we assume that the parameter space P is itself a Banach space and the dependence of f on p is similar to its dependence on x.

Theorem 3. *Assume in addition to the hypotheses of either Theorem 1 or Theorem 2 that P is itself a Banach space and $T(p)$ and $f(x,p)$ are $C^{1,1}$, the latter jointly in (x,p).*

Let $x(p)$ be the solution of the equation

$$Tx + \nabla\eta\ (x,p) = 0$$

and let $\Gamma_p(x)$ and $g(p)$ be defined as in Theorems 1 and 2. Then:

(a) g is $C^{1,1}$ and

$$g_p'(p) = (1/2)(T_p'(p)x(p)|x(p)) + \eta_p'(x(p),p),$$

and

(b) $(x,p) \to (\Gamma_p(x),p)$ is a Lipschitz homeomorphism from either $H \times P$ (in case of Theorem 1) or $B_r \times U$ (in case of Theorem 2) into $H \times P$ and $(x(p),p)$ is a critical point of f (as a function of (x,p)) if and only if p is a critical point of g.

Proof. We first observe that under the assumptions, $x(p)$ and $\Gamma_p(x)$ satisfy the Lipschitz condition, the latter jointly in (x,p). Therefore the equality

$$\begin{aligned} g(p+q) - g(p) &= f(x(p+q),p+q) - f(x(p),p) \\ &= (1/2)(T_p'(p)x(p)|x(p))q + \eta_p'(x(p),p)q + (T(p)x(p) \\ &\qquad +\nabla\eta(x(p),p))(x(p+q) - x(p)) + o(||q||) \end{aligned}$$

proves (a) (taking into account that by construction $T(p)x(p)+\nabla\eta(x(p),p) = \nabla f(x(p),p) = 0$).

To prove (b) we note that $(0,p)$ is a critical point of $\psi(x,p) = f(\Gamma_p(x),p) = (1/2)(T(p)x|x) + g(p)$ if and only if p is a critical point of g. The fact that $(x,p) \to (\Gamma_p(x),p)$ is a Lipschitz homeomorphism (on a corresponding domain) follows from the theorems as $\Gamma_p(x)$ satisfies the Lipschitz condition. It remains to take into account that the property of a function to have a zero derivative at a point is invariant under Lipschitz homeomorphisms.

In a more general context, the property of being a critical point in the sense of "metric critical point theory" [5],[8],[9] is invariant under Lipschitz homeomorphisms.

As an easy corollary of the local version of Theorem 3 we get the following $C^{1,1}$ extension of the splitting theorem of Gromol and Meyer [6] (see also [4]).

Corollary 3. *Let $F(x)$ be a $C^{1,1}$ function in a neighborhood of zero. We assume that zero is a critical point of F and that F is twice Fréchet differentiable at zero with $A = \nabla^2 F(0)$ being a strict Fréchet derivative of $\nabla f(\cdot)$ at zero. Assume further that the set $[-\lambda, \lambda] \setminus \{0\}$ does not contain points of the spectrum of A. Let $Y = Ker A$. Then there are an $r > 0$, a zero preserving Lipschitz diffeomorphism Φ from B_r into H and a Lipschitz mapping $\Psi : B_r \cap Y \to Z = Y^{\perp}$ such that for $x \in B_r$*

$$F(\Phi(x)) = (1/2)(Az|z) + F(y + \Psi(y)),$$

where y and z stand for the projections of x onto Y and Z respectively, and $F(y + \Psi(y))$ is a $C^{1,1}$ function on $B_r \cap Y$.

Proof. By the assumptions

$$||\nabla F(x + h) - \nabla F(x) - Ah|| = \rho(x, h)||h||,$$

where $\rho(x, h) \to 0$ as $x \to 0$, $h \to 0$. It follows that the mapping $\nabla F(x) - Ax$ satisfies on every ball B_r the Lipschitz condition with constant $K_r \to 0$ as $r \to 0$.

Consider the function $f(z, y) = F(y + z)$ on $Z + Y$, and let T be the restriction of A to Z. Then setting

$$\eta(z, y) = f(z, y) - (1/2)(Tz|z)$$

we conclude that η is $C^{1,1}$ with $\nabla \eta$ having Lipschitz constant on B_r going to zero as $r \to 0$. On the other hand, as A is a self-adjoint, T maps Z into Z and by the assumptions, T is invertible with $||T^{-1}||^{-1} \geq \lambda$. This means that we can choose R so small that the Lipschitz constant of η on B_r is strictly smaller than λ.

Now we apply the local version of Theorem 3 (that is corresponding to Theorem 2) to $P = Y$, $\bar{p} = 0$ and H replaced by Z. It follows that there are $r > 0$ and a continuous family $\Gamma_y(\cdot)$ of Lipschitz homeomorphisms from $B_r \cap Z$ into Z defined on $B_r \cap Y$ which satisfy the Lipschitz condition jointly in (z, y) on B_r and a Lipschitz mapping $\Psi(y) = z(y)$ from $B_r \cap Y$ into Z such that

$$f(\Gamma_y(z), y) = (1/2)(Tz|z) + f(\Psi(y), y),$$

$(z, y) \to (\Gamma_y(z), y)$ is a Lipschitz homeomorphism, and $y \to f(\Psi(y), y)$ is a $C^{1,1}$-function. But then the mapping $\Phi(x) = \Phi(z, y) = \Gamma_y(z) + y$ is a Lipschitz homeomorphism and it remains to note that $f(\Gamma_y(z), y) = F(\Gamma_y(z) + y) = F(\Phi(x))$ and $f(\Psi(y), y) = F(\Psi(y) + y)$.

References

[1] Arnold, V.I., Varchenko, A.N. and Gussein–Zade, S.M., *Singularities of differentiable mappings*, Nauka, Moscow 1982 (Russian).

[2] Aubin, J.P. and I. Ekeland, I., *Applied nonlinear analysis*, Wiley Interscience, 1984.

[3] Cambini, A., Sul lemma di Morse, Boll. Un. Mat. It. **7** (1973), 87-93.

[4] Chang, K., *Infinite dimensional Morse theory and multiple solution problems*, Birkhäuser, 1993.

[5] Corvellec J.M., Degiovanni, M. and Marzocchi, M., Deformation properties of continuous functionals and critical point theory, Topol. Methods in Nonlinear Analysis **1** (1993),151–171.

[6] Gromol, D. and Meyer, W., On differentiable functions with isolated critical points, *Topology* **8** (1969), 361-369.

[7] Hofer, H., The topological degree at a critical point of mountain pass type, Proc. Symp. Pure Math. **45** (1986), 501-509.

[8] Ioffe, A. and Schwartzman, E., Metric critical point theory I. Morse regularity and homotopic stability of a minimum, J. Math. Pures. Appl. **75** (1996), 125-153.

[9] Katriel, G., Mountain pass theorem and a global homeomorphism theorem, Ann. Inst. Henri Poincaré – Analyse Non–linéaire **11** (1994), 189–211.

[10] Kuiper, N., C^1-equivalence of functions near isolated critical points, Symp. on Infinite-dimensional Topology, *Annals of Math. Studies* **69**, Princeton Univ. Press, 1972.

[11] Mahwin, J. and Willem, M.,On the generalized Morse lemma, Bull. Soc. Math. Belgique **37** (1985), 23-29.

[12] Nirenberg, L., *Topics in nonlinear functional analysis*, Courant Institute Lecture Notes, New York, 1974.

[13] Palais, R.S., Morse theory on Hilbert manifolds, Topology **2** (1963), 299–340.

[14] Schwartz, J.T., *Nonlinear functional analysis*, Gordon and Breach, 1969.

[15] Takens, F., A note on sufficiency of jets, Inventions Math. **13** (1971), 225–231.

Contemporary Mathematics
Volume **204**, 1997

THE FRACTIONAL-LINEAR TRANSFORMATIONS OF THE OPERATOR BALL AND DICHOTOMY OF SOLUTIONS TO EVOLUTION EQUATIONS

VICTOR KHATSKEVICH AND LEONID ZELENKO

ABSTRACT. In this paper we consider the ball of angular operators determined by an indefinite metric on a Hilbert space. The weak compactness of the preimage of this ball by a fractional-linear transformation is established. This transformation is generated by a continuous linear operator, which is not continuously invertible in general. We apply the above nonlinear result to the study of dichotomous behavior of solutions of a nonautonomous linear differential equation with an unbounded operator coefficient in a Hilbert space. The evolution operator of this equation is not continuously invertible in general.

1. INTRODUCTION

We consider a fractional-linear transformation of the ball $\mathfrak{K}^-$ of all angular operators corresponding to the set of all maximal negative subspaces of a Hilbert space $\mathfrak{H}$ with an indefinite metric. This transformation is generated by a continuous linear operator U in $\mathfrak{H}$ (so called plus-operator: see for example [1]-[4]). We do not suppose U to be continuously invertible. The weak compactness of the preimage of $\mathfrak{K}^-$ by U is established (Theorem 2.1). The above result is a development of corresponding results of V. Khatskevich [1], [2] concerning fractional-linear transformations generated by plus-operators in Krein spaces [3], [4]. We apply this

1991 *Mathematics Subject Classification.* Primary 34G10, 35B40; Secondary 47N20.
The second author was supported by the Ministry of Science of Israel, Grant 3991-1-91.

result to the study of dichotomous behavior of solutions for a nonautonomous linear differential equation

$$\frac{dy}{dt} = A(t)y. \tag{1.1}$$

This behavior means that solutions with initial values belonging to some subspace of the phase space stabilize themselves to zero at infinity, but all the other solutions grow infinitely. This property was studied in the papers and books of O. Perron [5] and of J.L.Massera and J.J. Schaffer [6] for the case of a bounded operator coefficient A(t).

The study of the similar question for equation (1.1) with a unbounded operator coefficient $A(t)$ is of particular interest. This interest is stimulated by applications to the qualitative theory of partial differential equations [7].

We obtain a condition of dichotomous behavior in terms of a family of indefinite metrics (Theorem 3.1)) and use the method of imbedded bicones ([8],[9]). We obtain local and integral estimates for the velocity of dichotomy. In the case when the evolution operator of (1.1) is not continuously invertible, the application of the method of imbedded bicones is connected with certain topological difficulties. To overcome these difficulties we use Theorem 2.1, mentioned above. As an application of our results, we have obtained in the paper [11] some constructive conditions of dichotomous behavior of solutions of the boundary value problem for a nonautonomous linear diffusion equation.

Let us notice that in the paper of V. Khatskevich and V. Shulman [10] the convexity of the above mentioned preimage of the ball $\mathfrak{K}^-$ was established under the assumption that the operator U is continuously invertible. This result permit them to investigate in [10] the dichotomy for evolution equations which describe invertible processes only.

2. Plus-operators and fractional-linear transformations

2.1^o. Let $\mathfrak{H}$ be a complex or real Hilbert space with the inner product (x, y) and with the norm $\|x\|$. Define on $\mathfrak{H}$ an indefinite metric by means of an indefinite inner product $[x, y]$ generated by a selfadjoint bounded operator V, i.e. $[x, y] = (Vx, y)$. Assume V to be continuously invertible. We will denote the space $\mathfrak{H}$ with the indefinite metric $[x, y]$ by $(\mathfrak{H}, V)$. Let $\mathfrak{H}^+$, $\mathfrak{H}^-$ be the invariant subspaces of V, corresponding to the positive and negative parts of its spectrum respectively. Set $d_+ = \dim(\mathfrak{H}^+)$, $d_- = \dim(\mathfrak{H}^-)$. One has:

$$\mathfrak{H} = \mathfrak{H}^+ \bigoplus \mathfrak{H}^-. \tag{2.1}$$

Denote by P^+, P^- the orthogonal projections on $\mathfrak{H}^+$, $\mathfrak{H}^-$ respectively. It is well known that one can reduce the indefinite square form to the form:

$$[x, x] = \|P^+x\|^2 - \|P^-x\|^2,$$

passing to an equivalent norm in the space $\mathfrak{H}$ [3]. In the sequel we will suppose such passage to be realized. We define the following "bicones" connected with above indefinite metric:

$$C^+ = \{x \in \mathfrak{H} : [x, x] \geq 0\},$$

$$C^- = \{x \in \mathfrak{H} : [x,x] \leq 0\}.$$

Consider two spaces $(\mathfrak{H}, V_i)$ $(i = 1,2)$ with indefinite metrics of the form described above. Assume that the signatures of the square forms $[x,x]_i = (V_i x, x)$ are the same, namely $\{d_+, d_-\}$, where $d_+ < \infty$. Such spaces with indefinite metric are called in the literature *Pontryagin spaces* (a special case of Krein spaces) [3]. Let U be a linear bounded operator acting from $(\mathfrak{H}, V_1)$ into $(\mathfrak{H}, V_2)$, such that

$$U(C_1^+) \subset C_2^+,$$

i.e. U is a plus-operator [3]. Consider the block matrix representation of the operator U, corresponding to the decomposition (2.1) of the spaces $(\mathfrak{H}, V_i)$ $(i = 1,2)$:

$$U = \{U_{i,j}\}_{i,j=1}^2 : \mathfrak{H}_1^+ \bigoplus \mathfrak{H}_1^- \to \mathfrak{H}_2^+ \bigoplus \mathfrak{H}_2^-, \tag{2.2}$$

where the subspaces $\mathfrak{H}_i^+$, $\mathfrak{H}_i^-$ correspond to the decomposition of the form (2.1), associated with the Pontryagin space $(\mathfrak{H}, V_i)$ $(i = 1,2)$. We denote by $\mathfrak{K}^-$ the set of all angular operators K^-, which correspond to the maximal closed subspaces L_- lying in the "bicone" C_1^-. Let us recall that they are the linear operators $K^- : \mathfrak{H}_1^- \to \mathfrak{H}_1^+$, for which $L_- = \text{graph}(K^-)$, i.e. $L_- = \text{Im}(U(P^- + K^-))$. Let $\mathfrak{M}$ be the subset of $\mathfrak{K}^-$, corresponding to the subspaces L_-, for which $U(L_-) \subset C_2^-$. In other words:

$$\mathfrak{M} = \{K^- \in \mathfrak{K}^- : U(P^- + K^-)\mathfrak{H}_1^- \subset C_2^-\}. \tag{2.3}$$

2.2^o. We now turn to the main assertion of this section.

Theorem 2.1. *Assume that the operator U maps the set $C_1^+ \setminus \{0\}$ into another one $C_2^+ \setminus \{0\}$. Then the set $\mathfrak{M}$ defined by (2.3) is non-empty and compact with respect to the weak operator topology.*

Proof. . First we show that $\mathfrak{M}$ is non-empty. The assumptions on U show that $U_{1,1}$ has zero kernel (see the representation (2.2)). But $\dim(\mathfrak{H}_1^+) = \dim(\mathfrak{H}_)^+ < \infty$, so there exists the inverse $U_{1,1}^{-1}$. The operator $K^- = -U_{1,1}^{-1}U_{1,2}$ belongs to $\mathfrak{M}$, since the operator $U_{1,1} + U_{1,2}K^+$ is invertible for any contraction $K^+ : \mathfrak{H}_1^+ \to \mathfrak{H}_1^-$. So the set $\mathfrak{M}$ is nonempty.

Now we establish the compactness of $\mathfrak{M}$. Notice that the set $\mathfrak{K}^-$ is weakly compact, because this is the set of all contractions $\mathfrak{H}_1^- \to \mathfrak{H}_1^+$. Thus, it is enough to show that $\mathfrak{M}$ is weakly closed in $\mathfrak{K}^-$. But if $K_\alpha^- \to K^-$ is a weakly convergent net in K^- with $K_\alpha^- \in \mathfrak{M}(\forall \alpha)$, then for each $x \in \mathfrak{H}_1^-$: $K_\alpha^- x \to K^- x$ weakly and hence strongly in the finite dimensional space $\mathfrak{H}_1^+$. So

$$C_2^- \ni U(P^- + K_\alpha^-)x \to U(P^- + K^-)x$$

strongly, hence

$$U(P^- + K^-)x \in C_2^- \quad \forall x \in \mathfrak{H}_1^-.$$

This shows that $\mathfrak{M}$ is weakly closed in $\mathfrak{K}^-$. ■

Concerning Theorem 2.1 see also [3], Subsection 8.20, Chapter 1 (p. 59).

3. Dichotomy of solutions of an evolution equation

3.1^o Consider a linear differential equation of the form (1.1) in a Hilbert space $\mathfrak{H}$, where $t \in \mathbf{R}_+$ and $A(t)$ are in general linear closed unbounded operators with a common domain of definition $\mathrm{Dom}(A(t)) \equiv D_0$, which is dense in $\mathfrak{H}$. Further we impose the following restriction:

(A) *The Cauchy problem for (1.1) is uniformly correct on the half-line* $\mathbf{R}_+$ [12].

The condition (A) means that for (1.1) there exists the continuous on D_0 evolution operator $U(t,s)$ $(0 \le s \le t)$ associating with each $y_0 \in D_0$ the value $y(t,s)$ of the solution of (1.1), which satisfies the initial condition: $y(s,s) = y_0$. Then for each fixed t, s the operator $U(t,s)$ may be continuously extended to all of $\mathfrak{H}$; for this extension we conserve the same notation $U(t,s)$. For any $y_0 \in \mathfrak{H}$ the function $y(t,s) = U(t,s)y_0$ is said to be *the generalized solution* of the equation (1.1), satisfying the initial condition $y(s,s) = y_0$.

3.2^0. Consider a family of selfadjoint bounded operators $\{V(t)\}_{t\in\mathbf{R}_+}$, which define in $\mathfrak{H}$ the family of indefinite metrics

$$[x,y]_t = (V(t)x, y). \tag{3.1}$$

Further we will impose the following conditions on these metrics:

(B) *$V(t)$ is a strongly continuously differentiable operator function and for any $t \in \mathbf{R}_+$ the derivative $\frac{d}{dt}_{(1.1)}[x,x]_t$ of the square form $[x,x]_t$ with respect to (1.1) is positive definite with a lower bound $k(t)$, which is a positive locally integrable function on $\mathbf{R}_+$;*

(C) *For each fixed $t \in \mathbf{R}_+$ $V(t)$ is a continuously invertible operator;*

(D) *For each operator $V(t)$ the dimension d_+ of its invariant subspace corresponding to the positive spectrum is finite and does not depend on t;*

(E) *The function $\|V(t)\|$ is bounded on $\mathbf{R}_+$ and moreover:*

$$\sup_{t\in\mathbf{R}_+} \|V(t)\| = 1.$$

The following important statement is based on the condition (B):

Lemma 3.1. . *Assume that the conditions (A) and (B) are fulfilled. Then the following relation is valid for all $0 \le \tau \le t$ and for any $y_0 \in \mathfrak{H}$:*

$$[U(t,\tau)y_0, U(t,\tau)y_0]_t - [y_0, y_0]_\tau \ge \int_\tau^t k(s)\|U(t,\tau)y_0\|^2 ds. \tag{3.2}$$

Consider the following "bicones", connected both with the metrics (3.1) and the evolution operator of (1.1):

$$C_t^- = \{y_0 \in \mathfrak{H} : [U(t)y_0, U(t)y_0]_t \le 0\} \quad (t \in \mathbf{R}_+). \tag{3.3}$$

It is easy to show, using (3.2), that for the family of these "bicones" the property of monotonicity holds:

$$C_t^- \subset C_\tau^-, \text{ if } t > \tau. \tag{3.4}$$

We set

$$C_\infty^- = \bigcap_{t\in\mathbf{R}_+} C_t^-. \tag{3.5}$$

Applying Theorem 2.1 to the Pontryagin spaces $(\mathfrak{H}, V(0))$, $(\mathfrak{H}, V(t))$ and to the operator U = U(t,0) and taking into account (3.4), we obtain the following statement:

Lemma 3.2. *Assume that the conditions (A)-(D) are fulfilled. Then the set* C_{∞}^{-} *contains a closed subspace* $L_{\infty}^{-} \subset \mathfrak{H}$ *lying in the "bicone"* C_0^{-}, *which satisfies the condition:*

$$codim(L_{\infty}^{-}) = d_{+}. \tag{3.6}$$

Remark. It is easy to see that L_{∞}^{-} is a maximal closed subspace lying in C_0^{-}.

Using Lemmas 3.1, 3.2 and the Bellman-Gronwall inequality [13], we obtain the following assertion on dichotomy of solutions of the equation (1.1):

Theorem 3.1. *Assume that the conditions (A)-(E) are fulfilled and the function* $k(t)$ *(see (B)) satisfies the condition:*

$$\int_0^{\infty} k(t)dt = \infty.$$

Then the following statements are true:

(1) the set C_{∞}^{-} *(3.5) coincides with a closed subspace* L_{∞}^{-}, *which satisfies the condition (3.6);*

(2) for any $y_0 \in L_{\infty}^{-}$ *the generalized solution* $y(t) = U(t,0)y_0$ *of (1.1) satisfies the condition:*

$$I(y) = \int_0^{\infty} k(s)\|y(s)\|^2 ds < \infty$$

and the following estimate is valid for it:

$$\int_t^{\infty} k(s)\|y(s)\|^2 ds \leq I(y)exp(-2\int_0^t k(s)ds);$$

(3) for any $y_0 \notin C_0^{-}$ *the generalized solution* $y(t) = U(t,0)y_0$ *of (1.1) satisfies the inequality:*

$$\|y(t)\|^2 \geq [y_0, y_0]_0 exp(2\int_0^t k(s)ds)$$

References

1. V. Khatskevich. *On the symmetry of properties of plus-operator and its conjugate operator*, Funct. analysis (Ulyanovsk) 14 (1980), 177-186.

2. V. Khatskevich. *On characteristic spectral properties of focused operators*, Doklady Acad. Nauk Arm. S.S.R., 79 (1984), 102-105.

3. T. Ya. Azizov and I. S. Iohvidov. *Foundations of the linear operator theory in spaces with indefinite metric*, Nauka, Moscow, 1986.

4. M. G. Krein. *On one new application of fixed point principle to the operator theory in spaces with indefinite metric*, Dokl. Acad. Nauk S.S.S.R., 154 (1964), 1023- 1026.

5. O. Perron. *Die Stabilitatsfrage bei Differentialgleichungen*, Math. Z., 32 (1930), 703- 728.

6. J. L. Massera and J. J. Shaffer. *Linear differential equations and function spaces.* Acad. Press, New York, London, 1966.

7. S. Agmon, L. Nirenberg *Properties of solutions of ordinary differential equations in Banach space.* Comm. Pure and Appl. Math., 16, 2 (1963) 121-239.

8. A. D. Maizel. *On stability of solutions of systems of differential equations*, Ural. Politehn. Instit. Trudy, 51 (1954), 20-54 (Russian).

9. L. Zelenko. *On manifold of square integrable solutions of nonlinear systems of differential equations*, Differential equations and theory of functions, Saratov University (1984), 19-26 (Russian).

10. V. Khatskevich and V. Shulman. *Operator fractional-linear transformations: convexity and compactness of image; applications*, Studia Math. 116 (2) (1995), 189- 195.

11. V. Khatskevich and L. Zelenko. *Indefinite metrics and dichotomy of solutions of linear differential equations in Hilbert spaces.* Chinese J. of Math., No 2 (1996)- to appear.

12. S. G. Krein. *Linear differential equations in Banach space*, Transl. of Math. Monogr., vol 29, A.M.S. Providence, R.I., 1971.

13. L. Cesary. *Asymptotic behavior and stability problems in ordinary differential equations*, Springer Verlag, Berlin, Gottingen, Heidelberg, 1959.

Department of Applied Mathematics, International College of Technology, Karmiel, P.O.B. 78, Karmiel 20101, Israel.

Department of Mathematics and Computer Science, University of Haifa, Haifa 31905, Israel. *Email:* rsmaf05@haifauvm.bitnet

Contemporary Mathematics
Volume **204**, 1997

Stable Approximation of Nondifferentiable Optimization Problems with Variational Inequalities

M. ZUHAIR NASHED AND OTMAR SCHERZER

Abstract

This paper is devoted to optimization theory (existence, uniqueness, and stability) in nonreflexive Banach spaces. A general method is developed for the stable approximate solution of a class of optimization problems for which approximate minimizers can be characterized as solutions of variational inequalities. The functional to be minimized is not assumed to be differentiable, and the minimizers are not required to satisfy a variational inequality. Applications to an inverse source problem and a minimal surface problem are presented.

1 Introduction

In this paper we study minimization problems in nonreflexive Banach space for which approximate minimizers can be characterized as solutions of variational inequalities. We do not assume any differentiability condition on the functional to be minimized and we do not require the minimizers themselves to satisfy a variational inequality.

Many important applications (some are classical such as minimal surface problems; some are modern such as image restoration problems) lead to optimization problems in nonreflexive Banach spaces of the type considered in this paper.

Optimization theory (existence, uniqueness, and stability) in *nonreflexive* Banach spaces is more subtle than the theory in reflexive Banach spaces, which is partly

1991 Mathematics Subject Classification: Primary 49J40, 49Q05, 65J15, 65J20, 65K10, 86A22; Secondary 26A45, 46N10, 49J45, 49N60.

Key words and phrases: Nondifferentiable optimization problems, variational inequalities, nonreflexive spaces, regularization, inverse source problems, minimal surface problem, bounded variation norm.

The work of the second author is supported by the Austrian Fonds zur Förderung der wissenschaftlichen Forschung, Grant J01088-TEC; O.S. is on leave from Institut für Mathematik, Universität Linz, Altenberger Str. 69, A–4040 Linz, Austria

due to the fact that the weak topology of a Banach space does not coincide with the weak–star topology of its second dual. Another marked difference arises in consideration of stability of minimizing sequences. We illustrate this by the following situation which is relevant to the framework of this paper: For a function $f \in \mathrm{BV}(\Omega)$ (the space of functions of bounded variation) it cannot be expected (in general) to find functions $f_j \in C^\infty(\Omega)$ such that $f_j \to f$ in $L^1(\Omega)$ and $\int_\Omega |D(f_j - f)| \to 0$, where D denotes "differentiation". This is due to the fact that the closure of the C^∞–functions with respect to the BV–norm is the Sobolev space $W^{1,1}(\Omega)$ [4]. This shows that functions of bounded variation, which are not in $W^{1,1}(\Omega)$ (like piecewise constant functions) cannot be approximated by a family of smooth functions with respect to the bounded variation norm. Ramification and implications of this situation for an inverse source problem are analyzed in Section 3.1.

Existence and uniqueness of solutions of various classes of optimization problems in nonreflexive Banach spaces have been studied in the literature. However, questions of stability with respect to the input data have not been satisfactorily addressed, partly due to the difficulties described above. To circumvent these difficulties stability questions can be considered in weaker norms, which is sometimes quite unsatisfactory. For example such an approach would not be meaningful for the solution of the inverse source problem. A more satisfactory approach to overcome these difficulties is to consider a family of modified functionals on a Hilbert space (which are appropriately embedded in the nonreflexive Banach space) and to consider the minimizers of the modified optimization problems as approximate solutions of the original problem. This idea is used in this paper to study stable methods for the approximate solution of optimization problems of the type

$$\min_{v \in \mathcal{D}} \mathcal{J}(v), \text{ where } \mathcal{J}(v) := \phi(v, z^\delta) + k(v) - l(v); \tag{1.1}$$

here $\mathcal{D}$ is a (nonreflexive) Banach space, z^δ is given perturbed input data of the exact data z, l is a linear functional from $\mathcal{D}$ into $I\!R$, and finally ϕ, k are nonlinear (not necessarily differentiable) functionals from $\mathcal{D}$ into $I\!R$, satisfying certain conditions described in Section 2.

The thrust of this paper is a study of approximate solutions of a family of "regularized" optimization problems, for the optimization problem (1.1), of the form

$$\min_{v \in \mathcal{D}_\varepsilon} \mathcal{J}_\varepsilon(v), \text{ where } \mathcal{J}_\varepsilon(v) := \phi_\varepsilon(v, z^\delta) + k_\varepsilon(v) - l_\varepsilon(v) + \frac{1}{2} j_\varepsilon(v)\,, \tag{1.2}$$

for which the minimizers can be characterized as solutions of variational inequalities. A typical situation, which we have in mind is $\mathcal{D} = \mathrm{BV}(\Omega)$, the space of functions of bounded variation, and

$$\begin{aligned} k(u) &= \int_\Omega |Du| \\ &:= \sup\Big\{ -\int_\Omega u\nabla.g \, : \, g=(g_1,...,g_d) \in C_0^1(\Omega, I\!R^d)\,, \\ &\qquad\qquad |g(x)|^2 = \sum_{i=1}^d g_i^2(x) \le 1, x \in \Omega \Big\} \end{aligned}$$

the bounded variation seminorm. In order to make an attempt to characterize a directional derivative of k to obtain a variational inequality for a solution of the problem (1.1) we consider

$$\begin{aligned} &\lim_{t\to 0} t^{-1}\,[k(u+tv) - k(u)] \\ &= \lim_{t\to 0} t^{-1}\Big[\sup\Big\{-\int_\Omega (u+tv)\nabla.g : g \in C_0^1(\Omega, I\!R^d)\,,\ |g(x)| \le 1\Big\} - \\ &\qquad \sup\Big\{-\int_\Omega u\nabla.g : g \in C_0^1(\Omega, I\!R^d)\,,\ |g(x)| \le 1\Big\}\Big]\,. \end{aligned}$$

Without smoothness assumptions on u, v the right hand side cannot be characterized via a variational inequality. To avoid this situation we restrict attention to a subset $\mathcal{D}_\varepsilon$ of $\mathcal{D}$, consisting of "smoother" functions.

In Section 2 we develop general principles for the construction of a family of modified minimization problems, for which solutions provide *stable* approximations and which can be *efficiently* constructed by a computational procedure. We also give a convergence analysis of the approximate solutions. In Section 3 we apply the results of this analysis to an inverse source problem and to a minimal surface problem.

2 Regularization techniques for variational inequalities with nondifferentiable terms

Let Y be a real Banach space, and let $\mathcal{D}_\varepsilon$ be a family of real Hilbert spaces, where ε denotes a vector of positive numbers. In this section we consider the minimization of the functional $\mathcal{J}_\varepsilon(v)$ in (1.2) over $\mathcal{D}_\varepsilon$, under the following assumptions on the constituent functionals in (1.2):

1. The functionals

$$\begin{aligned} \phi_\varepsilon : \mathcal{D}_\varepsilon \times Y &\to \overline{I\!R} \\ (v,z) &\to \phi_\varepsilon(v,z) \end{aligned}$$

and

$$k_\varepsilon : D_\varepsilon \quad \to \quad I\!R$$

satisfy the following conditions:

(a) For fixed $z \in Y, \phi_\varepsilon(.,z)+k_\varepsilon(.)$ is a proper convex functional on $\mathcal{D}_\varepsilon$, which is weakly lower semicontinuous (with respect to the norm on $\mathcal{D}_\varepsilon$). Recall that a functional $\Psi : \mathcal{D}_\varepsilon \to I\!R \cup \{\infty\}$ is called proper if $\Psi(.) \not\equiv \infty$ (see e.g. [5]).

(b) Let $z, z^\delta \in Y$, $\|z^\delta - z\| \le \delta$. For fixed ε, there exists a constant C_{1,ϕ_ε} such that for all $u \in \mathcal{D}_\varepsilon$

$$\left|\phi_\varepsilon(u,z) - \phi_\varepsilon(u,z^\delta)\right| \le C_{1,\phi_\varepsilon}\delta.$$

(c) For fixed z, the functional $-\phi_\varepsilon(.,z)$ is uniformly bounded from above, in the sense that

$$-\phi_\varepsilon(u,z) \le \max\left\{C_{3,\phi_\varepsilon}\|u\|_{\mathcal{D}_\varepsilon}, C_{4,\phi_\varepsilon}\right\}.$$

2. The functional $-k_\varepsilon$ is uniformly bounded from above, so for all $u \in \mathcal{D}_\varepsilon$

$$-k_\varepsilon(u) \le \max\left\{C_{3,k_\varepsilon}\|u\|_{\mathcal{D}_\varepsilon}, C_{4,k_\varepsilon}\right\}.$$

3. The functional $l_\varepsilon : \mathcal{D}_\varepsilon \to I\!R$ is continuous and linear, so there exists a constant such that for all $u \in \mathcal{D}_\varepsilon$

$$|l_\varepsilon(u)| \le C_{2,l_\varepsilon}\|u\|_{\mathcal{D}_\varepsilon}.$$

4. $j_\varepsilon(u) = a_\varepsilon(u,u)$, where a_ε is a continuous symmetric bilinear form that is strongly elliptic, in the sense that there exists $e(\varepsilon) > 0$ such that for all $u \in \mathcal{D}_\varepsilon$

$$a_\varepsilon(u,u) \ge e(\varepsilon)\|u\|^2_{\mathcal{D}_\varepsilon}.$$

We emphasize that it is not assumed that $e(\varepsilon) > a > 0$ uniformly for all $\varepsilon > 0$. Indeed in our setting $e(\varepsilon) \to 0$ as $\varepsilon \to 0$.

Before we develop existence, uniqueness, and stability results, we emphasize the nature of the constituent terms in (1.2). The term ϕ_ε contains the data z, k_ε is a nonlinear functional independent of the data, l_ε is a linear functional, and j_ε is a penalty term.

Existence and Uniqueness: It is well known that under assumptions 1a, 3, and 4, the minimization problem (1.2), for fixed ε, has a unique solution, which can be characterized as the solution of the following variational inequality:

$$a_\varepsilon(v, v-u) + \phi_\varepsilon(v,z^\delta) + k_\varepsilon(v) - \phi_\varepsilon(u,z^\delta) - k_\varepsilon(u) \ge l_\varepsilon(v-u), \text{ for all } v \in \mathcal{D}_\varepsilon.$$

For a proof see e.g., Section 4 of Chapter 1 in [5].

Stability: In the next theorem it is shown that the optimization problem (1.2), for each fixed ε, is weakly stable with respect to perturbations in the data y:

Theorem 2.1 *Let ε be a vector of positive numbers. Then each minimizing sequence $\{u_\varepsilon^{\delta_n}\}$ of $\mathcal{J}_\varepsilon$ (where in (1.2) z^δ is replaced by z^{δ_n} and $\|z^{\delta_n} - z\| \leq \delta_n$) has a subsequence in $\mathcal{D}_\varepsilon$ which converges weakly as $\delta_n \to 0$ to a minimizing element of $\mathcal{J}_\varepsilon$ (where now in (1.2) z^δ is replaced by z).*

Proof: From the definition of a minimizing element u_ε^δ of $\mathcal{J}_\varepsilon$ it follows that for any $\eta_\varepsilon \in \mathcal{D}_\varepsilon$

$$\begin{aligned} &\phi_\varepsilon(u_\varepsilon^\delta, z^\delta) + k_\varepsilon(u_\varepsilon^\delta) - l_\varepsilon(u_\varepsilon^\delta) + \tfrac{1}{2} j_\varepsilon(u_\varepsilon^\delta) \\ &\quad \leq \quad \phi_\varepsilon(\eta_\varepsilon, z^\delta) + k_\varepsilon(\eta_\varepsilon) - l_\varepsilon(\eta_\varepsilon) + \tfrac{1}{2} j_\varepsilon(\eta_\varepsilon). \end{aligned}$$

From assumption 1 we have $\phi_\varepsilon(v, z) \leq \phi_\varepsilon(v, z^\delta) + C_{1,\phi_\varepsilon}\delta$. Using the abbreviation

$$\rho_\varepsilon(\eta_\varepsilon, z^\delta) := \phi_\varepsilon(\eta_\varepsilon, z^\delta) + k_\varepsilon(\eta_\varepsilon) - l_\varepsilon(\eta_\varepsilon) + \frac{1}{2} j_\varepsilon(\eta_\varepsilon),$$

and invoking assumption 1 we obtain

$$\begin{aligned} \frac{1}{2} j_\varepsilon(u_\varepsilon^\delta) &\leq \rho_\varepsilon(\eta_\varepsilon, z^\delta) - \phi_\varepsilon(u_\varepsilon^\delta, z^\delta) - k_\varepsilon(u_\varepsilon^\delta) + l_\varepsilon(u_\varepsilon^\delta) \\ &\leq \rho_\varepsilon(\eta_\varepsilon, z^\delta) - \phi_\varepsilon(u_\varepsilon^\delta, z) - k_\varepsilon(u_\varepsilon^\delta) + l_\varepsilon(u_\varepsilon^\delta) + C_{1,\phi_\varepsilon}\delta \\ &\leq \rho_\varepsilon(\eta_\varepsilon, z^\delta) + \max\left\{C_{3,\phi_\varepsilon}\|u_\varepsilon^\delta\|_{\mathcal{D}_\varepsilon}, C_{4,\phi_\varepsilon}\right\} \\ &\qquad + \max\left\{C_{3,k_\varepsilon}\|u_\varepsilon^\delta\|_{\mathcal{D}_\varepsilon}, C_{4,k_\varepsilon}\right\} + \\ C_{1,\phi_\varepsilon}\delta + C_{2,l_\varepsilon}\|u_\varepsilon^\delta\|_{\mathcal{D}_\varepsilon}, \end{aligned}$$

which together with assumption 4 implies that $\{u_\varepsilon^\delta\}$ is bounded in $\mathcal{D}_\varepsilon$. Therefore it has a weakly convergent subsequence in $\mathcal{D}_\varepsilon$ (which again will be denoted by $\{u_\varepsilon^{\delta_n}\}$), i.e.,

$$u_\varepsilon^{\delta_n} \rightharpoonup \underline{u} \text{ in } \mathcal{D}_\varepsilon.$$

Since $\phi_\varepsilon(., z) + k_\varepsilon(.)$ is weakly lower semicontinuous, it follows from assumption 1, for $\delta_n \to 0$, that

$$\begin{aligned} \phi_\varepsilon(\underline{u}, z) + k_\varepsilon(\underline{u}) &\leq \liminf_{n \in \mathbb{N}} \left\{\phi_\varepsilon\left(u_\varepsilon^{\delta_n}, z\right) + k_\varepsilon\left(u_\varepsilon^{\delta_n}\right)\right\} \\ &\leq \limsup_{n \in \mathbb{N}} \left\{\phi_\varepsilon\left(u_\varepsilon^{\delta_n}, z^{\delta_n}\right) + k_\varepsilon\left(u_\varepsilon^{\delta_n}\right) + C_{1,\phi_\varepsilon}\delta_n\right\} \\ &\leq \limsup_{n \in \mathbb{N}} \left\{\phi_\varepsilon\left(u_\varepsilon^{\delta_n}, z^{\delta_n}\right) + k_\varepsilon\left(u_\varepsilon^{\delta_n}\right)\right\}. \end{aligned}$$

From the preceding inequality and the weak semicontinuity of l_ε we obtain for all $v \in \mathcal{D}_\varepsilon$:

$$\begin{aligned}
&\phi_\varepsilon(\underline{u}, z) + k_\varepsilon(\underline{u}) - l_\varepsilon(\underline{u}) + \frac{1}{2} j_\varepsilon(\underline{u}) \\
\leq\ & \liminf_{n \in \mathbb{N}} \left\{ \phi_\varepsilon\left(u_\varepsilon^{\delta_n}, z^{\delta_n}\right) + k_\varepsilon\left(u_\varepsilon^{\delta_n}\right) - l_\varepsilon\left(u_\varepsilon^{\delta_n}\right) + \frac{1}{2} j_\varepsilon\left(u_\varepsilon^{\delta_n}\right) \right\} \\
\leq\ & \limsup_{n \in \mathbb{N}} \left\{ \phi_\varepsilon\left(u_\varepsilon^{\delta_n}, z^{\delta_n}\right) + k_\varepsilon\left(u_\varepsilon^{\delta_n}\right) - l_\varepsilon\left(u_\varepsilon^{\delta_n}\right) + \frac{1}{2} j_\varepsilon\left(u_\varepsilon^{\delta_n}\right) \right\} \\
\leq\ & \limsup_{n \in \mathbb{N}} \left\{ \phi_\varepsilon(v, z^{\delta_n}) + k_\varepsilon(v) - l_\varepsilon(v) + \frac{1}{2} j_\varepsilon(v) \right\} \\
\leq\ & \phi_\varepsilon(v, z) + k_\varepsilon(v) - l_\varepsilon(v) + \frac{1}{2} j_\varepsilon(v) ,
\end{aligned}$$

which shows that $\underline{u}$ is a minimizing element of (1.2). □

If the minimizer of $\mathcal{J}_\varepsilon$ for the unperturbed data z is unique, then it follows by a subsequence–subsequence argument that $\{u_\varepsilon^{\delta_n}\}$ is weakly convergent.

Theorem 2.1 guarantees that the minimization problem (1.2) is weakly stable. There are at least two practical important situations under which the method is strongly stable:

- If $\mathcal{D}_\varepsilon$ is a Sobolev space, it can be compactly embedded in each Sobolev space of lower order (see for example Chapter 1, Section 16 in [9]), i.e., the method (1.2) is strongly stable in the Sobolev space of lower order.
- If $\mathcal{D}_\varepsilon$ is a Hilbert space, $j_\varepsilon(v) = \varepsilon \|v\|^2_{\mathcal{D}_\varepsilon}$, and $0 \leq \phi_\varepsilon(., z^\delta) + k_\varepsilon(.) - l_\varepsilon(.)$ is weakly lower semicontinuous on $\mathcal{D}_\varepsilon$ for any z^δ, and $|\phi_\varepsilon(., z^\delta) - \phi_\varepsilon(., z)| \leq C\|z^\delta - z\|$, then it follows from the stability result in [2] (together with some minor modifications) that the method is also strongly stable in $\mathcal{D}_\varepsilon$.

(Weak) Convergence: We next prove that the minimizers of the problems (1.2), which have been shown to be stable with respect to perturbations in the data, also provide approximate solutions of the original minimization problem (1.1). To this end, we need to link the spaces $\mathcal{D}_\varepsilon$ to the space $\mathcal{D}$, and the functionals $\mathcal{J}_\varepsilon$ to the functional $\mathcal{J}$. Up to this point the spaces $\mathcal{D}_\varepsilon$ are not related to the space $\mathcal{D}$; moreover, the existence, uniqueness, and stability are all in the framework of the space $\mathcal{D}_\varepsilon$, for fixed ε. We shall prove that the family of minimization problems (1.2) is a regularization method for the problem (1.1) under the following assumptions:

5. The Hilbert spaces $\mathcal{D}_\varepsilon$ have continuous embeddings in the (nonreflexive) Banach space $\mathcal{D}$, and $\mathcal{D}$ has a continuous embedding in some reflexive Banach space $\mathcal{Z}$.

6. A minimum of $\mathcal{J}$ is attained at $\underline{u} \in \mathcal{D}$.

7. There exist sequences $\{\varepsilon_n\} \to 0$, $\{\eta_n\} \in \mathcal{D}_{\varepsilon_n}$ with the properties:

$$
\begin{array}{rcl}
j_{\varepsilon_n}(\eta_n) & \to & 0, \\
l_{\varepsilon_n}(\eta_n) & \to & l(\underline{u}), \\
\limsup_{n \in \mathbb{N}} \{k_{\varepsilon_n}(\eta_n) + \phi_{\varepsilon_n}(\eta_n, z)\} & \leq & k(\underline{u}) + \phi(\underline{u}, z).
\end{array} \tag{2.1}
$$

8. (1b) holds uniformly in ε, i.e., for every $\varepsilon > 0$ and $z^\delta \in Y$, which satisfies $\|z^\delta - z\| \leq \delta$ and for all $v \in \mathcal{D}_\varepsilon$

$$
\left|\phi_\varepsilon(v, z) - \phi_\varepsilon(v, z^\delta)\right| \leq C_1 \delta. \tag{2.2}
$$

9. The functional $\mathcal{J}_\varepsilon(.)$, defined in (1.2), is $\mathcal{D}$-coercive, uniformly in ε, i.e., there exist positive constants C_5, C_6 such that for all $u \in \mathcal{D}$

$$
\|u\|_{\mathcal{D}} \leq C_5 \mathcal{J}_\varepsilon(u) + C_6. \tag{2.3}
$$

10. For given data z^δ, the functional

$$
\phi(v, z^\delta) + k(v) - l(v) \tag{2.4}
$$

is weakly lower semicontinuous on $\mathcal{D}$, with respect to the norm on $\mathcal{Z}$, i.e., if $v_n \in \mathcal{D}$ and $\{v_n\}$ converges weakly to v in $\mathcal{Z}$, then $v \in \mathcal{D}$ and

$$
\phi(v, z^\delta) + k(v) - l(v) \leq \liminf_{n \in \mathbb{N}} \left\{\phi(v_n, z^\delta) + k(v_n) - l(v_n)\right\} .
$$

Theorem 2.2 *Let assumptions 5 – 10 be satisfied. Then each minimizing sequence* $\{u_{\varepsilon_n}^{\delta_n}\}$ *of* $\mathcal{J}_{\varepsilon_n}$ *has a subsequence in* $\mathcal{Z}$*, which is weakly convergent (in* $\mathcal{Z}$*) to a minimizing element of* $\mathcal{J}$*. If the minimizer of* $\mathcal{J}$ *is unique, then the sequence is itself weakly convergent.*
If additionally the embedding from $\mathcal{D}$ *into* $\mathcal{Z}$ *is compact, then weak convergence can be replaced by strong convergence; in this case, assumption 10 can be replaced by the following weaker assumption*

11. *(2.4) is lower semicontinuous on* $\mathcal{D}$*, with respect to the norm on* $\mathcal{Z}$*.*

Proof: From the definition of $u_{\varepsilon_n}^{\delta_n}$ and (2.2) it follows that

$$
\begin{array}{rl}
& \phi_{\varepsilon_n}(u_{\varepsilon_n}^{\delta_n}, z^{\delta_n}) + k_{\varepsilon_n}(u_{\varepsilon_n}^{\delta_n}) - l_{\varepsilon_n}(u_{\varepsilon_n}^{\delta_n}) + \frac{1}{2} j_{\varepsilon_n}(u_{\varepsilon_n}^{\delta_n}) \\
\leq & \phi_{\varepsilon_n}(\eta_{\varepsilon_n}, z^{\delta_n}) + k_{\varepsilon_n}(\eta_{\varepsilon_n}) - l_{\varepsilon_n}(\eta_{\varepsilon_n}) + \frac{1}{2} j_{\varepsilon_n}(\eta_{\varepsilon_n}) \\
\leq & \phi_{\varepsilon_n}(\eta_{\varepsilon_n}, z) + k_{\varepsilon_n}(\eta_{\varepsilon_n}) - l_{\varepsilon_n}(\eta_{\varepsilon_n}) + \frac{1}{2} j_{\varepsilon_n}(\eta_{\varepsilon_n}) + C_1 \delta_n.
\end{array}
$$

Consequently, since by assumption 6 there exists a global minimum $\underline{u} \in \mathcal{D}$ of the functional $\mathcal{J}$ (cf. (1.1)), the following inequality follows from assumption 7 as both δ_n and ε_n tend to zero:

$$
\begin{aligned}
&\liminf_{n\in I\!N} \mathcal{J}_{\varepsilon_n}\left(u_{\varepsilon_n}^{\delta_n}\right) \\
&\quad\le \limsup_{n\in I\!N} \mathcal{J}_{\varepsilon_n}\left(u_{\varepsilon_n}^{\delta_n}\right) \\
&\quad\le \limsup_{n\in I\!N} \left\{ \phi_{\varepsilon_n}(\eta_{\varepsilon_n}, z) + k_{\varepsilon_n}(\eta_{\varepsilon_n}) - l_{\varepsilon_n}(\eta_{\varepsilon_n}) + \frac{1}{2} j_{\varepsilon_n}(\eta_{\varepsilon_n}) + C_1 \delta_n \right\} \\
&\quad\le \mathcal{J}(\underline{u}).
\end{aligned}
$$

Together with (2.3) this shows that the sequence $\{u_{\varepsilon_n}^{\delta_n}\}$ is bounded in $\mathcal{D}$ and weakly compact in $\mathcal{Z}$, i.e., there exists a subsequence of $\{u_{\varepsilon_n}^{\delta_n}\}$, which will still be denoted by $\{u_{\varepsilon_n}^{\delta_n}\}$, which is weakly convergent, i.e., $u_{\varepsilon_n}^{\delta_n} \rightharpoonup w$ in $\mathcal{Z}$. From the weak lower semicontinuity of $\mathcal{J}$ on $\mathcal{D}$ with respect to the norm on $\mathcal{Z}$ it follows that

$$
\mathcal{J}(w) \le \liminf \mathcal{J}(u_{\varepsilon_n}^{\delta_n}) \le \mathcal{J}(\underline{u}),
$$

which shows that w is a minimizing element of $\mathcal{J}$. If the embedding from $\mathcal{D}$ into $\mathcal{Z}$ is also compact, then $\{u_{\alpha_n}^{\delta_n}\}$ is convergent in $\mathcal{Z}$. □

We note that it cannot be expected that we obtain (weak) convergence with respect to $\mathcal{D}_\varepsilon$, since a minimizer of $\mathcal{J}$ does usually not belong to $\mathcal{D}$. The assumption $\mathcal{D} \subseteq \mathcal{Z}$ with continuous embedding and assumption 10 are used to deduce that the limit of $\{u_{\varepsilon_n}^{\delta_n}\}$ is a minimizer of $\mathcal{J}$. This is the step of the proof that required the embedding of the (nonreflexive) Banach space $\mathcal{D}$ in the reflexive Banach space $\mathcal{Z}$.

3 Applications

In this section we present two applications which can be treated within the framework of the results of Section 2. Efficient numerical algorithms can be applied for the construction of the approximate minimizers.

3.1 An inverse source problem

We consider the inverse source problem of determining a function f on Ω satisfying

$$
A(f) = u \quad \text{in } \Omega \tag{3.1}
$$

from its indirect, noisy measurements $u^\delta \in L^1(\Omega)$ of the exact data u. It is assumed that $\Omega \subseteq I\!R^d, d \ge 1$ is a bounded domain with C^1–boundary $\partial\Omega$. The operator A is not assumed to be linear.

The source term f to be estimated is assumed to be a function of bounded variation, i.e., $f \in \mathrm{BV}(\Omega)$, where

$$\mathrm{BV}(\Omega) := \left\{ g \in L^1(\Omega) : \int_\Omega |Dg| < \infty \right\} . \tag{3.2}$$

Functions of bounded variation are of particular interest in practical situations, since they include piecewise continuous source terms.

In the sequel it will be assumed that the parameter-to-solution mapping A is compact from $L^1(\Omega)$ into $L^1(\Omega)$. This assumption can be weakened, but for the purpose of this paper it is general enough. The compactness of A is satisfied in many important practical applications in signal processing, for example when the available data of the original image is blurred and noisy and the original image has to be reconstructed.

Due to the compactness of the operator A, the problem (3.1) is ill–posed, in the sense that small perturbations in the data u may lead to significant distortions in the reconstruction of the source term. Regularization techniques are usually applied to cope with the ill–posedness. One possibility of a stabilizing algorithm is to approximate the solution of (3.1) by a minimizing element of the functional

$$\mathcal{R}(f) := \|A(f) - u^\delta\|_{L^1(\Omega)} + \alpha \|f\|^2_{L^2(\Omega)} , \quad f \in L^2(\Omega), \quad \alpha > 0. \tag{3.3}$$

The noise level is taken into consideration when regularization parameter α is chosen. This approach does not take into account that a source term to be reconstructed is a bounded variation function; therefore, it is more appropriate to consider the following regularization problem

$$\begin{gathered} \min_{f \in \mathrm{BV}(\Omega)} \mathcal{J}(f), \text{ where} \\ \mathcal{J}(f) := \|A(f) - u^\delta\|_{L^1(\Omega)} + \alpha \left(\|f\|_{L^1(\Omega)} + \int_\Omega |Df| \right) ; \end{gathered}$$

$\int_\Omega |Df|$ is as defined in the Introduction.

Since $\mathrm{BV}(\Omega)$ is not a reflexive Banach space, stability questions for these minimizers can only be proven in an L^p–norm (see e.g. [1]), which theoretically does not imply "more" stability than the approach (3.3). In particular, and more significantly in applications, we cannot infer stability in the bounded variation norm. In order to overcome this difficulty we consider an approach for constructing approximate solutions of the minimization problem (3.4) based on minimizing the following modified functional over the space $W^{1,2}(\Omega)$

$$\mathcal{J}_\varepsilon(f) := \mathcal{J}(f) + \alpha\varepsilon \|f\|^2_{W^{1,2}(\Omega)} . \tag{3.4}$$

Note that this minimization problem involves modifications of both the functional and the admissible domain for the minimization problem (3.4).

A similar algorithm has been considered in [6, 7] for image deblurring and denoising problems.

In order to apply the general results of Section 2, we formulate (3.4) in the setting of this paper by defining:

$$\mathcal{D}_\varepsilon = W^{1,2}(\Omega), \mathcal{D} = \mathrm{BV}(\Omega), \mathcal{Z} = L^p(\Omega), 1 < p < \frac{d}{d-1}, Y = L^1(\Omega),$$

$$\begin{aligned} \phi = \phi_\varepsilon : \mathcal{D}_\varepsilon \times Y &\to \mathbb{R} \\ (f, u) &\to \|A(f) - u\|_{L^1(\Omega)} \end{aligned}$$

$$\begin{aligned} k = k_\varepsilon &= \alpha \left\{ \|f\|_{L^1(\Omega)} + \int_\Omega |Df| \right\} \\ j_\varepsilon &= \alpha\varepsilon \|f\|^2_{W^{1,2}(\Omega)} \\ l_\varepsilon &= 0. \end{aligned}$$

For simplicity we assume that for fixed u, the functional $\|A(f) - u\|_{L^1(\Omega)}$ is convex, which is the case, for example, if A is a linear operator. Then assumptions 1,2,3, and 4 are satisfied, and thus Theorem 2.1 holds. Moreover, since $j_\varepsilon = \alpha\varepsilon\|f\|^2_{W^{1,2}(\Omega)}$, and $\|A(f) - u\|_{L^1(\Omega)}$ and $\|f\|_{\mathrm{BV}(\Omega)}$ are weakly lower semicontinuous on $\mathcal{D}_\varepsilon$, the minimizers of (3.4) are not only weakly stable, but also stable with respect to the $W^{1,2}$–norm (see the remark after the proof of Theorem 2.1).

Since the embedding of $\mathrm{BV}(\Omega)$ into $L^p(\Omega)$ $\left(1 \le p < \frac{d}{d-1}\right)$ is compact, and $\phi + k$ is lower continuous on $\mathrm{BV}(\Omega)$ with respect to the $L^p(\Omega)$–norm, assumptions 6 and 11 are satisfied.

It is known, but nontrivial to prove, that for any $\underline{f} \in \mathrm{BV}(\Omega)$, there exists a sequence $\left\{\underline{\eta}_n\right\} \in C^\infty(\Omega)$, which satisfies

$$\|\underline{\eta}_n - \underline{f}\|_{L^1(\Omega)} \to 0, \qquad \int_\Omega |D\underline{\eta}_n| \to \int_\Omega |D\underline{f}|.$$

For a proof see [4]. Especially from this result it follows that $\underline{\eta}_n \in W^{1,1}(\Omega)$, and it can therefore be approximated by a sequence $\{\psi_{n,m}\} \in C^\infty(\overline{\Omega})$ with respect to the

$W^{1,1}(\Omega)$-norm. For a proof see [3]. Thus any $\underline{f} \in \mathrm{BV}(\Omega)$ can be approximated by a sequence $\eta_n := \psi_{n,m(n)} \in C^\infty(\overline{\Omega})$, which satisfies

$$\|\eta_n - \underline{f}\|_{L^1(\Omega)} \to 0, \qquad \int_\Omega |D\eta_n| \to \int_\Omega |D\underline{f}|.$$

Hence assumption 7 is satisfied if ε_n is chosen such that $\varepsilon_n \|\eta_n\|^2_{W^{1,2}(\Omega)} \to 0$. Assumptions 8 and 9 are trivially satisfied.

Thus we have established the following theorems:

Theorem 3.1 (Stability) *Let ε be a fixed positive number. Then each minimizing sequence $\{f_\varepsilon^{\delta_n}\}$ of the functional $\mathcal{J}_\varepsilon$ (where in (3.4) u^δ is replaced by u^{δ_n} satisfying $\|u^{\delta_n} - u\|_{L^1(\Omega)} \le \delta_n$) converges in $W^{1,2}(\Omega)$ as $\delta_n \to 0$ to the minimizing element of $\mathcal{J}_\varepsilon$ (where now in (3.4) u^δ is replaced by u).*

Theorem 3.2 (Convergence) *Let ε_n and η_n be chosen as above. Then as $\delta_n \to 0$ each minimizing sequence $\{f_{\varepsilon_n}^{\delta_n}\}$ of $\mathcal{J}_{\varepsilon_n}$ has a subsequence that converges in $L^p(\Omega)$ $(1 \le p < \frac{d}{d-1})$ to a minimizing element of $\mathcal{J}$.*

3.2 A Minimal Surface Problem

In this subsection we give an overview of a method based on a variational inequality for the solution of the relaxed *Dirichlet problem for a minimal surface problem.*

We briefly describe the Dirichlet problem for the minimal surface problem: Let u be a function defined on some bounded domain $\Omega \subseteq \mathbb{R}^d$, $d \ge 2$, where the boundary satisfies certain smoothness assumptions. The area of the graph of u is defined by (see [4])

$$\mathcal{A}(u) := \int_\Omega \sqrt{1 + |Du|^2}\, dx\,, \tag{3.5}$$

where

$$\int_\Omega \sqrt{1+|Du|^2}\, dx \;:=\; \sup\Big\{ \int_\Omega (g_{n+1} + u\nabla . g)\, dx \;:$$
$$g = (g_1, ..., g_{d+1}) \in C_0^1(\Omega, \mathbb{R}^{d+1})\,,\ |g(x)|^2 := \textstyle\sum_{i=1}^{d+1} g_i^2(x) \le 1 \Big\}.$$

The *Dirichlet problem* for *the relaxed minimal surface problem* is to find a function u of minimal area (3.5) in the class of functions of bounded variation ($\mathrm{BV}(\Omega)$) with boundary data ϕ on $\partial\Omega$.

From the general results in [4] it can be shown that for a bounded domain the solution of the minimal surface problem can be reformulated in terms of an unconstrained minimization problem of the functional

$$
\begin{aligned}
\mathcal{J} : W^{1,1}(\Omega) &\to I\!R, \\
u &\to \mathcal{A}(u) + \int_{\partial\Omega} |u - \phi|,
\end{aligned} \tag{3.6}
$$

where $W^{1,1}(\Omega)$ is the space of all functions in $L^1(\Omega)$ whose first derivative exists almost everywhere and belongs to $L^1(\Omega)$. On the first glimpse the minimization of the functional in (3.6) and the Dirichlet problem for the minimal surface problem are rather different: A minimizer $\hat{u}$ of $\mathcal{J}$ over $W^{1,1}(\Omega)$ does not have to attain the boundary data ϕ – but by extending the boundary data ϕ to a function $\hat{\phi} \in W^{1,1}(\mathcal{B}_R - \Omega)$, where $\mathcal{B}_R$ is a sphere of radius R surrounding Ω, the corresponding function

$$\hat{u}\chi_\Omega + \hat{\phi}\chi_{B_R - \Omega}$$

can be interpreted as a function of $\mathrm{BV}(\mathcal{B}_R)$ which satisfies the boundary data $\hat{\phi}$. Here $\chi_\Omega, \chi_{B_R-\Omega}$ denote the characteristic functions of Ω, $B_R - \Omega$, respectively.

The reformulation in terms of an unconstraint minimization problem over $W^{1,1}(\Omega)$ is derived in two steps. From the fact that $\mathcal{A}(u))$ (defined in (3.5)) is lower semi–continuous with respect to weak L^1–convergence, one proves the existence of a minimizing element of (3.6) in $\mathrm{BV}(\Omega)$. Then analytical methods (see [4]) are used to verify step–by step that minimizing elements of (3.6) over $\mathrm{BV}(\Omega)$ must satisfy certain smoothness properties in Ω. The fact that a minimizing element of $\mathcal{J}$ is in $W^{1,1}(\Omega)$ is utilized below to show existence of a reference sequence which approximates the minimizer (cf. Lemma 3.4 below).

The minimization problem of the functional in (3.7) was solved in [8] with a finite element method, where the approximations converge with respect to the $W^{1,1}(\Omega)$–norm. In this section we are concerned with continuous dependence on the input data ϕ of the solution of the minimal surface problem.

In order to utilize the theoretical results in Section 2 we introduce modified functionals $\mathcal{J}_\varepsilon$ (defined on $\mathcal{D}_\varepsilon$) of $\mathcal{J}$ (defined on $\mathcal{D}$), which allow stable calculation of the minimizing elements, and for appropriately chosen parameter ε we verify that the approximate solutions converge to the solution of the minimal surface problem.

Let $\varepsilon = (\varepsilon_1, \varepsilon_2)$, where $\varepsilon_1, \varepsilon_2$ are fixed positive numbers. Consider the functional

$\mathcal{A}_\varepsilon(\phi^\delta) : \mathcal{D}_\varepsilon := W^{1,2}(\Omega) \to I\!R$ defined by

$$\mathcal{A}_\varepsilon(\phi^\delta) = \frac{\varepsilon_1\varepsilon_2}{2}\|u\|^2_{W^{1,2}(\Omega)} + \varepsilon_1\|u\|_{L^1(\Omega)} + \int_\Omega \sqrt{1+|\nabla u|^2} + \int_{\partial\Omega} |u-\phi^\delta| \,. \tag{3.7}$$

The term $\frac{\varepsilon_1\varepsilon_2}{2}\|u\|^2_{W^{1,2}(\Omega)}$ is used as a stabilization term, while $\varepsilon_1\|u\|_{L^1(\Omega)}$ is considered as penalty term in order to simulate the calculation of a minimizer of (3.6) with minimal L^1–norm.

Verifying the general assumptions of Theorem 2.1, it can be shown that the minimizers of $\mathcal{A}_\varepsilon$ are stable with respect to data perturbations (see [10]):

Theorem 3.3 *Let $\varepsilon = (\varepsilon_1, \varepsilon_2)$. For given $\phi^\delta \in L^1(\partial\Omega)$ there exists a unique element $u_\varepsilon(\phi^\delta) \in W^{1,2}(\Omega)$, which minimizes the functional $\mathcal{A}_\varepsilon$. If $\phi^\delta \to \phi$ in $L^1(\partial\Omega)$, then $u_\varepsilon(\phi^\delta) \to u_\varepsilon(\phi)$ in $W^{1,2}(\Omega)$, i.e., the minimizers of $\mathcal{A}_\varepsilon$ are stable with respect to data perturbations.*

Under the general assumptions of this subsection it is not known if the solution of the minimal surface problem is unique; however any two solutions can only differ by a constant (see Proposition 14.11 in [4]).

In the rest of this section we restrict our attention to a *"minimal" solution* of the minimal surface problem: Let

$$\text{dist} := inf\{\mathcal{J}(u) : u \in \text{BV}(\Omega)\} \,.$$

We seek a solution $\underline{u}$ of the minimal surface problem which satisfies

$$\begin{aligned} \mathcal{J}(\underline{u}) &= \text{dist} \\ \|\underline{u}\|_{L^1(\Omega)} &\leq \|u\|_{L^1(\Omega)} \text{ for all } u \text{ which satisfy } \mathcal{J}(u) = \text{dist} \,. \end{aligned}$$

The existence of a minimal solution follows from the fact that $\text{BV}(\Omega)$ can be compactly embedded in $L^p(\Omega)$, $1 \leq p < \frac{d}{d-1}$.

In order to apply Theorem 2.2 the existence of a reference sequence $\{\eta_n\} \subseteq \mathcal{D}_\varepsilon$, which approximates a minimal solution of the minimal surface problem, must be established.

Lemma 3.4 *Let $\underline{u} \in W^{1,1}(\Omega)$. Let $\{\rho_n\}$ be a monotonically decreasing sequence of positive numbers such that $\{\rho_n\} \to 0$. Then there exists a sequence $\{\eta_n\} \in W^{1,2}(\Omega)$ which satisfies:*

- $\int_\Omega |\eta_n - \underline{u}| \leq \frac{\rho_n}{2}$

- $\left| \int_{\Omega} \sqrt{1+|\nabla \eta_n|^2} - \int_{\Omega} \sqrt{1+|\nabla \underline{u}|^2} \right| \leq \frac{\rho_n}{2}$
- $\int_{\partial\Omega} |T\eta_n - T\underline{u}| \leq \frac{\rho_n}{2}$

Lemma 3.4 and Theorem 2.2 are the essential ingredients for proving that the minimizers of $\mathcal{A}_\varepsilon(\phi^\delta)$ are convergent.

Theorem 3.5 *Let ϕ^{δ_n} satisfy*

$$\|\phi^{\delta_n} - \phi\|_{L^1(\partial\Omega)} \leq \delta_n \to 0.$$

For $\underline{u}$, a minimal solution of the minimal surface problem (note that $\underline{u} \in W^{1,1}(\Omega)$), let $\eta_n \in W^{1,2}(\Omega)$ be chosen as in Lemma 3.4, and let $\{\varepsilon_n\} = \{(\varepsilon_{1,n}, \varepsilon_{2,n})\}$ be a sequence with positive numbers $\varepsilon_{i,n}, i = 1, 2$, such that $\varepsilon_{2,n}\|\eta_n\|^2_{W^{1,2}(\Omega)} \to 0$. Then, as

$$\frac{\rho_n}{\varepsilon_{1,n}} \to 0, \text{ and } \frac{\delta_n}{\varepsilon_{1,n}} \to 0,$$

we have

$$\int_{\Omega} \sqrt{1+|\nabla u^{\delta_n}_{\varepsilon_n}|^2} + \int_{\partial\Omega} |u^{\delta_n}_{\varepsilon_n} - \phi^{\delta_n}| \to \int_{\Omega} \sqrt{1+|\nabla \underline{u}|^2} + \int_{\partial\Omega} |\underline{u} - \phi^{\delta_n}| \, ,$$

where $\underline{u}$ is a minimal solution of the minimal surface problem. Moreover, each subsequence of $\{u^{\delta_n}_{\varepsilon_n}\}$ has a (weak) convergent subsequence and the limit of every (weak) convergent subsequence is a minimal solution, i.e., if we denote by $\{u^{\delta_n}_{\varepsilon_n}\}$ a (weak) convergent subsequence, then

$$\{u^{\delta_n}_{\varepsilon_n}\} \rightharpoonup \underline{u} \text{ in } L^{p_0}(\Omega), \quad \{u^{\delta_n}_{\varepsilon_n}\} \to \underline{u} \text{ in } L^p(\Omega), \quad 1 \leq p < p_0 = \frac{d}{d-1}\,.$$

The numerical minimization of $\mathcal{A}_\varepsilon$ can be performed with *Uzawa's algorithm*, which efficiently handles the nondifferentiable terms $\int_{\partial\Omega} |u - \phi^\delta|$ and $\int_{\Omega} |u|$. In [10] we prove (in an infinite dimensional setting) that *Uzawa's algorithm* is a stable and convergent method for minimizing $\mathcal{A}_\varepsilon$.

4 Remarks

We conclude with some remarks that highlight differences between optimization problems in reflexive and nonreflexive Banach spaces.

The traditional method in the calculus of variations for proving existence of a minimum of a functional is to establish precompactness of minimizing sequences and lower semicontinuity of the functional. It is true that a lower semicontinuous functional on a compact subset S in a topological space attains its infimum on S;

however, the norm–topology is not appropriate here, since a ball or a closed convex set is not compact in the norm–topology of infinite dimensional spaces. Such sets arise often in applications; this motivates introducing on a Banach space a topology relative to which such sets are compact. The weak topology does this job in *reflexive* Banach spaces. The role of the *weak topology* is derived from two important theorems. The first theorem, which was proved independently by Alaoglu, Bourbaki, and Kakutani, states that the unit ball in any dual space $\mathcal{B}^*$ is compact in the weak–star topology of $\mathcal{B}^*$. For reflexive spaces, the weak topology of $\mathcal{B}$ coincides with the weak–star topology of $\mathcal{B}^{**}$ and thus the ball is weakly compact. The second theorem, due to Mazur, states that every strongly closed convex set in a Banach space is weakly closed; hence in a reflexive Banach space every bounded closed convex set is weakly closed. On the basis of these two results, it follows that if f is a weakly lower semicontinuous real–valued functional on a weakly closed subset Ω (or in particular, on a bounded, closed, and convex subset Ω) of a reflexive Banach space, then f attains its infimum on Ω. The utility of this theorem depends on establishing sufficient conditions for f to be weakly lower semicontinuous. Often conditions on derivatives are imposed to guarantee weak lower semicontinuity.

The weak topology of a nonreflexive Banach space does not coincide with the weak-star topology of its dual. As a consequence the Alaoglu-Bourbaki-Kakutani theorem and the Mazur theorem, which are central to the role that the weak topology plays for optimization theory in reflexive Banach spaces, have no direct relevance for optimization theory in nonreflexive spaces.

Another marked difference arises in consideration of stability of minimizing sequences. A typical situation that we have illustrated in Section 3 involves minimization of a functional on the space $\mathrm{BV}(\Omega)$ of functions of bounded variations. In general such functions cannot be approximated by a family of smooth functions with respect to the bounded variation norm. The stable methods for approximating solutions of optimization problems in nonreflexive spaces, developed in this paper, are based on a family of regularized optimization problems involving modifications of both the functional and the domain. These methods are also applicable to image restoration problems where the total variation of the image is minimized subject to constraints. The use of the total variation norm enables one to reconstruct sharp edges or discontinuities in the signals and images. We have illustrated the applications of our theoretical results to an inverse source problem.

Acknowledgement

The authors would like to thank the referee for constructive comments which helped to improve the motivation and exposition of this paper.

References

[1] R. Acar and C.R. Vogel, *Analysis of bounded variation penalty methods for ill–posed problems*, Inverse Problems **10** (1994), 1217–1229.

[2] H.W. Engl, K. Kunisch and A. Neubauer, *Convergence rates for Tikhonov regularization of nonlinear ill-posed problems*, Inverse Problems **5** (1989), 523 – 540.

[3] L. C. Evans and R. F. Gariepy, *Measure Theory and Fine Properties of Functions*, CRC Press, Boca Raton, 1992.

[4] E. Giusti, *Minimal Surfaces and Functions of Bounded Variation*, Birkhäuser, Boston, 1984.

[5] R. Glowinski, *Numerical Methods for Nonlinear Variational Problems*, Springer–Verlag, Berlin–Heidelberg–New York, 1984.

[6] K. Ito and K. Kunisch, *Augmented Lagrangian methods for nonsmooth convex optimization in Hilbert spaces*, preprint.

[7] K. Ito and K. Kunisch, *An active set strategy based on the augmented Lagrangian formulation for image restoration*, preprint.

[8] C. Jouron, *Resolution numérique du probléme de surfaces minima*, Archive for Rational Mechanics and Analysis **59** (1975), 311 – 348.

[9] J. L.Lions and E. Magenes, *Non–Homogeneous Boundary Value Problems and Applications*, Springer–Verlag, Berlin–Heidelberg–New York, 1972.

[10] M. Z. Nashed and O. Scherzer, *Stable approximations of a minimal surface problem with variational inequalities*, preprint.

Department Of Mathematical Sciences,
University Of Delaware,
Newark, DE 19716, U. S. A.
E-mail Address: nashed@math. udel. edu

Institut fuer Mathematik
Johannes-Kepler-Universitaet
A-4040 Linz
Oesterreich / Austria
E-Mail: scherzer@indmath.uni-linz.ac.at

Contemporary Mathematics
Volume **204**, 1997

SOBOLEV GRADIENTS AND BOUNDARY CONDITIONS FOR PARTIAL DIFFERENTIAL EQUATIONS

J. W. NEUBERGER

ABSTRACT. Suppose F is a $C^{(3)}$ function from a Hilbert space H to a Hilbert space K. Using steepest descent we define a foliation of H so that each leaf contains one and only one zero of F. Defining G so that if $x \in H$ then $G(x)$ is the zero of F which is in the same leaf as x, we have that G is $C^{(1)}$.

1. INTRODUCTION

For a given system of partial differential equations, what side conditions may be imposed in order to specify a unique solution? For various classes of elliptic, parabolic or hyperbolic equations there are, of course, well established criteria in terms of boundary conditions. For many systems, however, there is some mystery concerning characterization of the set of all solutions to the system.

In this note we deal with systems specified by a function F from a Hilbert space H to a Hilbert space K. Under some rather strong (considerably stronger than we like) conditions we construct a foliation of H so that each leaf of the foliation contains one and only one element u so that $F(u) = 0$.

For a quick example of such a foliation relevant to boundary value problems, consider a linear, second order, non-singular differential operator F (with continuous coefficients) on the Sobolev space $H = H^{2,2}([a, b])$ into $K = L_2([a, b])$. Denote by G the orthogonal projection of H onto $N(F)$, the null space of F. The null space of G together with its translates (i.e., $H/N(G)$) provides a foliation of H each member of which contains precisely one element of $N(F)$. That $H/N(G)$ turns out to be two dimensional provides the classical basis for a study of boundary value problems associated with F (for example for two point boundary value problems).

For problems in which F is nonlinear, such orthogonal projections G are commonly not available. In this note we use a method of steepest descent to construct foliations in the following way: Under our hypotheses, if $x \in H$, then our continuous steepest descent starting at x converges to $G(x) \in N(F)$. For a given function F, the requirement that a solution u be in a specified level set of the corresponding function G provides a side condition which specifies a unique solution. Obtaining a characterization of the set of all solutions should place one in a good position to investigate other, perhaps more physically appealing, boundary conditions. The problem of determining whether a second set of boundary conditions led also to a unique solution could take the form of the establishment of an appropriate mapping between the two types of conditions.

Date: November 1, 1995.

1991 *Mathematics Subject Classification.* Primary: 35A35; Secondary: 35A15.

Key words and phrases. Sobolev Gradient, Foliation, Steepest Descent, Nonlinear Partial Differential Equations.

The present work seems applicable to systems of partial differential equations for which the set of all solutions is connected. In particular our hypothesis seem to preclude (some very interesting) cases in which the set of all solutions is denumerable. For such cases one might have sets of leaves separated by collections which contain no solution. Such a structure might be built using a local version of the present results.

In this paper we consider continuous steepest descent (2.3). The discrete version ($z_n = z_{n-1} - \delta_{n-1}(\nabla\phi)(z_{n-1}), \delta_{n-1}$ chosen optimally, $n = 1, 2, \dots$) dates back to Cauchy [3]. Discrete steepest descent is often used due to its numerical efficiency. Continuous steepest descent seems to be needed for present purposes. For additional background on steepest descent see [12],[11] (and references contained therein). See [1] and [2] which deal with descent-like processes without the differentiability hypothesis of the present note.

Our use of the term 'foliation' is not standard in that we do not prove that our 'leaves' are manifolds.

2. Statement of Results

Suppose that each of H and K is a Hilbert space and F is a $C^{(3)}$ function from $H \to K$. Define

$$\phi : H \to R$$

by

$$\phi(x) = \|F(x)\|_K^2/2, x \in H \tag{2.1}$$

and note that

$$\phi'(x)h = < F'(x)h, F(x) >_K = < h, F'(x)^* F(x) >_H, x, h \in H \tag{2.2}$$

where $F'(x)^* \in L(K, H)$ is the Hilbert space adjoint of $F'(x), x \in H$. In view of (2.2), we take $F'(x)^*F(x)$ to be $(\nabla\phi)(x)$, the gradient of ϕ at x.

It is shown in ([7], Lemma 2) that there is a unique function

$$z : [0, \infty) \times H \to H$$

such that

$$z(0, x) = x, z_1(t, x) = -(\nabla\phi)(z(t, x)), t \geq 0, x \in H \tag{2.3}$$

where the subscript in (2.3) indicates the partial derivative of z in its first argument.

In this note we have the following standing assumptions on F : If $r > 0$, there is $c > 0$ such that

$$\|F'(x)^*g\|_H \geq c\|g\|_K, \ \|x\| < r, \ g \in K \tag{2.4}$$

and if

$$x \in H, z(0, x) = x, z_1(t, x) = -(\nabla\phi)(z(t, x))t > 0,$$

then

$$\{z(t, x) : t \geq 0\} \text{ is bounded.} \tag{2.5}$$

Under these assumptions it follows from ([8], Theorem 2, essentially Lemma 1 below) that if $x \in H$, then

$$u = lim_{t\to\infty} z(t, x), exists \ and \ F(u) = 0. \tag{2.6}$$

Define $G: H \to H$ so that if $x \in H$ then $G(x) = u, u$ as in (2.6). Denote by Q the collection of all $g \in C^{(1)}(H, R)$ so that

$$g'(x)(\nabla\phi)(x) = 0, x \in H. \tag{2.7}$$

Main Theorem . *(a) G' exists, is continuous and has range in $L(H, H)$ and*

(b) $G^{-1}(G(x)) = \cap_{g \in Q}\, g^{-1}(g(x)), x \in H.$

3. Lemmas and Proofs

Lemma 1 below is a slightly weaker form of Theorem 10 from [8]; Lemma 2 is a slightly weaker version of Theorem 3 of [8].

Lemma 1. *Under our standing hypotheses, if $x \in H$,*

$$u = \lim_{t\to\infty} z(t, x) \text{ exists and } F(u) = 0. \tag{3.1}$$

Lemma 2. *Suppose that $x \in H, r > 0, c > 0,$*

$$\|F'(y)^* g\|_H \geq c\|g\|_K,\ g \in K,\ \|y - x\|_H \leq r,$$

and z satisfies (2.3). If $\|F(x)\|_K < rc$, then (3.1) holds and $\|x - u\| \leq r$.

Lemma 3. *Under the standing hypothesis suppose $x \in H$ and*

$$Q = \{z(t, x), t \geq 0\} \cup \{G(x)\}.$$

There are $\gamma, M, r, T > 0$ so that if

$$Q_\gamma = \cup_{w \in Q} B_\gamma(w),$$

then

$$|(\nabla\phi)'(w)| \leq M,\ |(\nabla\phi)''(w)| \leq M, w \in Q$$

and if $y \in H$, $\|y - x\|_H < r$, then

$$[z(t, y), z(t, x)] \subset Q_\gamma, t \geq 0 \text{ and } [z(t, y), G(y)] \subset Q_\gamma, t \geq T.$$

For $a, b \in H$, $[a, b] = \{ta + (1 - t)b : 0 \leq t \leq 1\}$ and for $w \in H$, $|(\nabla\phi)'(w)|$, $|(\nabla\phi)''(w)|$ denote the norms of $(\nabla\phi)'(w)$, $(\nabla\phi)''(w)$ as linear and bilinear functions on $H \to H$ respectively:

$$|(\nabla\phi)'(w)| = \sup_{h \in H, \|h\|_H = 1} |(\nabla\phi)'(w)h|$$

$$|(\nabla\phi)''(w)| = \sup_{h,k \in H, \|h\|_H = 1, \|k\|_H = 1} |(\nabla\phi)''(w)(h, k)|.$$

Proof. Since Q is compact and both $(\nabla\phi)', (\nabla\phi)''$ are continuous on H, there is $M > 0$ and an open subset α of H containing Q so that

$$|(\nabla\phi)'(w)|, |(\nabla\phi)''(w)| < M, w \in \alpha.$$

Pick $\gamma > 0$ such that $Q_\gamma \subset \alpha$. Then the first part of the conclusion clearly holds.

Note that Q_γ is bounded. Denote by c a positive number so that

$$\|F'(w)^* g\|_H \geq c\|g\|_K, g \in K, w \in Q_\gamma.$$

Pick $T > 0$ so that $\|z(T, x) - G(x)\|_H < \gamma/4$ and $\|F(z(T, x))\|_K < c\gamma/4$ (this is possible since $\lim_{t\to\infty} F(z(t, x)) = F(u) = 0$). Pick $v > 0$ such that $v exp(TM) < \gamma/4$ Suppose $y \in B_v(x)$ $(= \{w \in H : \|w - x\|_H < v)$. Then

$$z(t, y) - z(t, x) = y - x - \int_0^t ((\nabla\phi)(z(s, y)) - (\nabla\phi)(z(s.x)))ds$$

and so

$$z(t,y) - z(t,x) = y - x$$

$$-\int_0^t \int_0^1 ((\nabla\phi)'((1-\tau)z(s,x) + \tau z(s,y)))d\tau(z(s,y) - z(s,x))ds, t \geq 0.$$

Hence there is $T_1 > 0$ such that $[z(s,y), z(s,x)] \subset Q_\gamma$, $0 \leq s \leq T_1$, and so $|\int_0^1((\nabla\phi)'((1-\tau)z(s,x) + \tau z(s,y)))d\tau| \leq M$ and

$$\|z(t,y) - z(t,x)\|_H \leq \|y - x\|_H + M\int_0^t \|z(s,y) - z(s,x)\|_H ds, 0 \leq s \leq T_1.$$

But this implies that

$$\|z(t,y) - z(t,x)\|_H \leq \|y-x\|_H exp(tM) < vexp(tM)$$
$$\leq vexp(T_1 M) < \gamma, \|y - x\|_H < r$$

and so

$$[z(t,y), z(t,x)] \subset Q_\gamma, \|y - x\|_H < v, 0 \leq t \leq T_1.$$

Supposing that the largest such T_1 is less than T, we get a contradiction and so have that

$[z(t,y), z(t,x)] \subset Q_\gamma$, $0 \leq t \leq T$, $\|y - x\|_H < r$. Now choose $r > 0$ such that $r \leq v$ and such that if $\|y - x\|_H < r$, then

$$\|F(y)\|_K \leq 2\|F(x)\|_K, \|z(T,y) - z(T,x)\|_H < \gamma/4$$

and

$$\|F(z(T,y)) - F(z(T,x))\|_K < c\gamma/4.$$

Hence for $\|y - x\|_H < r$,

$$\|z(T,y) - G(x)\|_H \leq \|z(T,y) - z(T,x)\|_H + \|z(T,x) - G(x)\|_H < \gamma/2$$

and

$$\|F(z(T,y))\|_K \leq \|F(z(T,y)) - F(z(T,x))\|_K + \|F(z(T,x))\|_K < c\gamma/2.$$

According to Lemma 2, it must be that

$$\|z(t,y) - z(T,y)\|_H < \gamma/2, \ t \geq T$$

and so

$$\|G(y) - z(T,y)\|_H \leq \gamma/2, \ t \geq T$$

since $G(y) = \lim_{t\to\infty} z(t,y)$. Note also that Lemma 2 gives that

$$\|z(t,x) - z(T,x)\|_H < \gamma/2, \ t \geq T$$

and so we have that the convex hull of

$$G(x), G(y), \{z(t,x) : t \geq T\}, \{z(t,y) : t \geq T\}$$

is a subset of $B_\gamma(G(x)) \subset \alpha$. This gives us the second part of the conclusion since we already have that

$$[z(t,y), z(t,x)] \subset Q_\gamma, 0 \leq t \leq T.$$

□

Lemma 4. *Suppose* $B \in L(H, K)$, $c > 0$ *and*

$$\|B^* g\|_H \geq c\|g\|_K, g \in K. \tag{3.2}$$

Then

$$|\exp(-tB^*B) - (I - B^*(BB^*)^{-1}B)| \leq \exp(-tc^2), t \geq 0.$$

Note that the spectral theorem (cf [9]) gives that $\exp(-tB^*B)$ converges pointwise on H to $(I - B^*(BB^*)^{-1}B)$, the orthogonal projection of H onto $N(B)$, as $t \to \infty$. What Lemma 4 gives is exponential convergence in operator norm.

Proof. First note that (3.2) is sufficient for $(BB^*)^{-1}$ to exist and belong to $L(K, K)$. Note next the formula

$$\exp(-tB^*B) = I - B^*(BB^*)^{-1}B + B^*(BB^*)^{-1}\exp(-tBB^*)B \tag{3.3}$$

which is established by expanding $\exp(-tBB^*)$ in its power series and collecting terms, $t \geq 0$. Note also that

$$B^*(BB^*)^{-1}\exp(-tBB^*)B = B^*(BB^*)^{-1/2}\exp(-tBB^*)(BB^*)^{-1/2}B$$

and that

$$|B^*(BB^*)^{-1/2}| = |(BB^*)^{-1/2}B| \leq 1$$

and hence

$$|B^*(BB^*)^{-1}\exp(-tBB^*)B| \leq |\exp(-tBB^*)|. \tag{3.4}$$

Now denote by ξ a spectral family for BB^*. Since

$$< BB^*g, g >_K = \|B^*g\|_K^2 \geq c^2\|g\|_K^2, g \in K$$

it follows that c^2 is a lower bound to the numerical range of BB^*. Denote by b the least upper bound to the numerical range of BB^*. Then

$$BB^* = \int_{c^2}^{b} \lambda \, d\xi(\lambda)$$

and

$$\exp(-tBB^*) = \int_{c^2}^{b} \exp(-t\lambda)d\xi(\lambda), t \geq 0.$$

But this implies that $|\exp(-tBB^*)| \leq \exp(-tc^2)$, $t \geq 0$. This fact together with (3.2),(3.3) give the conclusion to the lemma. □

Lemma 5. *Suppose* x, γ, M, r, T, c *are as in Lemma 3. If* $\|x - w\| < r$, *then*

$$\|z(t, w) - G(w)\| \leq M_2 \exp(-tc^2), t \geq 0$$

where $M_2 = 2^{-1/2}\|F(x)\|/(1 - \exp(-c^2))$.

Proof. (From an argument for Theorem 10 in [8]). For $w \in H$, $\|w - x\|_H < r$, define $\eta(t) = \phi(z(t, w))$, $t \geq 0$. Then

$$\eta'(t) = \phi'(z(t, w)z_1(t, w) = < (\nabla\phi)(z(t, w)), z_1(t, w) >_H$$

$$= -\|(\nabla\phi)(z(t, w))\|_H^2 = -\|F'(z(t, w))^*F(z(t, w))\|_H^2$$

$$\leq -c^2\|F(z(t, w))\|_K^2 = -2c^2\eta(t), t \geq 0.$$

Thus $\eta'(t)/\eta(t) \leq -2c^2, t \geq 0$ and so $\eta(t) \leq \eta(0)\exp(-2tc^2), t \geq 0$, i.e.,

$$\|F(z(t, w))\|_K \leq \|F(w)\|_K \exp(-tc^2), t \geq 0.$$

Now if $a \geq 0$,

$$\|z(a+1,w) - z(a,w)\|_H^2 = \|\int_a^{a+1} z_1(s,w)ds\|_H^2 \leq (\int_a^{a+1} \|z_1(s,w)\|_H ds)^2$$

$$\leq \int_a^{a+1} \|z_1(s,w)\|_H^2 ds = \phi(z(a,w)) - \phi(z(a+1,w))$$

$$\leq \phi(z(a,w)) \leq \phi(w)\exp(-2c^2 a)$$

and so

$$\|z(a+1,w) - z(a,w)\|_H \leq 2^{-1/2}\|F(w)\|_K \exp(-c^2 a).$$

Hence for $b \geq 0$,

$$\sum_{a=b}^{\infty} \|z(a+1,w) - z(a,w)\|_H \leq (\|F(w)\|_K 2^{-1/2}) \sum_{a=b}^{\infty} \exp(-c^2 a)$$

$$= (\|F(w)\|2^{-1/2})\exp(-c^2 b) \sum_{a=0}^{\infty} \exp(-c^2)^a$$

$$= (\|F(w)\|_K 2^{-1/2})\exp(-c^2 b)/(1 - \exp(-c^2)) \leq M_2 \exp(-c^2 b).$$

Thus

$$\|z(b,w) - G(w)\|_H \leq M_2 \exp(-c^2 b)$$

and the lemma is established. □

From ([5], Theorem (3.10.5)) we have that $z_2(t,w)$ exists for all $t \geq 0, w \in H$ and that z_2 is continuous. Furthermore if $Y(t,w) = z_2(t,w), t \geq 0, w \in H$, then

$$Y(0,w) = I, Y_1(t,w) = -(\nabla\phi)'(z(t,w))Y(t,w), t \geq 0, w \in H.$$

Consult [5] for background on various techniques with differential inequalities used in this note.

Lemma 6. *Suppose x, γ, M, r, T, c are as in Lemma 3 and $\epsilon > 0$. There is $M_0 > 0$ so that if $t > s > M_0$ and $\|w - x\|_H < r$, then*

$$|Y(t,w) - Y(s,w)| < \epsilon.$$

Proof. First note that if $\|w - x\|_H < r$ then

$$|Y(t,w)| \leq \exp(Mt), t \geq 0 \textit{ since}$$

$$Y(t,w) = I - \int_0^t (\nabla\phi)'(z(s,w))Y(s,w)ds, t \geq 0 \tag{3.5}$$

and $|(\nabla\phi)'(z(s,w))| \leq M$, $0 \leq s$. In particular,

$$|Y(T,w)| \leq \exp(MT), \ \|w - x\|_H < r.$$

Suppose that $t > s \geq T$ and $\delta = t - s$. Then

$$|Y(t,w) - Y(s,w)| = \lim_{n\to\infty} |(\Pi_{k=1}^n (I - (\delta/n)(\nabla\phi)'(z(s + (k-1)\delta/n, w))))Y(s)| \tag{3.6}$$

(This is an expression that the Cauchy polygon methods works for solving (3.5) on the interval $[s,t]$). For n a positive integer and $\|w-x\|_H < r$,

$$|(\Pi_{k=1}^n (I-(\delta/n)(\nabla\phi)'(z(s+(k-1)\delta/n,w))))Y(s,w)|$$

$$\begin{aligned}\le |(\Pi_{k=1}^n (I-(\delta/n)(\nabla\phi)'(G(w))))Y(s,w)| \\ + |(\Pi_{k=1}^n (I-(\delta/n)(\nabla\phi)'(z(s+(k-1)\delta/n,w))))Y(s,w) \\ - (\Pi_{k=1}^n (I-(\delta/n)(\nabla\phi)'(G(w))))Y(s,w)|\end{aligned}$$

Now by Lemma 4,

$$|(\Pi_{k=1}^n (I-(\delta/n)(\nabla\phi)'(G(w))))Y(s,w)| \le \exp(-c^2 s)|Y(s,w)|. \tag{3.7}$$

Define

$$A_k = I-(\delta/n)(\nabla\phi)'(z(s+(k-1)\delta/n,w))$$

and

$$B_k = I-(\delta/n)(\nabla\phi)'(G(w)),$$

$k=1,2,...,n$, and denote $Y(s,w)$ by W. By induction we have that

$$|(\Pi_{k=1}^n A_k)W - (\Pi_{k=1}^n B_k)W| \le \sum_{k=1}^n |A_n \cdots A_{k+1}(A_k-B_k)B_{k-1}\cdots B_1 W|. \tag{3.8}$$

Now

$$\begin{aligned}|A_j| &\le |I-(\delta/n)(\nabla\phi)'(z(G(w)))| \\ &\quad + (\delta/n)|(\nabla\phi)'(G(w)) - (\nabla\phi)'(z(s+(j-1)\delta/n,w))| \\ &\le 1+(\delta/n)M|(\nabla\phi)'(G(w)) - (\nabla\phi)'(z(s+(j-1)\delta/n,w))| \\ &\le 1+(\delta/n)(\int_0^1 |(\nabla\phi)''((1-\tau)z(s+(j-1)\delta/n,w)+\tau G(w)|dr) \\ &\quad \|G(y)-z(s+(j-1)\delta/n,w)\|_H \\ &\le 1+(\delta/n)M\|G(y)-z(s+(j-1)\delta/n,w)\|_H \\ &\le 1+(\delta/n)MM_2\exp(-c^2(s+(j-1)\delta/n) \\ &= 1+(\delta/n)M_3\exp(-c^2 s)(\exp(-c^2\delta/n))^{j-1},\end{aligned}$$

$j=1,...,n$. We note that $|B_j| \le 1, j=1,...,n$. Note that

$$\begin{aligned}|A_n \cdots A_{k+1}| &\le |A_n|\cdots|A_{k+1}| \qquad (3.9) \\ &\le \Pi_{j=k+1}^n (1+(\delta/n)M_3\exp(-c^2 s)(\exp(-c^2\delta/n))^{j-1}) \\ &\le \Pi_{j=k+1}^n \exp((\delta/n)M_3\exp(-c^2 s)(\exp(-c^2\delta/n))^{j-1}) \\ &\le \exp(M_3\exp(-c^2 s)(\delta/n)\sum_{j=k+1}^n (\exp(-c^2\delta/n))^{j-1}) \\ &\le \exp(M_3\exp(-c^2 s)(\delta/n)/(1-\exp(-c^2\delta/n))) \\ &\le \exp(M_4\exp(-c^2 s))\end{aligned}$$

so long as $\delta/n \le 1$ where $M_4 = M_3 \sup_{\beta\in(0,1]} \beta/(1-\exp(-c^2\beta))$ and $M_3 = MM_2$. Note that

$$|A_k - B_k| = (\delta/n)|(\nabla\phi)'(z(s+(j-1)\delta/n, w)) - (\nabla\phi)'(G(w))|$$
$$= (\delta/n)|(\int_0^1 (\nabla\phi)''((1-\tau)z(s+(j-1)\delta/n, w) + \tau G(w))\, d\tau$$
$$(z(s+(j-1)\delta/n, w) - G(w))| \quad (3.10)$$
$$\le (\delta/n)M|(z(s+(j-1)\delta/n, w) - G(w))|$$
$$\le (\delta/n)MM_2\exp(-c^2 s)(\exp(-c^2\delta/n)^{j-1}_), k = 1, ..., n,$$

and so using (3.9),(3.10), we get that

$$|(\Pi_{k=1}^n A_k)W - (\Pi_{k=1}^n B_k)W|$$
$$\le \sum_{k=1}^n \exp(M_4\exp(-c^2 s))|A_k - B_k||W|$$
$$\le \exp(M_4\exp(-c^2 s))MM_2 exp(-c^2 s)\sum_{k=1}^n \exp(-c^2\delta/n)^{k-1})|W|$$
$$\le \exp(M_4\exp(-c^2 s))\exp(-c^2 s))M_4|W|.$$

Thus

$$|(\Pi_{k=1}^n(I - (\delta/n)(\nabla\phi)'(z(s+(k-1)\delta/n, w))))Y(s,w)| \quad (3.11)$$
$$\le |(\Pi_{k=1}^n(I - (\delta/n)(\nabla\phi)'(G(w))))Y(s,w)|$$
$$+ \exp(M_4\exp(-c^2 s))\exp(-c^2 s)M_4|Y(s,w)|$$

Taking the limit of both sides of (3.11) as $n \to \infty$, we have

$$|Y(t,w) - Y(s,w)| \le \exp(-c^2 s)|Y(s,w)|(1 + \exp(M_4\exp(-c^2 s))M_4) \quad (3.12)$$

$0 < T \le s < t$. Taking for the moment $s = T$ the above yields

$$|Y(t,w) - Y(T,w)| \le \exp(-c^2 T)|Y(T,w)|(1 + \exp(M_4\exp(-c^2 T))M_4). \quad (3.13)$$

Note $\{|Y(T,w)|, \|y - x\|_H < r\}$ is bounded, say by $M_5 > 0$. Hence

$$|Y(t,w) - Y(T,w)| \le$$
$$\exp(-c^2 s)M_5(1 + \exp(M_4\exp(-c^2 T))M_4),\ t > s \ge T,\ \|w - x\|_H < r, \quad (3.14)$$

and so the conclusion to Lemma 6 follows. □

Denote by U the function with domain $B_r(x)$ to which $\{Y(t,\cdot)\}_{t\ge 0}$ converges uniformly on $B_r(x)$. Note that $U : B_r(x) \to L(H,H)$. and U is continuous.

Lemma 7. *Suppose that $x \in H$, $r > 0$, each of $v_1, v_2, ...$ is a continuous function from $B_r(x) \to H$, q is a continuous function from H to $L(H,H)$ which is the uniform limit of $v_1', v_2', ...$ on $B_r(x)$. Then q is continuous and*

$$v'(y) = q(y), \|y - x\|_H < r.$$

Proof. Suppose $y \in B_r(x)$, $h \in H$, $h \neq 0$ and $y + h \in B_r(x)$. Then

$$\int_0^1 q(y+sh)h\,ds = \lim_{n\to\infty} \int_0^1 v_n'(y+sh)h\,ds$$

$$= \lim_{n\to\infty} (v_n(y+h) - v_n(y)) = v(y+h) - v(y).$$

Thus

$$\|v(y+h) - v(y) - q(y)h\|_H / \|h\|_H = \| \int_0^1 q(y+sh)h - q(y)h)ds\|$$

$$\leq \int_0^1 |q(y+sh) - q(y)|ds \to 0$$

as $\|h\| \to 0$. Thus v is Frechet differentiable at each $y \in B_r(x)$ and $v'(y) = q(y)$, $y \in B_r(x)$. □

Proof. To prove the Theorem, note that Lemmas 5,6,7 give the first conclusion. To establish the second conclusion, suppose that $g \in Q$. Suppose $x \in H$ and $\beta(t) = g(z(t,x)), t \geq 0$. Then

$$\beta'(t) = g'(z(t,x))z_1(t,x) = -g'(z(t,x))(\nabla\phi)(z(t,x)), t \geq 0.$$

Thus β is constant on $R_x = \{z(t,x) : t \geq 0\} \cup \{G(x)\}$. But if y is in $G^{-1}(G(x))$ then $g(\{z(t,y) : t \geq 0\} \cup \{G(y)\})$ must also be in $G^{-1}(G(x))$ since $G(y) = G(x)$. Thus $G^{-1}(G(x))$ is a subset of the level set $g^{-1}(g(x))$ of g. Therefore,

$$G^{-1}(G(x)) \subset \cap_{g\in Q} g^{-1}(g(x)).$$

Suppose now that $x \in H$, $y \in \cap_{g\in Q} g^{-1}(g(x))$ and $y \notin G^{-1}(G(x))$. Denote by f a member of H^* so that $f(G(x)) \neq f(G(y))$. Define $p : H \to R$ by $p(w) = f(G(w)), w \in H$. Then $p'(w) = f'(G(w))G'(w)$ and so

$$p'(w)(\nabla\phi)(w) = fG'(w)(\nabla\phi)(w) = 0,$$

$w \in H$, and hence $p \in Q$, a contradiction since

$$y \in \cap_{g\in Q} g^{-1}(g(x)) and \;\; p \in Q$$

together imply that $p(y) = p(x)$. Thus

$$G^{-1}(G(x)) \supset \cap_{g\in Q} g^{-1}(g(x))$$

and the second part of the theorem is established. □

We end this note with an example and some comments concerning related developments.

4. The Linear case; an example

Example. Take H to be the Sobolev space $H^{1,2}([0,1]), K = L_2([0,1])$,

$$F(y) = y' - y, y \in H.$$

We claim that the corresponding function G is specified by

$$(G(y))(t) = exp(t)(y(1)e - y(0))/(e^2 - 1), t \in [0,1], y \in H. \tag{4.1}$$

In this case

$$G^{-1}(G(x)) = \{w \in H : w(1)e - w(0) = y(1)e - y(0)\}.$$

This may be observed by noting that since F is linear,

$$\nabla(\phi)(y) = F^*Fy, y \in H.$$

The equation

$$z(0) = y \in H, z'(t) = -F^*Fz(t), t \geq 0$$

has the solution

$$z(t) = \exp(-tF^*F)y, t \geq 0.$$

But $\exp(-tF^*F)y$ converges to

$$(I - F^*(FF^*)^{-1}F)y, \tag{4.2}$$

the orthogonal projection of y onto $N(F)$, *i.e.*, the solution $w \in H$ to $F(w) = 0$ that is nearest (in H) to y. A little calculation shows that this nearest point is given by $G(y)$ in (4.1). The quantity $(y(1)e - y(0))/(e^2 - 1)$ provides an invariant for steepest descent for F relative to the Sobolev metric $H^{1,2}([0,1])$. Similar reasoning applies to all cases in which F is linear but invariants are naturally much harder to exhibit for more complicated functions F. In summary, for F linear, the corresponding function G is just the orthogonal projection of H onto $N(F)$. In general G is a projection on H in the sense that it is an idempotent transformation. We hope that a study of these idempotents in increasingly involved cases will give information about 'boundary conditions' for significant classes of partial differential equations.

5. Weighted Sobolev Spaces

For a given system of partial differential equations and within the restriction that $F \in C^{(3)}$, there is often considerable latitude in choice of an appropriate space H. It is suspected that for many systems, there is a natural 'weighted' Sobolev space (where the metric may change from point to point — making in effect an infinite dimensional Riemannian manifold of the set of all contenders for solutions). An example may be found in [10] in a minimal surface problem.

6. Semigroups of Steepest Descent; Linear Representations

The present setting fits naturally into some aspects of Lie theory. For a given F satisfying our standing hypothesis, define the one parameter semigroup T so that if $t \geq 0$, then

$$T(t)x = z(t, x), x \in H$$

where z satisfies (2.3). Denote by $CB(H)$ the Banach space (with sup norm) of all bounded continuous real valued functions on H. Define the linear representation S of T by

$$(S(t)g)(x) = g(T(t)x), g \in CB(H), t \geq 0, x \in H.$$

A recent result in [4] characterizes, in an appropriate topology, generators A of such semigroups S in case H is separable. Observe that our main result still holds if Q is replaced by QB, the set of all members of Q which are bounded (let g in Q correspond to $g^2/(1+g^2)$) In this setting, we have, for A the generator of S, $Ag = 0$ if and only if $g \in QG$. Thus for a nonlinear system specified by F, the construction of a foliation as described in the Theorem is reduced to a study of the kernel of the generator of the linear representation of the semigroup of steepest descent for R.

7. Closing Comments

The Sobolev gradient discussed in this note has been used as a basis for a number of numerical computations ([10], [8] and references contained in these). In a numerical study the spaces H and K of this note become finite dimensional versions of appropriate Sobolev spaces. The condition (2.4) is much more likely to be satisfied in such a finite dimensional setting; the 'trouble' is that, roughly speaking, it may be that $c \to 0$ as the computational mesh approaches zero. Nevertheless the theory of this note may have numerical content. The gradient inequality (2.4) may be considered a particularly strong version of part of some hypotheses of Moser ([6],p. 283). An attempt to obtain results similar to the present ones but using weaker hypotheses like Moser's might be an interesting line of investigation.

References

[1] H. Brezis, *Opérateurs maximaux monotones et semi-groupes nonlinéares*, Math Studies 5, North Holland, 1973.

[2] R. E. Bruck and S. Reich, *A General Convergence Principle in Nonlinear Functional Analysis*, Nonlinear Analysis 4 (1980), 939-950.

[3] Méthode générale pour la reésolution des systemes d'équations simultanées, C.R. Acad. Sci. Paris, 25 (1847), 536-538.

[4] J. R. Dorroh and J. W. Neuberger, *A Theory of Strongly Continuous Semigroups in Terms of Lie Generators*, J. Functional Analysis 136 (1996), 114-126.

[5] T. M. Flett, *Differential Analysis*, Cambridge University Press, Cambridge, 1980.

[6] Jurgen Moser, *A Rapidly Convergent Iteration Method and Non-linear Partial Differential Equations*, Annali Scuola Norm. Sup. Pisa 20 (1966),265-315.

[7] J. W. Neuberger, *Steepest Descent and Differential Equations*, J. Math. Soc. Japan 37 (1985), 187-195.

[8] J. W. Neuberger, *Constructive Variational Methods for Differential Equations*, Nonlinear Analysis, Theory, Meth. & Appl. 13 (1988), 413-428.

[9] F. Riesz and B. Sz.-Nagy, *Functional Analysis*, Ungar, 1955.

[10] R. Renka and J. W. Neuberger, *Minimal Surfaces and Sobolev Gradients*, Siam J. Sci. Comp. 16 (1995), 1412-1427.

[11] P. L. Rosenblum, *The Method of Steepest Descent*, Proc. Symp. Appl. Math. VI, Amer. Math. Soc.

[12] M. M. Vainberg, *On the Convergence of the Method of Steepest Descent for Nonlinear Equations*, Sibirsk. Math. A. 2 (1961), 201-220.

Department of Mathematics, University of North Texas, Denton, TX 76205
E-mail address: jwn@unt.edu

Contemporary Mathematics
Volume **204**, 1997

A NONLINEAR GENERALIZATION OF PERRON-FROBENIUS THEORY AND PERIODIC POINTS OF NONEXPANSIVE MAPS

ROGER D. NUSSBAUM

ABSTRACT. Let K^n denote the positive orthant in $\mathbb{R}^n$ and consider $\mathcal{F}_3(n)$, the collection of maps $f : K^n \to K^n$ which are nonexpansive in the ℓ_1-norm and satisfy $f(0) = 0$. We study the dynamics of iterates of maps $f \in \mathcal{F}_3(n)$. We describe exactly, in terms of combinatorial and number theoretical constraints, the finite set $Q(n)$ of positive integers p such that p is the minimal period of some periodic point of some $f \in \mathcal{F}_3(n)$: see Theorem 2.1 below. We also compute $Q(n)$ for $n \leq 50$. This paper provides a guide, with sketches of proofs, of joint work of the author with Sjoerd Verduyn Lunel and Michael Scheutzow.

1. Periodic Points of Nonexpansive Maps.

As usual, if a map g is defined on a subset D of a set X and $g : D \to X$, we say that $\xi \in D$ is a periodic point of g of period p if $g^p(\xi)$ is defined and $g^p(\xi) = \xi$. Here g^p denotes the composition of g with itself p times. If $g^j(\xi) \neq \xi$ for $1 \leq j < p$, we say that ξ is a periodic point of g of minimal period p.

To motivate our nonlinear results, we first recall some theorems concerning periodic points of nonnegative matrices. Let K^n denote the positive orthant in $\mathbb{R}^n$, so

$$K^n := \{x \in \mathbb{R}^n | x_i \geq 0 \text{ for } 1 \leq i \leq n\}.$$

Let $A = (a_{ij})$ be a nonnegative, $n \times n$, column stochastic matrix (so $\sum_{i=1}^{n} a_{ij} = 1$ for $1 \leq j \leq n$ and $a_{ij} \geq 0$ for $1 \leq i, j \leq n$). Perron-Frobenius theory [6,13] concerns the eigenvalues and eigenvectors of nonnegative matrices and yields a precise description of the behaviour of $A^k(x)$ as $k \to \infty$: For each $x \in \mathbb{R}^n$, there exists a positive integer $p_x = p$ and a point $\xi_x = \xi \in \mathbb{R}^n$ with (1) $A^p(\xi) = \xi$ and $A^j(\xi) \neq \xi$ for $1 \leq j < p$ and (2) $\lim_{k \to \infty} A^{kp}(x) = \xi$. Furthermore, p is the order of some element of the symmetric group on n letters, and every such number p is the minimal period of a periodic point of some $n \times n$, nonnegative column stochastic matrix A. A proof of a generalization of these results is given in Section 9 of [11].

To connect with our subsequent, nonlinear generalizations, we state the above result more formally. Let $\mathcal{S}(n)$ denote the set of $n \times n$, nonnegative, column stochastic matrices. Define $R(n)$ by $R(n) = \{p \in \mathbb{N} : \exists A \in \mathcal{S}(n)$ and a periodic point $\xi \in K^n$

1991 *Mathematics Subject Classification.* Primary 47H09, 58F20; Secondary 47H10.

The author was partially supported by NSF grant DMS 9401823

of A of minimal period $p\}$. We have described $R(n)$ precisely above. Alternately, $p \in R(n)$ if and only if there exists a set of positive integers $S = \{m_i | 1 \le i \le k\}$ such that $\sum_{i=1}^{k} m_i = n$ and $p = lcm(S)$, where $lcm(S)$ denotes the least common multiple of the elements of S.

Note that even though $\{R(n) : n \ge 1\}$ is described explicitly, it is a highly irregular and somewhat mysterious collection of sets. Thus, if $g(n)$ denotes the largest element of $R(n)$ (and hence the largest order of an element of the symmetric group on n letters) Landau [3] needed the prime number theorem to prove that

$$\log(g(n)) \sim \sqrt{n \log n};$$

and effective estimates for log $(g(n))$ were only obtained relatively recently [4].

We shall now describe precise nonlinear generalizations of the above facts. If S is a finite set of positive integers, $lcm(S)$ and $gcd(S)$ will always denote respectively the least common multiple and greatest common divisor of the elements of S. We shall always denote by $\|\cdot\|_1$ and $\|\cdot\|_\infty$ respectively the ℓ_1-norm and the ℓ_∞-norm (or sup norm) on $\mathbb{R}^n$:

$$\|x\|_1 = \sum_{i=1}^{n} |x_i| \text{ and } \|x\|_\infty = \sup\{|x_i| : 1 \le i \le n\}.$$

If a and b are real numbers we shall write

$$a \vee b = \max\{a, b\} \text{ and } a \wedge b = \min\{a, b\}.$$

If x and y are vectors in $\mathbb{R}^n$, $x \vee y = w \in \mathbb{R}^n$ and $x \wedge y = z \in \mathbb{R}^n$ are defined coordinate - wise:

$$w_i := (x \vee y)_i = x_i \vee y_i \text{ and } z_i := (x \wedge y)_i = x_i \wedge y_i.$$

The positive orthant K^n in $\mathbb{R}^n$ induces a partial ordering $\le$ on $\mathbb{R}^n$ by $x \le y$ if and only if $y - x \in K^n$. We shall also write $x < y$ if $x \le y$ and $x \ne y$. A map $f : D \subset \mathbb{R}^n \to \mathbb{R}^n$ is called "order-preserving" (with respect to the above partial ordering) if $f(x) \le f(y)$ whenever $x, y \in D$ and $x \le y$. The map f is called "nonexpansive in the ℓ_1-norm" if, for all $x, y \in D$,

$$\|f(x) - f(y)\|_1 \le \|x - y\|_1;$$

and f is called "integral-preserving" if, for all $x \in D$,

$$\sum_{i=1}^{n} f_i(x) = \sum_{i=1}^{n} x_i,$$

where $f_i(x)$ is the ith coordinate of $f(x)$. A useful result of Crandall and Tartar (see [2]) states that if $D = K^n$ or $D = \mathbb{R}^n$, $f(D) \subset D$ and f is integral-preserving, then f is order-preserving if and only if f is nonexpansive in the ℓ_1-norm. In particular, if $f : K^n \to K^n$ is integral-preserving and order-preserving, then f is nonexpansive in the ℓ_1-norm. Finally, we shall say that $f : D \subset \mathbb{R}^n \to \mathbb{R}^n$ is "sup norm decreasing" if, for all $x \in D$,

$$\|f(x)\|_\infty \leq \|x\|_\infty.$$

We are concerned with three classes of nonlinear maps:

$$\begin{aligned} &\mathcal{F}_1(n) = \{f : K^n \to K^n | f(0) = 0, \\ &f \text{ is order-preserving, integral-preserving and sup norm decreasing}\}, \end{aligned} \tag{1.1}$$

$$\begin{aligned} &\mathcal{F}_2(n) = \{f : K^n \to K^n | f(0) = 0, \\ &f \text{ is order-preserving and integral-preserving}\} \text{ and} \end{aligned} \tag{1.2}$$

$$\mathcal{F}_3(n) = \{f : K^n \to K^n | f(0) = 0 \text{ and } f \text{ is nonexpansive in the } \ell_1\text{-norm}\}. \tag{1.3}$$

Our remarks above show that

$$\mathcal{F}_1(n) \subset \mathcal{F}_2(n) \subset \mathcal{F}_3(n).$$

We are interested in understanding the behaviour of $f^k(x)$ as $k \to \infty$ for $x \in K^n$ and $f \in \mathcal{F}_j(n)$. The starting point of our work is a paper by Akcoglu and Krengel [1]. If $f \in \mathcal{F}_3(n)$ and $x \in K^n$, Ackoglu and Krengel proved that for every $x \in K^n$ there exists a periodic point $\xi_x = \xi$ of f of minimal period $p_x = p$ with

$$\lim_{k \to \infty} f^{kp}(x) = \xi_x. \tag{1.4}$$

Scheutzow [14] showed that the minimal period p of a periodic point ξ of a map $f \in \mathcal{F}_3(n)$ satisfies

$$p \leq lcm(\{1, 2, \cdots, n\}). \tag{1.5}$$

In [7] and [9], Nussbaum showed that the upper bound in eq. (1.5) is not, in general, optimal and determined, for $1 \leq n \leq 32$, the largest possible minimal period of a periodic point of a function $f \in \mathcal{F}_3(n)$.

The previous results suggest the importance of understanding the periodic points of $f \in \mathcal{F}_3(n)$ for understanding the dynamics of iterates of f. Thus, for $1 \leq j \leq 3$, we define

(1.6)

$$P_j(n) = \{p \geq 1 | \exists f \in \mathcal{F}_j(n) \text{ and a periodic point } \xi \text{ of } f \text{ of minimal period } p\}.$$

Our goal is to describe $P_j(n)$ precisely in terms of number theoretical and combinatorial constraints.

Before proceeding further it may be useful to give some examples of maps in $\mathcal{F}_j(n)$.

Example 1. Let σ and τ be one-one maps of $\{1, 2, \cdots, n\} = S$ onto itself. Define a map $h : K^n \to K^n$ by

$$h_i(x) = (x_{\sigma(i)} \vee 1) + (x_{\tau(i)} \wedge 1) - 1, \tag{1.7}$$

where $h_i(x)$ is the ith coordinate of $h(x)$. One can check that h is integral-preserving, order-preserving and sup norm decreasing, so $h \in \mathcal{F}_1(n)$. Even for this simple example, it is already nontrivial to determine the possible periods of periodic points of h.

Example 2. Our next example could be called "playing with sand by the seashore." For $1 \le i \le n$, suppose that C_i is a container of infinite volume and that C_i initially contains a volume of sand x_i. We write $x = (x_1, x_2, \cdots, x_n)$ and $S = \{1, 2, \cdots, n\}$ and suppose that a map $\gamma : S \times \mathbb{N} \to S$ is given ($\mathbb{N}$ denotes the natural numbers). For each i, we suppose that we have an infinite collection of containers $C_{i,k}, k \in \mathbb{N}$, and suppose that $C_{i,k}$ has volume $a_{i,k} > 0$, where

$$\sum_{k=1}^{\infty} a_{i,k} = \infty. \tag{1.8}$$

We now describe a procedure for shifting sand around in the containers C_i, $1 \le i \le n$. Pour sand from container C_i into container $C_{i,1}$ until container $C_{i,1}$ is full or there is no more sand left in C_i. Next pour sand into container $C_{i,2}$ until container $C_{i,2}$ is full or there is no more sand left in container C_i. Continue in this way. Equation (1.8) insures that after a finite number of steps, all the sand from container C_i will have been transfered from C_i to a finite number of the containers $C_{i,k}, k \in \mathbb{N}$. If $y^+ := \max(y, 0)$ and if $M_{ik}(x)$ denotes the volume of sand in container $C_{i,k}$, then one can see that

$$M_{ik}(x) = \left(x_i - \sum_{j=1}^{k-1} a_{ij} \right)^+ \wedge a_{ik}. \tag{1.9}$$

If we carry out the above procedure for each of the containers $C_i, 1 \le i \le n$, then all the containers C_i will be empty and all of the sand will have been transferred to the container $C_{i,k}$. Now we pour sand from the containers $C_{i,k}$ back into the containers C_j by the rule that the content of $C_{i,k}$ is poured into C_j if and only if $\gamma(i,k) = j$. If y_j is the total volume of sand in C_j and $y = (y_1, y_2, \cdots, y_n)$, then

$$y_j = \sum_{\gamma(i,k)=j} M_{ik}(x) := f_j(x). \tag{1.10}$$

Equation (1.10) defines a map $f : K^n \to K^n$, where $f_j(x)$ is the jth coordinate of $f(x)$. One can check that f is order-preserving and integral-preserving, so $f \in \mathcal{F}_2(n)$; but f is not necessarily sup norm decreasing. Even if $a_{ik} = 1$ for all i and k, it is nontrivial to determine the possible minimal periods of periodic points of f.

Example 3. Our previous examples were order-preserving maps. However, there are important instances of ℓ_1-norm nonexpansive maps $f : K^n \to K^n$ which need not be order-preserving. If $1 \le j < k \le n$, define $K^n(j,k)$ by

$$K^n(j,k) = \{x \in K^n : x_j x_k = 0\}.$$

We can define a retraction R_{jk} of K^n onto $K^n(j,k)$ by $R_{jk}(x) = y$, where $y_i = x_i$ for $i \neq j,k$ and

$$y_j = x_j - (x_j \wedge x_k), y_k = x_k - (x_j \wedge x_k).$$

The reader can verify that R_{jk} is nonexpansive in the ℓ_1-norm and $R_{jk}(0) = 0$, so $R_{jk} \in \mathcal{F}_3(n)$. However, R_{jk} is not order-preserving.

If $n = 2m$ is even, K^n can be identified with $K^m \times K^m$. With this identification we define $E^n \subset K^n$ by

$$E^n = \{(x,y) \in K^m \times K^m | x_j y_j = 0 \text{ for } 1 \le j \le m\}. \tag{1.11}$$

By composing different maps R_{jk}, we can obtain an ℓ_1-norm nonexpansive retraction R of K^n onto E^n by

$$R(x,y) = (\xi, \eta) \text{ and } \xi_j = x_j - (x_j \wedge y_j) \text{ and } \eta_j = y_j - (x_j \wedge y_j). \tag{1.12}$$

Again, we have $R \in \mathcal{F}_3(n)$ but R is not order-preserving.

Scheutzow [15] has used the retraction R to make a useful observation. Note that $(\mathbb{R}^m, \|\cdot\|_1)$ is isometric to $(E^{2m}, \|\cdot\|_1)$ (see eq.(1.11)) by the map

$$J(x) = (x \vee 0, (-x) \vee 0).$$

If $f : \mathbb{R}^m \to \mathbb{R}^m$ is nonexpansive in the ℓ_1-norm and has a periodic point ξ of minimal period p, it is known that f has a fixed point x_o; for the reader's convenience, we sketch the proof. If $S := \{f^j(\xi) | j \ge 0\}$ note that $f(S) = S$ and that diameter $(S) := d := \sup\{\|x-y\|_1 : x, y \in S\} < \infty$.

Define a set C by

$$C = \bigcap_{x \in S} V_d(x),$$

where $V_d(x) = \{y | \|y - x\|_1 \le d\}$. By our definition of d, we have $C \supset S$; and C is closed, bounded and convex, because it is the intersection of closed, bounded convex sets. If $y \in C$, we see that $\|y-x\|_1 \le d$ for all $x \in S$, so the nonexpansiveness of f implies that $\|f(y) - f(x)\|_1 \le d$ for all $x \in S$. Since $f(S) = S$, we conclude that $\|f(y) - z\|_1 \le d$ for all $z \in S$, so $f(y) \in C$. It follows that f is a continuous map with $f(C) \subset C$, so the Brouwer fixed point theorem implies that f has a fixed point $x_o \in C$.

Now by replacing f by

$$g(x) = f(x + x_o) - x_o,$$

we see that $g(0) = 0$, g is nonexpansive in the ℓ_1-norm and g has a periodic point of minimal period p. It follows that if

$$h = JgJ^{-1}R,$$

$h \in \mathcal{F}_3(2m)$ and h has a periodic point of minimal period p, so

$$p \in P_3(2m). \tag{1.13}$$

At present, nothing more precise than eq. (1.13) is known.

2. Lattices and Admissible Arrays.

If W is a subset of $\mathbb{R}^n, W$ will be called a "lower semilattice" if $x \wedge y \in W$ whenever $x, y \in W; W$ will be called a "lattice" if $x \wedge y \in W$ and $x \vee y \in W$ whenever $x, y \in W$. Here $\wedge$ and $\vee$ denote the standard lattice operations on $\mathbb{R}^n$. If W is a lower semilattice, a map $h : W \to W$ will be called a "lower semilattice homomorphism" if, for all $x, y \in W$,

$$h(x \wedge y) = h(x) \wedge h(y);$$

a lower semilattice homomorphism $h : W \to W$ will be called a lower semilattice automorphism if h is one-one and onto. Similarly, if W is a lattice, a map $h : W \to W$ will be called a "lattice homorphism" if h preserves the lattice operations $\wedge$ and $\vee$. We shall say that $h \in \mathcal{G}_1(n)$ (respectively, $h \in \mathcal{G}_2(n)$) if there is a lattice (respectively, lower semilattice) $W \subset \mathbb{R}^n$ such that domain $(h) = W$, range $(h) \subset W$ and $h : W \to W$ is a lattice homorphism (respectively, lower semilattice homomorphism). For $j = 1, 2$ we define a set of positive integers $Q_j(n)$ by

$$Q_j(n) = \{p | \exists h \in \mathcal{G}_j(n) \text{ and a periodic point } \xi \text{ of } h \text{ of minimal period } p\}. \tag{2.1}$$

The motivation for studying $G_j(n)$ and $Q_j(n)$ comes from an observation by Scheutzow [14]. Suppose that $f \in \mathcal{F}_3(n)$ and $\xi \in K$ is a periodic point of f of minimal period p. Define $A = \{f^j(\xi) | j \geq 0\}$ and let W denote the smallest lower semilattice which contains A. Then $f(W) \subset W, f^p(x) = x$ for all $x \in W$ and $f|W$ is a lower semilattice homomorphism. Thus we immediately obtain that

$$P_1(n) \subset P_2(n) \subset P_3(n) \subset Q_2(n) \text{ and } Q_1(n) \subset Q_2(n). \tag{2.2}$$

Further motivation for studying $Q_1(n)$ comes from a theorem in [8]. A norm $\|\cdot\|$ on $\mathbb{R}^n$ is called strictly monotonic in [8] if $\|x\| < \|y\|$ whenever $0 \leq x \leq y$ and $x \neq y$. The ℓ_p-norms on $\mathbb{R}^n$ are strictly monotonic for $1 \leq p < \infty$, but the sup norm is not strictly monotonic. A map $f : D \subset \mathbb{R}^n \to \mathbb{R}^n$ is called nonexpansive with respect to a norm $\|\cdot\|$ if, for all $x, y \in D$,

$$\|f(x) - f(y)\| \leq \|x - y\|.$$

Now suppose that $\|\cdot\|$ is a strictly monotonic norm on $\mathbb{R}^n$ and that $f : K^n \to K^n$ is order-preserving and nonexpansive with respect to $\|\cdot\|$ and $f(0) = 0$. Then it is proved in [8] that for every $x \in K^n$ there exists a periodic point $\xi_x = \xi$ of f of minimal period $p_x = p$ and

$$\lim_{k \to \infty} f^{kp}(x) = \xi_x.$$

Furthermore if $A = \{f^j(\xi) | j \geq 0\}$ and W is the smallest lattice which contains A, then $f(W) = W, f^p(y) = y$ for all $y \in W$ and $f|W$ is a lattice homomorphism. It follows that $p \in Q_1(n)$ and that $P_2(n) \subset Q_1(n)$.

To describe $Q_2(n)$ more precisely we need to recall some elementary facts about lower semilattices. Let $W \subset \mathbb{R}^n$ be a finite set and a lower semilattice. If $A \subset W$ is a nonempty set, there exists $b \in W$ such that (1) $b \leq a$ for all $a \in A$ and (2) if $\beta \leq a$ for all $a \in A$ and $\beta \in W$, then $\beta \leq b$. We write

$$b = \inf{}_W(A)$$

and note that b is actually independent of $W \supset A$. If A is a nonempty subset of W, we shall say that A is bounded above in W if there exists $\gamma \in W$ with $\gamma \geq a$ for all $a \in A$. If A is bounded above in W, we define $B = \{\gamma \in W | \gamma \geq a$ for all $a \in A\}$ and define

$$\sup{}_W(A) := \inf{}_W(B).$$

If $g : W \to W$ is a lower semilattice automorphism, there is a positive integer p such that g^p is the identity. If $S \subset W$ is bounded above, then we have

$$g(\sup{}_W(S)) = \sup{}_W(g(S)). \tag{2.3}$$

If $W \subset \mathbb{R}^n$ is a finite lower semilattice and $x \in W$, we define $h_W(x)$, "the height of x in W", by

$$\begin{aligned} h_W(x) = \sup\{k \geq 0 :\ &\text{there exist } y^i \in W, 0 \leq i \leq k, \\ &\text{with } y^k = x \text{ and } y^i < y^{i+1} \text{ for } 0 \leq i < k\}. \end{aligned} \tag{2.4}$$

If there does not exist $y \in W$ with $y < x$, we define $h_W(x) = 0$; and we note that there is a unique element $x \in W$ (the "minimal element of W") with $h_W(x) = 0$. If $x \in W$, we define $S_x = \{y \in W | y < x\}$ and we say that "x is irreducible in W" if S_x is empty or if

$$x > z := \sup{}_W(S). \tag{2.5}$$

If x is irreducible in W, S_x is nonempty and $z = \sup_W(S_x)$, we define

$$I_W(x) = \{i | x_i > z_i\}. \tag{2.6}$$

If x is the minimal element of W, we make the convention that

$$I_W(x) = \{i | 1 \leq i \leq n\}. \tag{2.7}$$

If $x \in W$ one can easily prove (use induction on the height of elements of W) that

$$x = \sup{}_W(\{z \in W | z \leq x \text{ and } z \text{ is irreducible in } W\}). \tag{2.8}$$

Our goal is to describe a set $Q(n)$, defined solely by certain combinatorial and number theoretical constraints, and show eventually that $Q(n) = Q_1(n) = Q_2(n) = P_2(n) = P_3(n)$. However, to motivate the definition of $Q(n)$ we need to recall some propositions obtained by Nussbaum and Scheutzow in [10]. In the following, recall that vectors $x, y \in \mathbb{R}^n$ are "not comparable" if it is not true that $x \leq y$ and it is not true that $y \leq x$.

Proposition 2.1. *(See [10]). Let W be a finite lower semilattice in $\mathbb{R}^n$ and $f : W \to W$ a lower semilattice homomorphism. Assume that $\xi \in W$ is a periodic point of f of minimal period p and that $f^p(x) = x$ for all $x \in W$ (the latter condition can be insured by replacing W by the lower semilattice generated by $A = \{f^j(\xi)|j \geq 0\}$). Then there exist elements $y^i \in W, 1 \leq i \leq m$, with the following properties:*

(1) *$y^i \leq \xi$ for $1 \leq i \leq m$.*
(2) *y^i is an irreducible element of W and a periodic point of f with minimal period p_i, where $1 \leq p_i \leq n$.*
(3) *$p =$lcm $(\{p_i|1 \leq i \leq m\})$.*
(4) *$h_W(y^i) \leq h_W(y^{i+1})$ for $1 \leq i < m$, where $h_W(\cdot)$ is defined by eq. (2.4).*
(5) *For $1 \leq i < j \leq m$, the sets $\{f^k(y^j) : k \geq 0\}$ and $\{f^k(y^i) : k \geq 0\}$ are disjoint.*
(6) *For $1 \leq i < j \leq m$, the elements y^i and y^j are not comparable.*

Outline of proof. By using eq. (2.8) we see that there are irreducible elements $z^i, 1 \leq i \leq \mu$, with

$$\xi = \sup{}_W(\{z^i|1 \leq i \leq \mu\}). \tag{2.9}$$

We can assume that μ is minimal in the sense that eq. (2.9) is not satisfied for any set of irreducible elements in W with fewer than μ elements. The elements y^i are obtained by relabelling the z^i and possibly removing some of the z^i to insure that the fifth condition of the proposition is satisfied.

With regard to condition 2 in Proposition 2.1, it is known (see [10,14]) that if y is any irreducible element in a finite lower semilattice W in $\mathbb{R}^n$ and $f : W \to W$ is a lower semilattice automorphism, then y is a periodic point of f of minimal period $q \leq n$. Furthermore, $f^j(y)$ is also an irreducible element of W for all $j \geq 0$.

The next proposition, which is also proved in [10], will lead directly to the definition of $Q(n)$.

Proposition 2.2. *(See [10].) Let W be a finite lower semilattice in $\mathbb{R}^n$ and $f : W \to W$ a lower semilattice automorphism. Assume that $y^i, 1 \leq i \leq m$, are elements of W and satisfy*

(a) y^i is an irreducible element of W.

(b) $h_W(y^i) \leq h_W(y^{i+1})$, where $h_W(\cdot)$ is the height function of eq.(2.4).

(c) For $1 \leq i < j \leq m$, the sets $\{f^k(y^i) : k \geq 0\}$ and $\{f^k(y^j) : k \geq 0\}$ are disjoint.

(d) For $1 \leq i < j \leq m$, the elements y^i and y^j are not comparable.
Let p_i denote the minimal period of y^i as a periodic point of f, so $1 \leq p_i \leq n$; and for $0 \leq j < p_i$ select $a_{ij} \in I_W(f^j(y^i))$ and extend the map $j \to a_{ij}$ to $\mathbb{Z}$ by making $j \to a_{ij}$ periodic of period p_i. (Here $I_W(x)$ is defined by eq. (2.6)). Define $\sum = \{i \in \mathbb{Z} : 1 \leq i \leq n\}$, so $\sum$ is a set with n elements. Then the semi-infinite array $a_{ij}, 1 \leq i \leq m, j \in \mathbb{Z}$, has the following properties:

(1) *$a_{ij} \in \sum$ for $1 \leq i \leq m$ and $j \in \mathbb{Z}$ and $a_{ij} \neq a_{ik}$ for $0 \leq j < k < p_i$ and $1 \leq i \leq m$. The map $j \to a_{ij}$ is periodic of minimal period p_i and $p_i \leq n$.*

(2) *If* $1 \leq m_1 < m_2 < \cdots < m_{r+1} \leq m$ *is any increasing sequence of* $(r+1)$ *integers between* 1 *and* m *and if*

$$a_{m_i s_i} = a_{m_{i+1} t_i} \tag{2.10}$$

for $1 \leq i \leq r$, *then*

$$\sum_{i=1}^{n} (s_i - t_i) \not\equiv 0, \quad \text{mod } (\rho), \tag{2.11}$$

where $\rho = gcd(\{p_{m_i} | 1 \leq i \leq r+1\})$ *and "gcd" denotes "greatest common divisor".*

We wish to **define** an admissible array on n symbols as a semi-infinite array $a_{ij}, i \in L := \{i \in \mathbb{N} : 1 \leq i \leq m\}$ and $j \in \mathbb{Z}$, which satisfies properties (1) and (2) in Proposition 2.2. It is convenient to give a slightly more general definition.

Definition 2.3. Let $(L, <)$ be a finite, totally ordered set, let $\mathbb{Z}$ denote the integers, and let $\sum$ denote a set with precisely n elements. For $i \in L$, suppose that $\varphi_i : \mathbb{Z} \to \sum$ is a map. We shall say that $\{\varphi_i : \mathbb{Z} \to \sum | i \in L\}$ is "an admissible array on n symbols" if the following conditions are satisfied:

(1) The map φ_i is a periodic map of minimal period p_i, where $p_i \leq n$. Furthermore, $\varphi_i(j) \neq \varphi_i(k)$ for $0 \leq j < k < p_i$.
(2) If $m_1 < m_2 < \cdots < m_{r+1}$ is any increasing sequence of $(r+1)$ elements of L and $\varphi_{m_i}(s_i) = \varphi_{m_{i+1}}(t_i)$ for $1 \leq i \leq r$, then

$$\sum_{i=1}^{r} (s_i - t_i) \not\equiv 0, \quad \text{mod } (\rho),$$

where $\rho = gcd(p_{m_1}, p_{m_2}, \cdots, p_{m_{r+1}})$.

In the notation of Proposition 2.2, $L = \{i \in \mathbb{N} | 1 \leq i \leq m\}$, $\sum = \{i \in \mathbb{N} | 1 \leq i \leq n\}$ and $\varphi_i(j) = a_{ij}$.

If $\{\varphi_i : \mathbb{Z} \to \sum | i \in L\}$ is an admissible array on n symbols and φ_i has minimal period p_i, we shall say that $p = lcm(\{p_i | i \in L\})$ is the minimal period of the admissible array $\{\varphi_i : \mathbb{Z} \to \sum | i \in L\}$. If $L_0 \subset L$ and L_0 inherits its ordering from L, we shall say that $\{\varphi_i : \mathbb{Z} \to \sum | i \in L_0\}$ is a subarray of $\{\varphi_i : \mathbb{Z} \to \sum | i \in L\}$; and we shall call it a proper subarray if $L_0 \neq L$. If $\{\varphi_i : \mathbb{Z} \to \sum | i \in L\}$ is an admissible array on n symbols, we shall call it a "minimal admissible array" if there is no proper subarray of $\{\varphi_i | i \in L\}$ whose minimal period equals the minimal period of $\{\varphi_i | i \in L\}$. With this terminology we can define $Q(n)$:

$$Q(n) = \{p | \text{ there exists an admissible array on } n \text{ symbols of minimal period } p\} \tag{2.12}$$

If $S = \{q_i | 1 \leq i \leq m\}$ is a set of m integers with $1 \leq q_i \leq n$ for $1 \leq i \leq m$ and $q_i \neq q_j$ for $1 \leq i < j \leq m$, we shall say that "S is array-admissible for n" if there exist an admissible array on n symbols $\{\varphi_i : \mathbb{Z} \to \sum | i \in L\}$ and a one-one map σ

of $\{i \in \mathbb{N} | 1 \leq i \leq m\}$ onto L such that $q_i = p_{\sigma(i)}$, where p_j is the minimal period of φ_j. In this notation we have

$$Q(n) = \{lcm(S) | S \text{ is array admissible for } n\}$$

In general, the concept of admissible array on n symbols depends on the ordering $<$ on L. However, it is easy to show that we can always assume that $L = \{i \in \mathbb{N} | 1 \leq i \leq m\}$ with the usual ordering and $\sum = \{i \in \mathbb{N} | 1 \leq i \leq n\}$, if so desired.

We can now state our first big theorem.

Theorem 2.4. *(See [10], [12].) If $P_j(n)$ is defined by eq.(1.6), $Q_j(n)$ by eq.(2.1) and $Q(n)$ by eq. (2.12), we have for all $n \geq 1$ that $P_1(n) \subset P_2(n)$ and*

$$P_2(n) = P_3(n) = Q_1(n) = Q_2(n) = Q(n).$$

Outline of proof. Propositions 2.1 and 2.2, eq. (2.2) and the fact that $P_2(n) \subset Q_1(n)$ imply that

$$P_1(n) \subset P_2(n) \subset P_3(n) \subset Q_2(n) \subset Q(n) \text{ and } P_2(n) \subset Q_1(n) \subset Q(n).$$

Thus it suffices to prove that if $\{\varphi_i : \mathbb{Z} \to \sum | i \in L\}$ is an admissible array on n symbols of minimal period p, then there exist $f \in \mathcal{F}_2(n)$ and a periodic point ξ of f of minimal period p. In [12] it is shown how to associate to $\{\varphi_i | i \in L\}$ a map $f \in \mathcal{F}_2(n)$ of the type given in Example 2 of Section 1 and a periodic point ξ of f of minimal period p.

In fact a sharper result than Theorem 2.4 is obtained in [12]. We mention one aspect of this sharper result. Suppose that $\{\varphi_i : \mathbb{Z} \to \sum | i \in L\}$ is a minimal admissible array on n symbols and that φ_i has minimal period p_i. Then there exists $f \in \mathcal{F}_2(n)$ and a periodic point ξ of f of minimal period $p = lcm(\{p_i | i \in L\})$. Furthermore, if W is the lower semilattice generated by $\{f^j(\xi) | j \geq 0\}$, W contains irreducible elements $y^i \leq \xi, i \in L$, such that y^i has minimal period p_i.

3. Computing $Q(n)$.

The set $Q(n)$ is determined solely by certain combinatorial and number theoretical constraints, but it is by no means clear how to compute $Q(n)$. Indeed, the problem of computing $Q(n)$ appears highly nontrivial. In this section we shall describe some results from [11] which, in particular, allow the computation of $Q(n)$ for $n \leq 50$.

We begin with some general definitions. Suppose that for each $n \geq 1$, $S(n)$ is a finite collection of positive integers and that $1 \in S(n)$.

Definition 3.1. We shall say that $\{S(n) : 1 \leq n \leq N\}$ satisfies Rule A for $n \leq N$ if whenever $p_1 \in S(m_1)$ and $p_2 \in S(m_2)$ and $m_1 + m_2 \leq N$, then $lcm(p_1, p_2) \in S(m_1 + m_2)$. We shall say that $\{S(n) : n \geq 1\}$ satisfies Rule A if $\{S(n) : 1 \leq n \leq N\}$ satisfies Rule A for $n \leq N$ for all $N \geq 1$.

Definition 3.2. We shall say that $\{S(n) : 1 \leq n \leq N\}$ satisfies Rule B for $n \leq N$ if whenever $p_1, p_2, \cdots, p_r \in S(m)$ and $rm \leq N$, it follows that $rlcm(p_1, p_2, \cdots, p_r) \in S(rm)$. We shall say that $\{S(n) : n \geq 1\}$ satisifies rule B if $\{S(n) : 1 \leq n \leq N\}$ satisfies Rule B for $n \leq N$ for all $N \geq 1$.

If one is given a set of positive integers $S(n)$ for $1 \leq n \leq N_0$ with $1 \in S(1)$ and if $\{S(n) : 1 \leq n \leq N_0\}$ satisfies Rule A and Rule B for $n \leq N_0$, then one can define a unique, minimal set of positive integers $S(N_0+1)$ such that $\{S(n) : 1 \leq n \leq N_0+1\}$ satisfies Rules A and B for $n \leq N_0 + 1$. Thus one can define inductively a minimal collection $\{S(n) : n \geq 1\}$ such that $S(n)$ is a set of positive integers for each $n \geq 1$ and $\{S(n) : n \geq 1\}$ satisfies Rules A and B.

With the above remarks in mind, we now define for each $n \geq 1$ an important collection of positive integers $P(n)$.

Definition 3.3. Define $P(1) = \{1\}$. If $P(n)$ denotes a set of positive integers and $P(n)$ has been defined for $n \leq N$, define $P(N+1)$ to be the smallest set of positive integers such that $\{P(n) : 1 \leq n \leq N+1\}$ satisfies Rules A and B.

The sets $P(n)$ are defined inductively for $n \geq 1$, and, as is discussed in [11], $P(n)$ can be computed relatively efficiently. If $1 \in S(1)$ and $\{S(n) : n \geq 1\}$ satisfies Rules A and B, then $P(n) \subset S(n)$ for all n. The reader can verify that $S(n) \subset S(n+1)$ for $n \geq 1$ and that

$$\{lcm(m_1, m_2, \cdots, m_k) | \sum_{i=1}^{k} m_i \leq n, k \geq 1\} \subset P(n).$$

In general, $P(n)$ is strictly larger than the collection of the orders of elements of the symmetric group on n letters, e.g., $12 \in P(6)$.

The relevance of these ideas to our problems is indicated by the following theorem.

Theorem 3.4. *(See [7] and [11]). The family of sets $\{P_1(n) : n \geq 1\}$ satisfies Rules A and B; $\{P_2(n) : n \geq 1\} = \{Q(n) : n \geq 1\}$ also satisfies Rules A and B.*

The fact that $\{P_1(n) : n \geq 1\}$ satisfies Rules A and B is proved in [7], and a slight modification of the argument in [7] (see [11]) proves that $\{P_2(n) : n \geq 1\}$ satisfies Rules A and B. Theorem 2.4 implies that $P_2(n) = Q(n)$ for all n, so $\{Q(n) : n \geq 1\}$ satisfies Rules A and B.

It follows from Theorem 3.4 and our previous remarks that

$$P(n) \subset P_1(n) \subset P_2(n) = P_3(n) = Q(n). \tag{3.1}$$

If we use equations (3.1) and (2.12), we see that

$$Q(n) = P(n) \cup \{lcm(S) | S \text{ is array-admissible for } n \text{ and } lcm(S) \notin P(n)\}. \tag{3.2}$$

It follows from eq. (3.2) that computing $Q(n)$ is equivalent to computing the second set on the right hand side of eq. (3.2).

The following theorems, when used in conjunction with eq. (3.2), are useful in determining $Q(n)$.

Theorem 3.5. *(See [11]). Suppose that $S \subset \{j \in \mathbb{N} | 1 \leq j \leq n\}$ and that S is array-admissible for n. If $|S| \leq 3$ or if $|S| = 4$ and $gcd(S) > 1$, it follows that $lcm(S) \in P(n)$.*

Our next theorem is intensively exploited in [11] to find conditions which insure that sets $S \subset \{j \in \mathbb{N} : 1 \leq j \leq n\}$ are not array-admissible for n.

Theorem 3.6. *(See Theorem 3.1 in [11]). Let $\hat{L} = \{i \in \mathbb{N} : 1 \leq i \leq m+1\}$ with the usual ordering and let $\sum$ denote a set with n elements. Assume that $\{\hat{\theta}_i : \mathbb{Z} \to \sum | i \in \hat{L}\}$ is an admissible array on n symbols. Let $\hat{B}_i = \{\hat{\theta}_i(j) | j \in \mathbb{Z}\}$ and write $p_i = |\hat{B}_i|$ (so $\hat{\theta}_i$ is periodic with minimal period p_i). Assume that $\hat{B}_i \cap \hat{B}_{i+1} \neq \phi$ for $1 \leq i \leq m$. Suppose that $S_j \subset \hat{L}, 1 \leq j \leq \mu$, are pairwise disjoint sets with $\hat{L} = \bigcup_{j=1}^{\mu} S_j$. For each $j, 1 \leq j \leq \mu$, assume that there is an integer $r_j > 1$ such that for all $i \in S_j$,*

$$gcd(p_{i-1}, p_i) | r_j \text{ and } gcd(p_i, p_{i+1}) | r_j. \tag{3.3}$$

(If $i = 1$, the equation $gcd(p_{i-1}, p_i)|r_i$ is vacuous; and if $i = m+1$, the equation $gcd(p_i, p_{i+1})|r_i$ is vacuous). If $s_j = |S_j|$, we have

$$\sum_{j=1}^{\mu} s_j = m+1 \tag{3.4}$$

and

$$\sum_{j=1}^{\mu} \frac{s_j}{r_j} \leq 1. \tag{3.5}$$

If all hypotheses of Theorem 3.6 are satisfied except possibly the condition that $\hat{B}_i \cap \hat{B}_{i+1} \neq \phi$ for $1 \leq i \leq m$ and if $\sum_{j=1}^{\mu} \frac{s_j}{r_j} > 1$, then Theorem 3.6 implies that $\hat{B}_i \cap \hat{B}_{i+1} = \phi$ for some $i, 1 \leq i \leq m$.

Typically, Theorem 3.6 is not applied directly. Rather one starts with an admissible array on n symbols $\{\theta_p : \mathbb{Z} \to \sum | p \in L\}$, where $\prec$ denotes the total ordering on L, one considers a subarray $\{\theta_p : \mathbb{Z} \to \sum | p \in L_o\}$, where $L_o = \{p_j | 1 \leq j \leq m+1\}$ and $p_i \prec p_{i+1}$ for $1 \leq i \leq m$, and one tries to apply Theorem 3.6 to the admissible array $\{\hat{\theta}_i : \mathbb{Z} \to \sum | i \in \hat{L}\}$ where $\hat{\theta}_i := \theta_{p_i}$ and $\hat{L} = \{i \in \mathbb{N} | 1 \leq i \leq m+1\}$.

Given a set $S = \{p_i | 1 \leq i \leq m+1\} \subset \{1, 2, \cdots, n\}$, a variety of directly verifiable conditions are given in [11] which are sufficient to show that S is not array-admissible for n. For reasons of length, we give only one of these conditions.

Definition 3.7. A set $S \subset \{1, 2, \cdots, n\}$ satisfies the generalized condition C for the integer n if S does not contain disjoint subsets Q and R with the following properties:

(1) $gcd(\alpha, \beta) = 1$ for all $\alpha \in Q$ and $\beta \in Q \cup R$ with $\alpha \neq \beta$.

(2) $\alpha + \beta > n^\star := n - \sum_{\gamma \in Q} \gamma$ for all $\alpha, \beta \in R$ with $\alpha \neq \beta$.

(3) there exist pairwise disjoint sets $S_j \subset R, 1 \leq j \leq \mu$, with $R = \bigcup_{j=1}^{\mu} S_j$ and such that for each $j, 1 \leq j \leq \mu$, there is an integer $r_j > 1$ such that $gcd(p, q) | r_j$ for all $p \in S_j$ and all $q \neq p, q \in R$.

(4) $\sum_{j=1}^{\mu} \frac{s_j}{r_j} > 1$, where $s_j = |S_j|$.

In the definition 3.7, the set Q is allowed to be empty. If $Q = \phi$, condition (1) in Definition 3.7 is vacuous and condition (2) means that $\alpha + \beta > n$ for all $\alpha, \beta \in R$ with $\alpha \neq \beta$.

Proposition 3.8. *If $S \subset \{1, 2, \cdots, n\}$ is array-admissible for n, then S satisfies the generalized condition C for n.*

It follows from Definition 3.7 and Proposition 3.8 that if $S \subset \{1, 2, \cdots, n\}$ contains disjoint sets Q and R which satisfy conditions (1) - (4) of Definition 3.7, then S is not array-admissible for n.

As an example of applying the generalized condition C, for $n = 45$ consider $S = \{24, 30, 33, 36, 39, 42\}$. Define $R = \{24, 30, 33, 39, 42\}, S_1 = \{33, 39\}$ and $S_2 = \{24, 30, 42\}$, so $s_1 = |S_1| = 2$ and $s_2 = |S_2| = 3$. Observe that $p + q > 45$ for all $p, q \in R$ with $p \neq q$. Note also that $gcd(p, \alpha)|3$ for all $p \in S_1$, and all $\alpha \neq p, \alpha \in R$ and $gcd(p, \alpha)|6$ for all $p \in S_2$ and all $\alpha \neq p, \alpha \in R$. If we set $r_1 = 3$ and $r_2 = 6$,

$$\sum_{j=1}^{2} \frac{s_j}{r_j} = \frac{2}{3} + \frac{3}{6} > 1.$$

It follows that S and R are not array-admissible for $n = 45$.

There are sets S for which $lcm(S) \notin P(n)$ and for which it is more difficult to determine whether S is array-admissible for n. For example, if $S = \{12, 14, 16, 20, 26, 34\}$, then $lcm(S) \notin P(49)$, and one can prove with the aid of Theorem 3.6 that S is not array-admissible for $n = 49$. However, the proof does not follow from generalized condition C.

With the aid of the above ideas and some other lemmas in [11], the following result is obtained in Theorem 6.3 of [11].

Theorem 3.9. *For $1 \leq n \leq 50, P(n) = P_1(n) = Q(n)$.*

The proof of Theorem 3.9 is computer assisted. For $n \leq 21$, a direct proof (not computer assisted) is given that $P(n) = Q(n)$. With little difficulty, this proof can be extended to $n \leq 25$; and with considerably more effort a direct proof can be given for $n \leq 32$. In general the computer considers various sets $S \subset \{1, 2, \cdots, n\}$ such that $lcm(S) \notin P(n)$ and checks whether the sets satisfy certain sufficient conditions (like generalized condition C) which imply that S is not array-admissible for n. After this sieving procedure one is left with what are called "candidate exceptional sets", i.e., sets $S \subset \{1, 2, \cdots, n\}$ such that $lcm(S) \notin P(n)$ and for which the sieving procedure leaves the possibility that S is array-admissible for n. Our particular sieving procedure finds no candidate exceptional sets S for $n \leq 41$, and a small number of candidate exceptional sets S for $42 \leq n \leq 50$. One of these candidate exceptional sets is the previously mentioned example $S = \{12, 14, 16, 20, 26, 34\}$ for $n = 49$. We then use subtler arguments to show directly that none of these candidate exceptional sets is actually array-admissible for n.

The number 50 in Theorem 3.9 has no particular significance. It is possible that $P(n) = Q(n)$ for a larger range of n. However, as is suggested by the following two theorems, it is almost certainly true that $P(n) \neq Q(n)$ for infinitely many n.

Thoerem 3.10. *(See Corollary 7.2 in [11]). Suppose that λ_4 and $\lambda_1 := \lambda_4 + 2$ are prime numbers with $\lambda_4 \geq 11$ and $\lambda_4 \neq 41$. Then it follows that* $q := 2^3 \times 7^2 \times \lambda_1 \times \lambda_4 \in Q(56 + 2\lambda_4)$ *and* $q \in P(57 + 2\lambda_4)$, *but* $q \notin P(56 + 2\lambda_4)$. *In particular, we have* $q = 2^3 \times 7^2 \times 11 \times 13 \in Q(78)$ *and* $q \in P(79)$, *but* $q \notin P(78)$.

Theorem 3.11. *(See Corollary 7.3 in [11]). Let λ_1 and λ_4 be prime numbers with* $97 \leq \lambda_4 < \lambda_1 \leq \lambda_4 + 6$. *Then it follows that* $q := 2^5 \times 3^4 \times \lambda_1 \times \lambda_4 \in Q(96 + 2\lambda_4)$ *and* $q \in P(96 + \lambda_1 + \lambda_4)$ *but* $q \notin P(96 + 2\lambda_4)$. *In particular, we have* $q := 2^5 \times 3^4 \times 97 \times 101 \in Q(290)$ *but* $q \notin P(290)$.

Remark 3.12. At present we do not know the smallest n such that $P(n) \neq Q(n)$, though Theorems 3.9 and 3.10 imply that $50 < n \leq 78$.

Remark 3.13. Until now all examples of integers q for which $q \in Q(n)$ but $q \notin P(n)$ also have the property that $q \in P_1(n)$. Thus it remains possible that $P_1(n) = Q(n)$ for all n.

Remark 3.14. Although presumably $P(n) \neq Q(n)$ for large n, perhaps $P(n)$ is, in some sense, a good approximation to $Q(n)$.

Remark 3.15. If $\gamma(n)$ denotes the largest element of $Q(n)$, can one determine the asymptotics of $\log(\gamma(n))$? The table below lists the largest element of $Q(n)$ for $n \leq 50$.

n	largest element of $Q(n)$	n	largest element of $Q(n)$
2	2	26	$2^4\cdot 3\cdot 5\cdot 13$
3	3	27	$2^4\cdot 3^2\cdot 5\cdot 7$
4	2^2	28	$2^3\cdot 3\cdot 5\cdot 7\cdot 11$
5	$2\cdot 3$	29	$2^3\cdot 3\cdot 5\cdot 7\cdot 11$
6	$2^2\cdot 3$	30	$2^3\cdot 3\cdot 5\cdot 7\cdot 11$
7	$2^2\cdot 3$	31	$2^4\cdot 3\cdot 5\cdot 7\cdot 11$
8	$2^3\cdot 3$	32	$2^5\cdot 3\cdot 5\cdot 7\cdot 11$
9	$2^3\cdot 3$	33	$2^5\cdot 3\cdot 5\cdot 7\cdot 11$
10	$2^2\cdot 3\cdot 5$	34	$2^5\cdot 3\cdot 5\cdot 7\cdot 11$
11	$2^2\cdot 3\cdot 5$	35	$2^5\cdot 3\cdot 5\cdot 7\cdot 11$
12	$2^3\cdot 3\cdot 5$	36	$2^4\cdot 3^2\cdot 5\cdot 7\cdot 11$
13	$2^3\cdot 3\cdot 5$	37	$2^4\cdot 3^2\cdot 5\cdot 7\cdot 11$
14	$2^3\cdot 3\cdot 7$	38	$2^4\cdot 3^2\cdot 5\cdot 7\cdot 11$
15	$2^2\cdot 3^2\cdot 5$	39	$2^5\cdot 3\cdot 5\cdot 7\cdot 17$
16	$2^4\cdot 3\cdot 7$	40	$2^5\cdot 3^2\cdot 5\cdot 7\cdot 13$
17	$2^2\cdot 3\cdot 5\cdot 7$	41	$2^5\cdot 3^2\cdot 5\cdot 7\cdot 13$
18	$2^2\cdot 3\cdot 5\cdot 7$	42	$2^2\cdot 3^2\cdot 5\cdot 7\cdot 11\cdot 13$
19	$2^3\cdot 3\cdot 5\cdot 7$	43	$2^2\cdot 3^2\cdot 5\cdot 7\cdot 11\cdot 13$
20	$2^4\cdot 3\cdot 5\cdot 7$	44	$2^4\cdot 3\cdot 5\cdot 7\cdot 11\cdot 13$
21	$2^4\cdot 3\cdot 5\cdot 7$	45	$2^5\cdot 3\cdot 5\cdot 7\cdot 11\cdot 13$
22	$2^4\cdot 3\cdot 5\cdot 7$	46	$2^5\cdot 3\cdot 5\cdot 7\cdot 11\cdot 13$
23	$2^4\cdot 3\cdot 5\cdot 7$	47	$2^5\cdot 3\cdot 5\cdot 7\cdot 11\cdot 13$
24	$2^4\cdot 3\cdot 5\cdot 11$	48	$2^4\cdot 3^2\cdot 5\cdot 7\cdot 11\cdot 13$
25	$2^4\cdot 3\cdot 5\cdot 11$	49	$2^4\cdot 3^2\cdot 5\cdot 7\cdot 11\cdot 13$
		50	$2^4\cdot 3^2\cdot 5\cdot 7\cdot 11\cdot 13$

References

1. M.A. Akcoglu and U. Krengel, *Nonlinear models of diffusion on a finite space*, Prob. Th. Rel. Fields **76** (1987), 411-420.
2. M.G. Crandall and L. Tartar, *Some relations between nonexpansive and order preseving mappings*, Proc. Amer. Math. Soc. **78** (1980), 385-391.
3. E. Landau, *Über die Maximal ordnung der Permutation gegebenen Grades*, Archiv. der. Math. und Phys., Ser. 3,5 (1903), 92–103.
4. J. Massias, *Majoration explicite de l'ordre maximum d'un élément du group symétrique*, Ann. Fac. Sci. Toulouse Math., ser. 5,6 (1984), 269–281.
5. W. Miller, *The maximum order of an element of a finite symmetric group*, Amer. Math. Monthly **94** (1987), 497–506.
6. H. Minc, *Nonnegative Matrices*, John Wiley and Sons, New York, (1988).
7. R.D. Nussbaum, *Estimates of the periods of periodic points for nonexpansive maps*, Israel J. Math. **76** (1991), 345–380.
8. ———, *Lattice isomorphisms and iterates of nonexpansive maps*, Nonlinear Analysis, T., M., and A. **22** (1994), 945–970.
9. ———, *Periodic points of nonexpansive maps*, in Optimization and Nonlinear Analysis, edited by A. Ioffe, M. Marcus and S. Reich, Pitman Research Notes in Math. Series vol. 244, Longman Scientic and Technical, Harlow, United Kingdom, 1992, 214–226.
10. R.D. Nussbaum and M. Scheutzow, *Admissible arrays and a nonlinear generalization of Perron-Frobenius theory*, to appear in J. London Math Society.
11. R. D. Nussbaum and S. Verduyn Lunel, *Generalizations of the Perron-Frobenius theorem for nonlinear maps*, submitted for publication.
12. R. D. Nussbaum, M.Scheutzow and S. Verduyn Lunel, in preparation.
13. H. Schaefer, *Banach Lattices and Positive Operators*, New York, Springer-Verlag, 1974.
14. M. Scheutzow, *Periods of nonexpansive operators on finite ℓ_1-spaces*, European J. Combinatorics **9** (1988), 73–78.
15. ———, *Corrections to Periods of nonexpansive operators on finite ℓ_1-spaces*, European J. Combinatorics **12** (1991), 183.

Mathematics Department, Rutgers University, New Brunswick, NJ 08903
E-mail address: nussbaum@math.rutgers.edu

Contemporary Mathematics
Volume **204**, 1997

On Generalized Quasiconvex Conjugation

A. M. Rubinov and B. M. Glover

ABSTRACT. This paper discusses various concepts of conjugation that have been developed in the literature for quasiconvex and generalized quasiconvex functions. We discuss an important subclass of quasiconvex functions for which a conjugation scheme is presented involving fewer parameters than in previous approaches to conjugation. An application to solvability of generalized quasiconvex inequality systems is presented where the dual condition characterizing solvability is expressed using conjugates. Finally we consider dual problems for certain quasiconvex maximization problems.

1. Introduction

Quasiconvex functions have found important applications particularly in economics and mathematical programming. These functions are characterized by the convexity of their level sets - an important property in economics. Many authors (see [**7**] for a review of the literature) have studied the problem of introducing a dual problem to a quasiconvex extremal problem. Since the class Q of all quasiconvex functions is extremely broad it is necessary to use an extra parameter in the definition of a general quasiconvex conjugate function. In this paper we restrict our attention to a special subclass of Q for which it is possible to define and study quasiconvex conjugation using fewer parameters.

Conjugacy operators play a very important role in both convex and quasiconvex analysis. These operators are very convenient tools for the study of extremal problems involving convex and quasiconvex functions. The Fenchel-Moreau conjugacy operator, $*$, plays a pivotal role in Convex Analysis. The well known Fenchel-Moreau conjugate, f^*, of a function f is defined as follows:

$$f^*(v) = \sup_{x \in X}(v(x) - f(x)) \qquad (\forall v \in V) \tag{1}$$

where X is a locally convex Hausdorff topological vector space, V is the dual space of X, and $f : X \to \mathbb{R} \cup \{\infty\}$. This conjugacy operator (1) is also used in the study of functions which are (generalized) convex with respect to a given class of functions H. By definition a function f is a *generalized convex function with respect to* H (H-convex) if it can be represented as the upper envelope of a subset of H; thus

1991 *Mathematics Subject Classification.* Primary 90C26; Secondary 26B25.

there is a set $U \subseteq H$ such that, for all x,

$$f(x) = \sup_{v \in H} v(x).$$

We will use the conjugacy operator (1) for generalized convex functions where the class H comprises certain special two-step functions.

Many conjugacy operators have been suggested in the study of quasiconvex functions. As a rule they are applied to minimization problems involving quasiconvex functions. Thach [**15, 16**] (see also [**18**]) has recently used a version of one of such operator to study problems of quasiconvex maximization. Such problems fall within the broad area of (multiextremal) global optimization. In fact Thach considered only the set of functions f such that $f(0) = \inf\{f(x) : x \neq 0\}$ and his conjugacy operator is a special case of more general approaches to conjugation developed by several authors including [**19, 7, 6, 2, 13**].

The principal result, which allows us to apply duality to the study of various problems (in particular for study global quasiconvex maximization) is the description of functions where *the second conjugate (or biconjugate) function coincides with the given function.* This description was given by various authors for a variety of conjugation schemes. (See [**7, 19, 6**] for a survey of results in this direction.) Thach described two classes of functions where the biconjugate coincides with the initial function under his conjugation scheme. In fact Thach proved this result for functions with the above mentioned property $f(0) = \inf\{f(x) : x \neq 0\}$, and which are either upper semicontinuous (u.s.c) or lower semicontinuous (l.s.c) and coercive.

We study generalized quasiconvex functions. Our approach is based on a representation of the functions as generalized convex functions with respect to a special class of upper semicontinuous (u.s.c.) quasiaffine two-step functions. We study a type of conjugacy operator for these functions. Following Thach we give a special definition of the value of the conjugate function at the point zero. However we show that it is more convenient in the context of this paper to give an alternative definition, from that of Thach, for this value. We will show that both definitions coincide for generalized quasiconvex functions. *In particular we give a full description of the set of all functions with the property $f(0) = \inf\{f(x) : x \neq 0\}$ for which the second conjugate function coincides with the given function.* For the proof of this result we use the conjugacy operator (1). It is interesting to observe that a duality result for quasiconvex functions follows from the usual convex duality theory applied under generalized convexity assumptions. We also consider a version of the conjugacy operator which arises if we take l.s.c two-step functions instead of u.s.c ones. The restriction of this version to the set of nonnegative functions with the property $f(0) = \inf\{f(x) : x \neq 0\}$ coincides with the restriction of a version of *conjugaison par tranches* [**19**]. A multiplicative version of the operator under consideration defined on the set of all nonnegative l.s.c quasiconvex functions vanishing at zero is studied in [**8**]. Finally, we study the solvability of certain inequality systems involving generalized quasiconvex functions utilizing this conjugacy operator and certain dual problems for quasiconvex maximiation problems.

Some results of this paper can be obtained from corresponding results due to I. Singer [**14**]. Here we provide direct proofs of these results for the sake of completeness. Our approach sheds new light on the links between the classical Fenchel-Moreau conjugation scheme and quasiconvex conjugation schemes that have been suggested recently in the literature.

The structure of the paper is as follows. In section 2 we present some preliminary results concerning generalized convex sets and functions and their basic properties. In section 3 we establish connections between generalized convex and generalized quasiconvex functions using special classes of two-step functions. In section 4 the conjugation scheme is studied and compared with the Fenchel-Moreau scheme under generalized convexity assumptions. Section 5 discusses conjugation schemes for generalized quasiconvex functions generated by l.s.c two-steps quasi-affine functions. In section 6 we present some solvability results for quasiconvex inequality systems and in the final section we discuss dual problems of quasiconvex maximization.

2. Preliminaries

Let us consider a pair (X, V) of sets with a coupling functional $[\,,\,] : V \times X \to \bar{\mathbb{R}}$ where $\bar{\mathbb{R}}$ denotes the extended real line. This function allows us to consider an element $v \in V$ as a function defined on X and, similarly, an element $x \in X$ as a function on V. Sometimes we will write either $v(x)$ of $x(v)$ interchangably with $[v, x]$ in the sequel. We shall say a function f minorizes a function g, where f and g are extended real-valued functions defined on a set X if $f(x) \leq g(x)$ for all $x \in X$.

Let us recall some definitions regarding concepts of generalized convexity (see [**5, 10**]).

For a function $f : X \to \bar{\mathbb{R}}$ we denote the *support set of* f *with respect to the pair* (X, V) by $s(f, X, V)$ where

$$s(f, X, V) = \{v \in V : (\forall x \in X)\, v(x) \leq f(x)\}.$$

A function $f : X \to \bar{\mathbb{R}}$ is called V*-convex* if there is a set $U \subseteq V$ such that

$$f(x) = \sup\{v(x) : v \in U\}.$$

Clearly f is V-convex if and only if

$$f(x) = \sup\{v(x) : v \in s(f, X, V)\}.$$

In the same way we can define X-convex functions defined on the set V.

A set $Z \subseteq X$ is called (V, X)-convex if there is a function $f : V \to \bar{\mathbb{R}}$ such that $Z = s(f, V, X)$. It follows immediately from the definition that Z is (V, X)-convex if and only if for each $x \in X \backslash Z$ there is a $v \in V$ such that $v(x) > \sup\{v(z) : z \in Z\}$.

REMARK 2.1. (V, X)-convex sets are called V-convex in [**4**] and X-convex in [**5, 9**].

A function $f : X \to \bar{\mathbb{R}}$ is called *quasiconvex with respect to the pair* (X, V) (or, simply, (V, X)-qc) if its level sets

$$S_c(f) = \{x : f(x) \leq c\}$$

are (V, X)-convex for all $c \in \mathbb{R}$.

A nonempty set $Z \subseteq X$ is called *evenly-*(V, X)*-convex* if for each $x \in X \backslash Z$ there exists $v \in V$ such that $v(x) > v(z)$, for all $z \in Z$. The empty set is evenly-(V, X)-convex by definition.

A function $f : X \to \bar{\mathbb{R}}$ is called *evenly quasiconvex* with respect to (X, V) (or, simply, ev-(V, X)-qc) if its level sets $S_c(f)$ are evenly (V, X)-convex for each $c \in \mathbb{R}$.

PROPOSITION 2.1. *Let* $f : X \to \bar{\mathbb{R}}$ *be a* (V, X)*-qc function (respectively an ev-*(V, X)*-qc function),* $Y \supseteq f(X)$ *and* $\tau : Y \to \bar{\mathbb{R}}$ *be a strictly increasing function. Then the function* $g = \tau \circ f$ *is also* (V, X)*-qc (respectively ev-*(V, X)*-qc).*

PROOF. For $c \in \bar{\mathbb{R}}$ we have

$$S_c(g) = \{x : \tau(f(x)) \leq c\} = \{x : f(x) \leq \tau^{-1}(c)\} = S_{\tau^{-1}(c)}(f).$$

Thus the result follows. □

It is easy to check that the following is valid.

PROPOSITION 2.2. *Every V-convex function is (V, X)-qc.*

We will consider in this paper only pairs of sets (X, V) such that there is an element $0 \in X$ with the property $[v, 0] = 0$ for all $v \in V$ and there is an element $0 \in V$ with the property $[0, x] = 0$ for all $x \in X$. Assume that the coupling functional $[\,,\,]$ maps $X \times V$ into $\mathbb{R}$. Let us denote

$$\tilde{V} = \{\tilde{v} : X \to \mathbb{R} : (\forall x \in X)\, \tilde{v}(x) = v(x) - c,\ v \in V,\ c \in \mathbb{R}\}.$$

$$\tilde{X} = \{\tilde{x} : V \to \mathbb{R} : (\forall v \in V)\, \tilde{x}(v) = x(v) - c,\ x \in X,\ c \in \mathbb{R}\}.$$

The sets $\tilde{V}$ and $\tilde{X}$ are useful in the study of the Fenchel-Moreau conjugacy operator $*$. Let $f : X \to \mathbb{R}_{+\infty}$ where $\mathbb{R}_{+\infty} = \mathbb{R} \cup \{+\infty\}$. The function f^* defined by the formula

$$f^*(v) = \sup_{x \in X} (v(x) - f(x))$$

is called the *conjugate function* of f (with respect to the pair (X, V)). It is easy to check that f^* is an $\tilde{X}$-convex function. The following can be shown to hold (see, for example, [**5**]):

THEOREM 2.1. *Let $f : X \to \mathbb{R}_{+\infty}$, then f^{**} coincides with the greatest $\tilde{V}$-convex function which minorizes f.*

We now consider two examples.

EXAMPLE 1. Let X be a locally convex Hausdorff topological vector space and let $V = X'$ be the dual space. We shall consider a pair (X, V) with the coupling functional $[v, x] = v(x)$. It is well known that a function $f : X \to \mathbb{R}_{+\infty}$ is V-convex if and only if f is sublinear and l.s.c., a function $f : X \to \mathbb{R}_{+\infty}$ is $\tilde{V}$-convex if and only if f is convex and l.s.c. A set $Z \subseteq X$ is (V, X)-convex if and only if it is closed and convex; a set Z is evenly (V, X)-convex if and only if Z is evenly convex (i.e. it is an intersection of open half-spaces (see [**3, 7**])). A function $f : X \to \bar{\mathbb{R}}$ is (V, X)-qc if and only if f is quasiconvex and l.s.c.; a function $f : X \to \bar{\mathbb{R}}$ is ev-(V, X)-qc if and only if f is evenly quasiconvex . Recall (see, for example, [**7**]) that a function $f : X \to \bar{\mathbb{R}}$ is evenly quasiconvex if its level sets $S_c(f)$ are evenly convex for all $c \in \mathbb{R}$. It is easy to check that both l.s.c. and u.s.c. (upper semicontinuous) quasiconvex functions are evenly quasiconvex.

EXAMPLE 2. Let $X = V = \mathbb{R}^n_{++} = \{x = (x_1, x_2, \dots, x_n) : (\forall i)\, x_i > 0\} \cup \{0\}$ with the coupling functional defined by

$$[v, x] = \min_i v_i x_i.$$

Then, see [**10, 9**], a function $f : X \to \mathbb{R}_{+\infty}$ is V-convex if and only if f is increasing, $f(\lambda x) = \lambda f(x)$ for $\lambda > 0$, $x \in X$ and $f(0) = 0$; a function $f : X \to \mathbb{R}_{+\infty}$ is $\tilde{V}$-convex (see [**9**]) if and only if f is increasing and convex along rays (i.e. $f_x(\lambda) = f(\lambda x)$, $(\lambda > 0)$, is convex for all $x \in X$). A set $Z \subseteq X$ is (V, X)-convex if and only if it is closed (in the topological space $X = \mathbb{R}^n_{++}$) and normal (i.e. the inclusions $z \in Z$, $z' \in X$ along with the relation $z' \leq z$ imply $z' \in Z$). A function

$f : X \to \bar{\mathbb{R}}$ is (V, X)-qc if and only if f is increasing and l.s.c. By definition a function $f : \mathbb{R}^n_{++} \to \bar{\mathbb{R}}$ is ev-(V, X)-qc if, for all $c \in \mathbb{R}$ and $x \notin S_c(f)$, there exists $v \in V = \mathbb{R}^n_{++}$ such that $\min_i v_i x_i > \min_i v_i z_i$ for all $z \in S_c(f)$. It is easy to check that an ev-(V, X)-qc function is increasing and both l.s.c and u.s.c increasing functions are ev-(V, X)-qc.

3. Generalized Convexity and Quasiconvexity

Let us consider a pair (X, V) with a coupling functional $[\,,] : V \times X \to \bar{\mathbb{R}}$. We will denote by K_o the set of all two-step functions $\kappa : X \to \mathbb{R}$ which have the following structure:

$$\kappa(x) = \begin{cases} c & [v, x] \geq d \\ c' & [v, x] < d \end{cases} \tag{2}$$

with $v \in V$, $c, c' \in \mathbb{R}$, $c \geq c'$, $d \in \mathbb{R}$.

THEOREM 3.1. *A function $f : X \to \mathbb{R}_{+\infty}$ is K_o-convex if and only if f is bounded from below and ev-(V, X)-qc.*

PROOF. 1. Assume that f is K_o-convex. Then there is a family $(\kappa_\alpha)_{\alpha \in A} \subseteq K_o$ such that

$$f(x) = \sup_\alpha \kappa_\alpha(x) \tag{3}$$

Clearly f is bounded from below. It follows easily from (3) that

$$S_c(f) = \cap_{\alpha \in A} S_c(\kappa_\alpha).$$

Since $\kappa_\alpha \in K_o$ we can find $v_\alpha \in V$, $d_\alpha \in \mathbb{R}$, $c_\alpha, c'_\alpha \in \mathbb{R}$ with $c_\alpha \geq c'_\alpha$ such that

$$\kappa_\alpha(x) = \begin{cases} c_\alpha & [v_\alpha, x] \geq d_\alpha \\ c'_\alpha & [v_\alpha, x] < d_\alpha \end{cases}$$

Therefore $S_c(\kappa_\alpha) = \{x : \kappa_\alpha(x) \leq c\}$ either coincides with X, $\emptyset$ or with the set $\{x : [v_\alpha, x] < d_\alpha\}$. Choose $c \in \mathbb{R}$ such that $S_c(f)$ is nonempty. Assume that $\bar{x} \notin S_c(f)$. Then there is an index α such that $\bar{x} \notin S_c(\kappa_\alpha)$. Also $S_c(f) \subseteq S_c(\kappa_\alpha)$. Clearly $S_c(\kappa_\alpha) \neq \emptyset$, $S_c(\kappa_\alpha) \neq X$. Thus, for all $x \in S_c(f) \subseteq S_c(\kappa_\alpha)$

$$[v_\alpha, x] < d_\alpha$$

and $[v_\alpha, \bar{x}] \geq d_\alpha$. Thus $S_c(f)$ is an evenly (V, X)-convex set.

2. Let f be an bounded from below ev-(V, X)-qc function. Denote $m = \inf_{x \in X} f(x)$. Clerly $m > -\infty$. Let $x_o \in X$ and $f(x_o) = c$. If $c = m$ then take a constant function $\bar{\kappa} \in K_o$ such that $\bar{\kappa}(x) = m$ for all $x \in X$. Clearly $\bar{\kappa} \leq f$ and $\bar{\kappa}(x_o) = f(x_o)$. So

$$\sup \{\kappa(x_o) : \kappa \in s(f, K_o, X)\} = f(x_o).$$

Assume $c > m$ and take an arbitrary number c' such that $m < c' < c$.

The set $S_{c'}(f) = \{x : f(x) \leq c'\}$ is not empty. Since f is ev-(V, X)-qc it follows that there is a function $v \in V$ such that $[v, x_o] > [v, x]$ for all $x \in S_{c'}(f)$, thus $S_{c'}(f) \subseteq \{x : [v, x] < d\}$ where $d = [v, x_o]$. Let $\bar{c} = \inf\{f(x) : [v, x] \geq d\}$. Then $\bar{c} \leq c$. We also have $c' \leq \bar{c}$ since the inequality $[v, x] \geq d$ implies $f(x) > c'$. Let

$$\bar{\kappa}(x) = \begin{cases} \bar{c} & [v, x] \geq d \\ m & [v, x] < d \end{cases}$$

Clearly $\bar{\kappa} \leq f$ and $\bar{\kappa}(x_o) = \bar{c} \geq c'$. Therefore

$$\sup \{\kappa(x_o) : \kappa \in s(f, K_o, X)\} \geq c'$$

where c' is an arbitrary number such that $m < c' < c = f(x_o)$. Thus $f(x_o) = \sup \{\kappa(x_o) : \kappa \in s(f, K_o, X)\}$. □

REMARK 3.1. Let us consider the set K of all functions of the form (2) with $c, c' \in \bar{\mathbb{R}}$. The same method of proof shows that the following is valid: a function $f : X \to \bar{\mathbb{R}}$ is K-convex if and only if it is ev-(V, X)-qc. A related result has been previously noted, see, for example, [**6, 14**].

Let us consider a pair (X, V) of sets with a coupling functional $[\,,\,] : X \times V \to \mathbb{R}$. Recall that we consider V as a set of functions on X and X as a set of functions on V.

We will call (X, V) a *conic pair* if

(i) V is a conic set of functions on X (that is $\lambda v \in V$ if $\lambda \geq 0$, $v \in V$) and X is a conic set of functions on V;
(ii) For each v_1, $v_2 \in V$, $v_1 \neq v_2$, there exists an element $x \in X$ such that $[v_1, x] \neq [v_2, x]$; for each x_1, $x_2 \in X$, $x_1 \neq x_2$, there exists an element $v \in V$ such that $[v, x_1] \neq [v, x_2]$.
(iii) For each $x \in X$ there is a $v \in V$ such that $[v, x] > 0$ and for each $v \in V$ there is a $x \in X$ such that $[v, x] > 0$.

It follows from (i), with $\lambda = 0$, that the zero function (i.e. the function identically equal to zero) belongs to V and to X. In other words there is an element $0 \in X$ (respectively $0 \in V$) such that $[v, 0] = 0$ for all $v \in V$ (respectively $[0, x] = 0$ for all $x \in X$). It follows from (ii) that there exists a unique element with this property belonging to X (respectively to V).

Clearly the pairs of sets in examples 1 and 2 are conic pairs.

Let (X, V) be a conic pair and denote by $K(V)$ the set of all functions κ defined by (2) with $d > 0$. Thus $\kappa \in K(V)$ if there are $v' \in V$, c, $c' \in \bar{\mathbb{R}}$ with $c' \leq c$ and $d > 0$ such that

$$\kappa(x) = \begin{cases} c & [v', x] \geq d \\ c' & [v', x] < d \end{cases}$$

Since V is a conic set it follows that $v = \frac{1}{d} v' \in V$. So $K(V)$ consists of all functions κ of the following form:

$$\kappa(x) = \begin{cases} c & [v, x] \geq 1 \\ c' & [v, x] < 1 \end{cases} \tag{4}$$

with $v \in V$, $c, c' \in \bar{\mathbb{R}}$ and $c' \leq c$. We will denote the function (4) by $\kappa = (v, c, c')$. Clearly if $\kappa \in K(V)$ then $\kappa + \tilde{c} \in K(V)$ for all $\tilde{c} \in \mathbb{R}$. In particular $K(V)$ contains all constant functions.

THEOREM 3.2. *A function $f : X \to \bar{\mathbb{R}}$ is $K(V)$-convex if and only if f is ev-(V, X)-qc and $f(0) = \inf_{x \in X} f(x)$.*

PROOF. 1. Let f be a $K(V)$-convex function. Then

$$f(x) = \sup \{\kappa(x) : \kappa \in s(f, K(V), X)\}. \tag{5}$$

Since $0 = [v, 0] < 1$ and $c' \leq c$ we have $\kappa(0) \leq \kappa(x)$ for all $\kappa = (v, c, c') \in K(V)$ and $x \in X$. It follows from (5) that $f(0) \leq f(x)$ for all $x \in X$. Since $K(V) \subseteq K$

we can conclude that f is a K-convex function. So it follows from Remark 3.1 that f is ev-(V, X)-qc.

2. Conversely let f be an ev-(V, X)-qc function with $f(0) = \inf\{f(x) : x \in X\}$. Applying Remark 3.1 we obtain the equality

$$f(x) = \sup\{\kappa(x) : \kappa \in s(f, K, X)\}. \tag{6}$$

Let U be the set of functions $\kappa = (v, c, c') \in s(f, K, X)$ such that $c' = \inf\{f(x) : x \in X\} = f(0)$; thus

$$\kappa(x) = \begin{cases} c & [v,x] \geq d \\ f(0) & [v,x] < d. \end{cases} \tag{7}$$

It easily follows from (6) that

$$f(x) = \sup\{\kappa(x) : \kappa \in U\}. \tag{8}$$

We shall now show that $U \subseteq K(V)$. Let $\kappa \in U$ with κ as in (7). If $c = f(0)$ then κ is a constant function and thus in $K(V)$. Assume $c > f(0)$. It follows from (7) that $0 \in \{x : [v, x] < d\}$, that is $[v, 0] = 0 < d$. Since V is a conic set and $d > 0$ we can suppose that $d = 1$. Thus $\kappa \in K(V)$ and $U \subseteq K(V)$. Equality (8) shows that f is a $K(V)$-convex function. □

REMARK 3.2. Consider the set $K_o(V)$ consisting of all functions $\kappa = (v, c, c')$ with $-\infty < c'$, $c < +\infty$. The same proof shows that a function $f : X \to \mathbb{R}_{+\infty}$ is $K_o(V)$-convex if and only if f is ev-(V, X)-qc and $f(0) = \inf_{x \in X} f(x)$.

4. A Conjugation Scheme

Consider a conic pair (X, V). For a function $f : X \to \bar{\mathbb{R}}$ define

$$f^H(v) = -\inf\{f(x) : [v, x] \geq 1\},\ v \in V,\ v \neq 0.$$

$$f^H(0) = -\sup_{v \neq 0} \inf_{[v,x] \geq 1} f(x).$$

(By definition $\inf \emptyset = +\infty$, $\sup \emptyset = -\infty$). Clearly $f^H(0) = \inf\{f^H(v) : v \in V,\ v \neq 0\}$.

We will call f^H the *conjugate function* with respect to (X, V).

REMARK 4.1. For a function f defined on $\mathbb{R}^n$, Thach [**15, 16, 17**], see also Tuy [**18**] proposed that the value of the conjugate function at zero be given by $f^H(0) = -\sup\{f(x) : x \in X\}$. It will be shown (see Remark 4.4) that our definition coincides with that of Thach.

REMARK 4.2. There is an evident connection between the given definition and the definitions of conjugacy investigated by various authors including Volle [**19**], Martinez-Legaz [**6**], Singer [**13, 14**] and others. See [**7, 6**] for a detailed review of the literature. For example, Volle [**19**] defines, for a function $f : X \to \bar{\mathbb{R}}$, the *conjugaison par tranches* function f^Δ by the formula $f^\Delta(v, d) = -\inf\{f(x) : [v, x] \geq d\}$. Clearly $f^H(v) = f^\Delta(v, 1)$.

The following is a summary of the properties of the conjugate function.

(i) $(\lambda f)^H = \lambda f^H$ for all $\lambda > 0$.
(ii) If c is a constant then $(f + c)^H = f^H - c$.
(iii) If $f_1 \geq f_2$ then $f_1^H \leq f_2^H$.
(iv) If $f(x) = \inf_{\alpha \in A} f_\alpha(x)$ then $f^H(v) = \sup_{\alpha \in A} f_\alpha^H(v)$, $v \neq 0$.

(v) $\inf_{x\in X\setminus\{0\}} f(x) = -\sup_{v\in V\setminus\{0\}} f^H(v)$.

We will prove only (v). Indeed we have

$$\begin{aligned} -\inf_{x\in X\setminus\{0\}} f(x) &= \sup_{x\in X\setminus\{0\}} (-f(x)) \\ &= \sup_{v\in V\setminus\{0\}} \sup_{[v,x]\geq 1} (-f(x)) \\ &= \sup_{v\in V\setminus\{0\}} f^H(v). \end{aligned}$$

REMARK 4.3. It should be noted that Singer [**11, 12**] calls an operation satisfying properties (ii) and (iv) (for all v) a *conjugation*. Since, in this paper, the property (iv) holds only for $v \neq 0$ it follows that the operation $f \mapsto f^H$ is not a conjugation in the sense of Singer [**11**].

PROPOSITION 4.1. *Let $\tau : \bar{\mathbb{R}} \to \bar{\mathbb{R}}$ be a strictly increasing function, $f : X \to \bar{\mathbb{R}}$ and $g = \tau \circ f$. Then, for all $v \in V$,*

$$g^H(v) = -\tau(-f^H(v)). \tag{9}$$

PROOF. At first assume that $v \neq 0$. Since $f(x) = \tau^{-1}(g(x))$ we have

$$\begin{aligned} f^H(v) &= -\inf_{[v,x]\geq 1} f(x) \\ &= -\inf_{[v,x]\geq 1} \tau^{-1}(g(x)). \end{aligned}$$

Let $\inf\{\tau^{-1}(g(x)) : [v,x] \geq 1\} = u$. Then $f^H(v) = -u$ and

$$g^H(v) = -\inf_{[v,x]\geq 1} g(x) = -\tau(u) = -\tau(-f^H(v)).$$

We now check (9) for the case $v = 0$. Since

$$g^H(0) = \inf\{g^H(v) : v \neq 0\}, \quad f^H(0) = \inf\{f^H(v) : v \neq 0\}$$

we have

$$g^H(0) = \inf_{v\neq 0}(-\tau(-f^H(v))) = -\tau(-\inf_{v\neq 0} f^H(v)) = -\tau(-f^H(0)).$$

Thus the result follows. □

PROPOSITION 4.2. *Let τ, f, g be as in Proposition 4.1. Then, for all $x \in X$,*

$$g^{HH}(x) = \tau(f^{HH}(x)).$$

PROOF. Denote $\tau^o(u) = -\tau(-u)$. Clearly τ^o is a strictly increasing function. Proposition 4.1 shows that $g^H = \tau^o \circ f^H$. Applying this proposition to the functions τ^o, f^H, g^H we can conclude that $g^{HH}(x) = -\tau^o(-f^{HH}(x)) = \tau(f^{HH}(x))$ for all $x \in X$. □

We shall now show that conjugate functions are ev-(X, V)-qc.

PROPOSITION 4.3. *For an arbitrary function $f : X \to \bar{\mathbb{R}}$, the function f^H is ev-(X, V)-qc.*

PROOF. Let $\bar{v} \in V$ and $\bar{v} \notin S_c(f^H) = \{v : f^H(v) \leq c\}$. Assume $S_c(f^H) \neq \emptyset$ Then $0 \in S_c(f^H)$ so $\bar{v} \neq 0$. We have

$$-\inf\{f(x) : [\bar{v}, x] \geq 1\} = f^H(\bar{v}) > c.$$

Therefore there is an $\bar{x} \in X$ such that $[\bar{v}, \bar{x}] \geq 1$ and $-f(\bar{x}) > c$. On the other hand if $v \in S_c(f^H)$, $v \neq 0$, then $f^H(v) = -\inf\{f(x) : [v, x] \geq 1\} \leq c$. That is $-f(x) \leq c$ for all $x \in X$ such that $[v, x] \geq 1$. Thus $[v, \bar{x}] < 1 \leq [\bar{v}, \bar{x}]$ for all $v \in S_c(f^H)$. Therefore $S_c(f^H)$ is evenly (X, V)-convex and f^H is an ev-(X, V)-qc function. □

We now consider functions f defined on X with the property $f(0) = \inf\{f(x) : x \in X \setminus \{0\}\}$. We shall establish connections between the second conjugate function f^{HH} and the second Fenchel-Moreau conjugate function f^{**} with respect to a special set, $L(V)$, of two-step functions.

Recall (see Remark 3.2) that $K_o(V)$ is the set of all functions $\kappa = (v, c, c')$ with $c' > -\infty$ and $c < +\infty$. We will denote by $L(V)$ the subset of $K_o(V)$ consisting of all functions (v, c, c') with $c' = 0$, $c < +\infty$. Thus $\ell \in L(V)$ if there are $v \in V$ and $0 \leq c < +\infty$ such that

$$\ell(x) = \begin{cases} c & v(x) \geq 1 \\ 0 & v(x) < 1 \end{cases} \tag{10}$$

We will denote the function described by (10) by $\ell = (v, c)$. If $v = 0$ then $\{x : [v, x] \geq 1\} = \emptyset$, thus $\ell = 0$. We suppose also that $c = 0$ in this case. So if $\ell = (v, c)$ with $v = 0$ then $c = 0$.

Clearly $\kappa \in K_o(V)$ if and only if there is $\ell \in L(V)$ and $c \in \mathbb{R}$ such that $\kappa = \ell + c$. If we apply the notation of section 2 then we can write $\widetilde{L(V)} = K_o(V)$.

We now consider the conic pair of sets $(X, L(V))$. We denote by $*$ the Fenchel-Moreau conjugacy operator with respect to this pair. Thus for a function $f : X \to \mathbb{R}_{+\infty}$ we have, by definition,

$$f^*(\ell) = \sup_{x \in X} (\ell(x) - f(x)) \ \ (\ell \in L(V));$$

$$f^{**}(x) = \sup_{\ell \in L(V)} (\ell(x) - f^*(\ell)) \ \ (x \in X).$$

Thus we have the conic pair (X, V) with the conjugacy operation $f \mapsto f^H$ and the conic pair $(X, L(V))$ with the Fenchel-Moreau conjugacy operation $f \mapsto f^*$. We shall compare these two conjugacy operations.

PROPOSITION 4.4. *Let $\ell = (v, c) \in L(V)$ and let $f : X \to \mathbb{R}_{+\infty}$ be a function such that $f(0) = m$ where $m = \inf\{f(x) : x \neq 0\}$. Then*

$$f^*(\ell) = \sup\{c + f^H(v),\ -m).$$

PROOF. At first assume that $v \neq 0$. We then have

$$\begin{aligned} f^*(\ell) &= \sup_{x \in X} (\ell(x) - f(x)) \\ &= \sup\{\sup_{[v,x] \geq 1} (c - f(x)),\ \sup_{[v,x] < 1} (-f(x))\} \end{aligned}$$

Since $[v, 0] = 0 < 1$ it follows that

$$\sup_{[v,x]<1} -f(x) = -\inf_{[v,x]<1} f(x) = -f(0) = -m.$$

Also we have

$$\sup_{[v,x]\geq 1}(c - f(x)) = c + \sup_{[v,x]\geq 1}(-f(x)) = c + f^H(v).$$

Thus the result follows for $\ell = (v, c)$ with $v \neq 0$.

Now let $v = 0$. Then also $\ell = 0$ and $c = 0$. We have

$$f^*(\ell) = \sup_{x\in X}(-f(x)) = -\inf_{x\in X} f(x) = -m;$$

$$f^H(v) = \inf_{v'\neq 0} f^H(v') = \inf_{v'\neq 0}\sup_{[v',x]\geq 1}(-f(x)) \leq \sup_{x\in X\setminus\{0\}}(-f(x)) = -m.$$

Thus $f^*(\ell) = \sup\{c + f^H(v),\ -m\}$ as required. □

THEOREM 4.1. *Let $f : X \to \mathbb{R}_{+\infty}$ be a function such that $f(0) = m = \inf_{x\neq 0} f(x)$. Then $f^{**} = f^{HH}$.*

PROOF. Let $x \in X$. We have by definition that

$$f^{**}(x) = \sup_{\ell=(v,c)\in L(V)}(\ell(x) - f^*(\ell)).$$

Consider the following sets

$$B_1 = \{(v,c) \in L(V) : [v,x] \geq 1,\ c + f^H(v) \geq -m\}$$

$$B_2 = \{(v,c) \in L(V) : [v,x] \geq 1,\ c + f^H(v) \leq -m\}$$

$$B_3 = \{(v,c) \in L(V) : [v,x] < 1,\ v \neq 0,\ c + f^H(v) \geq -m\}$$

$$B_4 = \{(v,c) \in L(V) : [v,x] < 1,\ v \neq 0,\ c + f^H(v) \leq -m\}$$

$$B_5 = \{(0,0)\}.$$

We have

$$f^{**}(x) = \sup_{i=1,2,\dots,5}\ \sup_{\ell=(v,c)\in B_i}(\ell(x) - f^*(\ell)).$$

We now calculate $b_i = \sup_{\ell=(v,c)\in B_i}(\ell(x) - f^*(\ell))$, for each $i = 1, 2, \dots 5$. We will apply Proposition 4.4 and the definition of the function $\ell = (v, c)$.

1. If $(v,c) \in B_1$ then $f^*(\ell) = c + f^H(v)$, $\ell(x) = c$ so that

$$b_1 = \sup_{(v,c)\in B_1} -f^H(v) = \sup\{-f^H(v) : [v,x] \geq 1,\ c + m \geq -f^H(v)\}.$$

2. If $(v,c) \in B_2$ then $f^*(\ell) = -m$, $c + m \leq -f^H(v)$, $\ell(x) = c$ so that

$$\begin{aligned} b_2 &= \sup_{(v,c)\in B_2}(c+m) \\ &= \sup_{[v,x]\geq 1,\, c+m\leq -f^H(v)}(c+m) \\ &= \sup_{[v,x]\geq 1,\, c+m\leq -f^H(v)} -f^H(v). \end{aligned}$$

3. If $(v,c) \in B_3$ then $f^*(\ell) = c + f^H(v) \geq -m$, $\ell(x) = 0$ so

$$b_3 = \sup_{(v,c)\in B_3} -(c + f^H(v)) \leq m.$$

4. If $(v,c) \in B_4$ then $f^*(\ell) = -m$, $\ell(x) = 0$ so

$$b_4 = \sup_{(v,c)\in B_4} m = m.$$

5. If $(v, c) \in B_5$ then $v = 0$, $c = 0$ and

$$b_5 = -f^*(0) = -\sup_{x \in X} -f(x) = \inf_{x \in X} f(x) = m.$$

Assume that $x \neq 0$. For all $v \in V$, $v \neq 0$, we have

$$-f^H(v) = \inf\{f(x) : [v, x] \geq 1\} \geq m.$$

Therefore $\sup(b_1, b_2) \geq b_3$, b_4, and b_5. So, for $x \neq 0$, we have

$$f^{**}(x) = \sup\{b_1, b_2\} = \sup_{[v,x] \geq 1} -f^H(v) = f^{HH}(x).$$

Now let $x = 0$. Then $B_1 = B_2 = \emptyset$ so $b_1 = b_2 = -\infty$ and $f^{**}(0) = \sup\{b_3, b_4, b_5\} = m$. We now calculate $f^{HH}(0)$. By definition $f^{HH}(0) = \inf_{x \neq 0} f^{HH}(x)$. Applying the equality $f^{**}(x) = f^{HH}(x)$ which has already been established for $x \neq 0$ we have $f^{HH}(0) = \inf_{x \neq 0} f^{**}(x)$. It follows from the definitions that for $x \neq 0$:

$$\begin{aligned} f^{**}(x) &= \sup_{\ell} (\ell(x) - f^*(\ell)) \\ &\geq -f^*(0) \\ &= -\sup_{x' \in X} -f(x') \\ &= \inf_{x' \in X} f(x') \\ &= m. \end{aligned}$$

Therefore $\inf_{x \neq 0} f^{**}(x) \geq m$. By applying the well known inequality $f \geq f^{**}$ (which follows from Theorem 2.1) we have

$$m = \inf_{x \neq 0} f(x) \geq \inf_{x \neq 0} f^{**}(x).$$

Therefore

$$f^{HH}(0) = \inf_{x \neq 0} f^{HH}(x) = \inf_{x \neq 0} f^{**}(x) = m = f^{**}(0),$$

as required. □

PROPOSITION 4.5. *Let $f : X \to \mathbb{R}_{+\infty}$ be a function which is bounded from below with $f(0) = \inf_{x \neq 0} f(x)$. Then f^{HH} is the greatest ev-(V, X)-qc function which minorizes f.*

PROOF. Recall that $K_o(V) = \widetilde{L(V)}$; that is for all $\kappa \in K_o(V)$ there is an $\ell \in L(V)$ and a $c \in \mathbb{R}$ such that $\kappa = \ell + c$. So we can apply Theorem 2.1 which shows that f^{**} is the greatest $K_o(V)$-convex function which minorizes f. In order to complete the proof, apply Remark 3.2 and Theorem 4.1. □

We now extend this result to arbitrary functions $f : X \to \bar{\mathbb{R}}$ with the property $f(0) = \inf_{x \neq 0} f(x)$. Consider a strictly increasing function $\tau : \bar{\mathbb{R}} \to [0, 1]$. Let $g = \tau \circ f$. It is clear that $g(0) = \inf_{x \neq 0} g(x)$ and the inequality $\tilde{g} \leq g$ is equivalent to the inequality $\tau^{-1} \circ \tilde{g} \leq \tau^{-1} \circ g = f$, for a function $\tilde{g} : X \to [0, 1]$. Proposition 2.1 shows that a function $\tilde{g}$ is ev-(V, X)-qc if and only if $\tilde{f} = \tau^{-1} \circ \tilde{g}$ is ev-(V, X)-qc. Since g is bounded from below we can apply Proposition 4.5 to this function. Thus g^{HH} is the greatest ev-(V, X)-qc minorant of g and $g^{HH}(x) \geq 0$ for all $x \in X$; so $\tau^{-1} \circ g^{HH}$ is the greatest ev-(V, X)-qc minorant of $\tau^{-1} \circ g = f$. Combining this with Proposition 4.2 shows that the following result is valid:

THEOREM 4.2. *Let $f : X \to \bar{\mathbb{R}}$ with $f(0) = \inf_{x \neq 0} f(x)$. Then f^{HH} is the greatest ev-(V, X)-qc function which minorizes f.*

COROLLARY 4.1. *A function $f : X \to \bar{\mathbb{R}}$ with $f(0) = \inf_{x \neq 0} f(x)$ is an ev-(V, X)-qc function if and only if $f = f^{HH}$.*

REMARK 4.4. Let $f : X \to \bar{\mathbb{R}}$ be an ev-(V, X)-qc function and assume that $f(0) = \inf_{x \neq 0} f(x)$. Then $f^H(0) = -\sup_{x \in X} f(x)$. Thus $f^H(0)$ coincides with the value of the conjugate function proposed by Thach at the origin.

Indeed we will apply property (v) for conjugate function to the function f^H. We have

$$f^{HH}(0) = \inf_{v \neq 0} f^H(v) = -\sup_{x \neq 0} f^{HH}(x).$$

It follows from Corollary 4.1 that $f^{HH} = f$. Since $f(0) = \inf_{x \neq 0} f(x)$ we have

$$f^{HH}(0) = -\sup_{x \neq 0} f(x) = -\sup_{x \in X} f(x).$$

REMARK 4.5. We can apply the method of proof in Theorem 4.2 to study a version of *conjugaison par tranches* for evenly quasiconvex functions. Let us consider a pair (X, V) as in Example 1, i.e X is a locally convex Hausdorff topological vector space and $V = X'$ is a dual space. For $f : X \to \bar{\mathbb{R}}$ and $(v, d) \in V \times \mathbb{R}$ denote

$$f^\Delta(v, d) = -\inf\{f(x) : [v, x] \geq d\}.$$

For $g : (V \times \mathbb{R}) \to \bar{\mathbb{R}}$ and $x \in X$ denote

$$g^\Delta(x) = \inf\{g(v, d) : [v, x] \geq d\}.$$

Let $L'(V)$ be the set of functions ℓ of the form

$$\ell = \begin{cases} c & [v, x] \geq d \\ 0 & [v, x] < d \end{cases}$$

with $v \in V$, $0 \leq c < +\infty$ and $d \in \mathbb{R}$. We will denote by f^* the Fenchel-Moreau conjugacy operator with respect to the pair of sets $(X, L'(V))$. The same reasoning as in proof of Proposition 4.4 show that for the function ℓ defined above and for a function $f : X \to \bar{\mathbb{R}}$ we have

$$f^*(\ell) = \sup\{c + f^H(v, d), -m(v, d)\}.$$

with $m(v, d) = \inf\{f(x) : [v, x] < d\}$. Following the method of Theorem 4.1 we can show that for a bounded from below function $f : X \to \mathbb{R}_{+\infty}$ the following equality holds:

$$f^{**}(x) = \sup(f^{\Delta\Delta}(x), \sup_{[v,x] < d} m(v, d)).$$

We have

$$f^{\Delta\Delta}(x) = -\inf_{(v,d):[v,x] \geq d} \left(-\inf_{x':[v,x'] \geq d} f(x')\right) = \sup_{[v,x] \geq d} \inf_{[v,x'] \geq d} f(x').$$

At the same time

$$\sup_{[v,x] < d} m(v, d) = \sup_{[v,x] < d} \inf_{[v,x'] < d} f(x') =$$

$$\sup_{[-v,x] < -d} \inf_{[-v,x'] < -d} f(x') = \sup_{[v,x] > d} \inf_{[v,x'] > d} f(x').$$

It is easy to check that

$$\sup_{[v,x] \geq d} \inf_{[v,x'] \geq d} f(x') \geq \sup_{[v,x] > d} \inf_{[v,x'] > d} f(x').$$

So

$$f^{**}(x) = f^{\Delta\Delta}(x).$$

Thus we can conclude by applying Theorem 4.1 and Remark 3.1 that $f^{\Delta\Delta}$ is the greatest ev-(V,X)-qc function which minorizes f (for a bounded from below function f). Following the method of proof of Theorem 4.2 shows that this conclusion holds for an arbitrary function $f: X \to \bar{\mathbb{R}}$.

5. Conjugation for (V,X)-qc Functions

Let (X,V) be a pair of sets with coupling functional $[\,,\,]$. For $v \in V$, $c, c' \in \bar{\mathbb{R}}$ with $c' \leq c$ and $d \in \mathbb{R}$ we construct the following functions κ and j:

$$\kappa(x) = \begin{cases} c & [v,x] \geq d \\ c' & [v,x] < d \end{cases} \tag{11}$$

$$j(x) = \begin{cases} c & [v,x] > d \\ c' & [v,x] \leq d \end{cases} \tag{12}$$

The set K of all functions of the form (11) is useful in studying ev-(V,X)-qc functions. We also consider the set J of all functions of the form (12) under the assumption that $v \in V$, c, $c' \in \bar{\mathbb{R}}$ and $c' \leq c$ and $d \in \mathbb{R}$. We will show that this set is useful for studying (V,X)-qc functions. The same method of proof as used in Theorem 3.1 establishes the following.

THEOREM 5.1. *A function $f: X \to \bar{\mathbb{R}}$ is (V,X)-qc if and only if it is J-convex.*

Let (X,V) be a conic pair of sets and let $J(V)$ be the set of all functions j defined by the formula (12) with $d > 0, v \in V$. Clearly $J(V)$ consists of all functions j which have the following form:

$$j(x) = \begin{cases} c & [v,x] > 1 \\ c' & [v,x] \leq 1 \end{cases}$$

THEOREM 5.2. *A function $f : X \to \bar{\mathbb{R}}$ is $J(V)$-convex if and only if f is (V,X)-qc and $f(0) = \inf_{x\in X} f(x)$.*

PROOF. Applying similar arguments as in the proof of Theorem 3.2 we can verify that a $J(V)$-convex function is (V,X)-qc and $f(0) \leq f(x)$ for all $x \in X$.

Let f be (V,X)-qc and $f(0) = \inf_{x\in X} f(x) = m$. Let $f(x_o) = c$. If $c = m$ then the constant function $\bar{\kappa}$ where $\bar{\kappa}(x) = m$ possesses the following properties: $\bar{\kappa} \leq f$ and $\bar{\kappa}(x_o) = f(x_o)$. Now assume that $c > m$ and take an arbitrary number c' such that $m < c' < c$. Clearly $S_{c'}(f) \neq \emptyset$. Since f is (V,X)-qc it follows that there is a function $v \in V$ such that $[v,x_o] > \sup\{[v,x] : x \in S_{c'}(f)\}$. Take a number d such that

$$\sup\{[v,x] : x \in S_{c'}(f)\} \leq d < [v,x_o].$$

Denote $\inf\{f(x) : [v,x] > d\} = \bar{c}$. Since $[v,x_o] > d$ we have $f(x_0) = c \geq \bar{c}$. On the other hand the inequality $[v,x] > d$ implies $f(x) > c'$ therefore $\bar{c} \geq c'$. Let

$$\bar{j}(x) = \begin{cases} \bar{c} & [v,x] > d \\ m & [v,x] \leq d \end{cases}$$

Since $m = f(0) < c' \leq \bar{c}$ it follows that $0 = [v,0] \leq d$. If $d = 0$ then we can take d' such that

$$\sup\{[v,x] : x \in S_{c'}(f)\} \leq d = 0 < d' < [v,x_o].$$

So we can assume without loss of generality that $d > 0$. Since V is a conic set we can assume also that $d = 1$, that is $\bar{j} \in J(V)$. Clearly $\bar{j} \leq f$ and $\bar{j}(x_o) = \bar{c} \geq c'$. Since c' is an arbitrary number such that $m < c' < c = f(x_o)$ we can conclude that

$$f(x_o) = \sup \{j(x_o) : j \leq f,\ j \in J(V)\},$$

as required. □

Let $f : X \to \bar{\mathbb{R}}$ and denote by f^N the following function defined on V:

$$f^N(v) = \begin{cases} -\inf \{f(x) : [v, x] > 1\} & v \neq 0 \\ \inf_{v' \neq 0} f^N(v') & v = 0 \end{cases}$$

Applying similar arguments as in the proofs of corresponding results in section 4 we can verify the following:

PROPOSITION 5.1. *For all functions $f : X \to \bar{\mathbb{R}}$ the function f^N is (X, V)-qc.*

THEOREM 5.3. *Let $f : X \to \bar{\mathbb{R}}$ be a function bounded from below with $f(0) = \inf_{x \neq 0} f(x)$. Then $f^{NN} = f^{**}$.*

The operator $*$ in Theorem 5.3 is the Fenchel-Moreau conjugacy operator with respect to the pair $(X, M(V))$ where $M(V)$ is the set of two-step functions of form

$$\ell(x) = \begin{cases} c & [v, x] > 1 \\ 0 & [v, x] \leq 1 \end{cases}$$

with $0 \leq c < +\infty$.

THEOREM 5.4. *Let $f : X \to \bar{\mathbb{R}}$ be a function with $f(0) = \inf_{x \neq 0} f(x)$. Then f^{NN} is the greatest (V, X)-qc function which minorizes f.*

COROLLARY 5.1. *Let $f : X \to \bar{\mathbb{R}}$ be a function with $f(0) = \inf_{x \neq 0} f(x)$. Then f is (V, X)-qc if and only if $f = f^{NN}$.*

REMARK 5.1. Let X be a locally convex Hausdorff topological vector space and $V = X'$ is a dual space. As noted in [**19**] the conjugate function f^o defined by the formula

$$f^o(v) = -\inf\{f(x) :\ [v, x] > 1\} \quad \text{for all } v \in V$$

has been studied previously by Atteia and El Qortobi [**1**]. It is proved in [**1**] that $f^{oo} = f$ if and only if f is l.s.c quasiconvex and $f(0) = -\infty$.

6. Quasiconvex Inequality Systems

Thach [**15, 16**] (see also [**18**]) has applied the conjugacy operator $f \mapsto f^H$ to the study of dual problems in quasiconvex maximization. We will study in the next section a different approach to this problem. At first we apply this operator to the study of dual conditions which characterize solvability for inequality systems involving quasiconvex functions.

Let (X, V) be a pair of conic sets. For functions $f,\ g : X \to \bar{\mathbb{R}}$ let us consider the following statements:

(i) $g(x) < 0 \implies f(x) < 0$;
(ii) $g(x) \leq 0 \implies f(x) \leq 0$;
(iii) $f^H(v) < 0 \implies g^H(v) < 0$;
(iv) $f^H(v) \leq 0 \implies g^H(v) \leq 0$;
(v) $f^H(v) < 0 \implies g^H(v) \leq 0$.

PROPOSITION 6.1. *Statement (i) implies statement (iv).*

PROOF. Assume that $f^H(v) \leq 0$. We have

$$\inf_{[v,x]\geq 1} f(x) \geq 0 \implies (\forall x : [v,x] \geq 1)\, f(x) \geq 0 \tag{13}$$
$$\implies (f(x) < 0 \implies [v,x] < 1).$$

Let $g(x) < 0$. It follows from (i) that $f(x) < 0$. Applying (14) we can deduce that $[v,x] < 1$. Therefore $[v,x] \geq 1 \implies g(x) \geq 0$ that is $g^H(v) = -\inf_{[v,x]\geq 1} g(x) \leq 0$. □

PROPOSITION 6.2. *Statement (ii) implies statement (v).*

PROOF. Assume $f^H(v) = -\inf_{[v,x]\geq 1} f(x) < 0$. Hence $f(x) > 0$ for all x with $[v,x] \geq 1$. That is

$$f(x) \leq 0 \implies [v,x] < 1. \tag{14}$$

Let $g(x) \leq 0$. It follows from (ii) that $f(x) \leq 0$ and (14) shows that $[v,x] < 1$. Therefore

$$[v,x] \geq 1 \implies g(x) > 0 \implies \inf_{[v,x]\geq 1} g(x) \geq 0 \implies g^H(v) \leq 0. \tag{15}$$

Thus the result follows. □

Let us now consider the following property of the function g:

$$\inf_{[v,x]\geq 1} g(x) \text{ is achieved for all } v \neq 0 \tag{16}$$

This property holds, for example, if X is finite dimensional, g is coercive (so that $\lim_{x\to\infty} g(x) = +\infty$) and continuous.

PROPOSITION 6.3. *If property (16) holds for the function g then statement (ii) implies statement (iii).*

PROOF. We can substitute (15) for the following formula

$$[v,x] \geq 1 \implies g(x) > 0 \implies \inf_{[v,x]\geq 1} g(x) > 0 \implies g^H(v) < 0,$$

and then apply the same arguments as in the proof of Proposition 6.2. □

THEOREM 6.1. *Let f, g be ev-(V,X)-qc functions with $f(0) = \inf_{x\neq 0} f(x)$ and $g(0) = \inf_{x\neq 0} g(x)$.*

1. *Suppose that property (16) holds for the function g. Then statement (ii) is equivalent to statement (iii).*
2. *Suppose that property (16) holds for the function f^H. Then statement (i) is equivalent to statement (iv).*

PROOF. 1. We have to prove only that (iii) implies (ii). Let us apply Proposition 6.1 substituting g for f^H and f for g^H. We have

$$\text{(iii)} \implies (g^{HH}(x) \leq 0 \implies f^{HH}(x) \leq 0).$$

Since $g^{HH} = g$, $f^{HH} = f$ it follows that (iii) implies (ii).

The same proof can be applied for part 2 using Proposition 6.3. □

7. Dual problems of quasiconvex maximization

Let (X, V) be a pair of conic sets. Let us consider the following problem of quasiconvex maximization:

$$f(x) \longrightarrow \max \text{ subject to } x \in X,\ g(x) \leq \beta \tag{17}$$

where f and g are ev-(V, X)-qc functions.

PROPOSITION 7.1. *Let f, g be ev-(V, X)-qc functions with $f(0) = \inf_{x \neq 0} f(x)$ and $g(0) = \inf_{x \neq 0} g(x)$. Let us denote*

$$\sup\{f(x) : g(x) \leq \beta\} = \alpha. \tag{18}$$

Assume that $\alpha < +\infty$, property (16) holds for the function g and, moreover,

$$\sup\{f(x) :\ g(x) \leq \gamma\} > \alpha \quad \text{for all } \gamma > \beta. \tag{19}$$

Then

$$\inf\{-g^H(v) : -f^H(v) > \alpha\} = \beta. \tag{20}$$

PROOF. Let us consider the functions

$$f_\alpha(x) = f(x) - \alpha \quad \text{and} \quad g_\beta(x) = g(x) - \beta.$$

It follows from (18) that

$$g_\beta(x) \leq 0 \Longrightarrow f_\alpha(x) \leq 0.$$

Applying Theorem 6.1 we can conclude that

$$(f_\alpha)^H(v) < 0 \Longrightarrow (g_\beta)^H(v).$$

Since $(f_\alpha)^H(v) = f^H(v) + \alpha$ and $(g_\beta)^H(v) = g^H(v) + \beta$ we have

$$\sup\{g^H(v) : f^H(v) < -\alpha\} \leq -\beta. \tag{21}$$

Assume that the inequality in (21) is strict. Then we have for some $\gamma > \beta$:

$$(f_\alpha)^H(v) < 0 \Longrightarrow (g_\gamma)^H(v) < 0$$

with $g_\gamma(x) = g(x) - \gamma$. Again applying Theorem 6.1 we get

$$g_\gamma(x) \leq 0 \Longrightarrow f_\alpha(x) \leq 0$$

So

$$\sup\{f(x) :\ g(x) \leq \gamma\} \leq \alpha$$

This inequality contradicts (19). Thus our assumption does not hold and there is equality in (21). Clearly we can represent this equality in the form (20). □

We will call the problem

$$-g^H(v) \longrightarrow \min \text{ subject to } - f^H(v) \leq \alpha$$

the *dual* to problem (17). Proposition 7.1 allows us to obtain the following result.

THEOREM 7.1. *Assume that all the conditions of Proposition 7.1 are fulfilled and g^H is a l.s.c. function and the function f possesses the property*

$$cl\{v :\ f^H(v) < \alpha\} = \{v :\ f^H(v) \leq \alpha\}. \tag{22}$$

Then the value of dual probiem coincides with the 'constraint constant' β.

Now assume that X is a locally convex Hausdorff topological vector space and $V = X'$. Let x_o be a solution of the problem (17) with f assumed continuous and $\alpha < +\infty$ is the value of this problem. Since $f(x_o) = \alpha$ it follows that x_o is a boundary point of the convex open set $\{x : f(x) < \alpha\}$. So there exists a vector $v_o \in X'$ such that

$$1 = v_o(x_o) = \max\{v_o(x) : x \in S_\alpha(f)\}. \tag{23}$$

THEOREM 7.2. *Assume that all the conditions of Theorem 7.1 are fulfilled and f is continuous. Let x_o be a solution of the problem (17) such that $g(x_o) = \beta$ and assume for some vector $v_o \in X'$ the equalities (23) hold. Then v_o is a solution of the dual problem.*

PROOF. Since $v_o(x) \geq 1 \Longrightarrow f(x) \geq \alpha$ we have

$$-f^H(v_o) = \inf\{f(x);\ v_o(x) \geq 1\} = \alpha = f(x_o).$$

Since $S_\beta(g) \subset S_\alpha(f)$ and $g(x_o) = \beta$ it follows that

$$-g^H(v_o) = \inf\{g(x) :\ v_o(x) \geq 1\} = \beta = g(x_o).$$

Theorem 7.1 now shows that v_o is a solution of the dual problem. □

ACKNOWLEDGEMENT. The authors wish to thank Professor I. Singer and Professor M. Volle for useful discussions on an early draft of this paper and the anonymous referees for constructive comments.

References

[1] M. Atteia and A. El Qortobi, *Quasiconvex duality*, Lecture Notes in Control, **30**, Springer-Verlag, Berlin, (1981) 3-8.

[2] J.-P. Crouzeix, *Conjugacy in quasiconvex analysis*, in *Convex Analysis and its Applications*, A. Auslender (ed.), Springer-Verlag, Berlin (1977), 66-99.

[3] W. Fenchel, *Convex Cones, Sets and Functions*, Princeton University Press, Princeton, 1951.

[4] K. Fan, *On the Krein-Milman theorem*, in *Convexity*, Proceedings of the Symposium in Pure Mathematics, **7** (1963), American Mathematical Society, Providence, 211 - 219.

[5] S. S. Kutateladze and A. M. Rubinov, *Minkowski duality and applications*, Russian Mathematical Surveys, **27** (1972), 137-192.

[6] J. E. Martinez-Legaz, *Quasiconvex duality theory by generalized conjugation methods*, Optimization, **19** (1988), 603-652.

[7] J.-P. Penot and M. Volle, *On quasiconvex duality*, Math. Oper. Research, **15** (1990), 597-625.

[8] A. M. Rubinov and B. Simsek, *Conjugate quasiconvex nonnegative functions*, Optimization, **35** (1995), 1-22.

[9] A. M. Rubinov and B. M. Glover, *Duality for increasing positively homogeneous functions and normal sets*, Working paper 95/12, School of Information Technology and Mathematical Sciences, University of Ballarat, 1995.

[10] A. M. Rubinov, B. M. Glover and V. Jeyakumar, *A general approach to dual characterization of solvability of inequality systems with applications*, J. Convex Analysis, **2** (1995), 309-344.

[11] I. Singer, *Conjugation operators*, in Selected Topics in Operations Research and Mathematical Economics, Lecture Notes in Economics and Mathematical Systems, **226**, Springer-Verlag, Berlin (1984), 80-97.

[12] I. Singer, *Some relations between dualities, polarities, coupling functionals, and conjugation*, J. Math. Anal. Appl., **115** (1986), 1-22.

[13] I. Singer, *The lower semicontinuous quasiconvex hull as a normalized second conjugate*, Nonlinear Analysis: Theory, Methods and Appl. **7** (1983), 1115-1121.

[14] I. Singer, *Generalized convexity, functional hulls and applications to conjugate duality in optimization*, Lecture Notes in Economics and Mathematical Systems, **226**, Springer-Verlag, Berlin (1984), 49-79.

[15] P. T. Thach, *Quasiconjugates of functions: Duality relationships between quasiconvex minimization under a reverse convex constraint and application*, J. Math. Anal. Appl., **159** (1991), 299-322.

[16] P. T. Thach, *Global optimality criterion and a duality with zero gap in nonconvex optimization*, SIAM J. Math. Anal., **24** (1993), 1537- 1556.

[17] P. T. Thach and M. Kojima, *A generalized convexity and variational inequality for quasiconvex minimization*, SIAM J. Optimization, **6(1)** (1996), 212-226.

[18] H. Tuy, *D. C. Optimization: Theory, methods and algorithms*, in *Handbook of Global Optimization*, edited by R. Horst and P. M. Pardalos, Kluwer Academic Publishers, (1995) 149-216.

[19] M. Volle, *Conjugaison par tranches*, Ann. Mat. Pura ed Appl., **139** (1985), 279-312.

Department of Mathematics and Computer Science, Ben Gurion University of the Negev, Beer-Sheva, Israel

Current address: School of Information Technology and Mathematical Sciences, University of Ballarat, Ballarat 3350, Victoria, Australia

E-mail address: amr@ballarat.edu.au

School of Information Technology and Mathematical Sciences, University of Ballarat, Ballarat 3350, Victoria, Australia

E-mail address: bmg@ballarat.edu.au

Contemporary Mathematics
Volume **204**, 1997

Subdifferentials of convex functions

S. Simons

This paper is dedicated to Robert R. Phelps on the occasion of his 70th birthday

ABSTRACT. In this paper, we discuss a number of results about the subdifferential of a proper, convex lower semicontinuous function (on a Banach space). We discuss various subclasses of the class of maximal monotone multifunctions which include all such subdifferentials. We also give some results of a more geometric flavor about the subtangents to such a convex function.

Introduction

We start this introduction by discussing the *maximal monotonicity theorem.* We suppose throughout that E is a real Banach space with adjoint E^*, and $f : E \to \mathbb{R} \cup \{\infty\}$ is a proper convex, lower semicontinuous function. (To say that f is *proper* means that $\mathrm{dom} f \neq \emptyset$, where $\mathrm{dom} f := \{x : x \in E,\ f(x) \in \mathbb{R}\}$). If $x \in E$, the *subdifferential* of f at x is defined by

$$\partial f(x) := \{x^* : x^* \in E^*, \text{ for all } y \in E,\ f(x) + \langle y - x, x^* \rangle \leq f(y)\}.$$

In order to simplify notation, we shall write $G(\partial f)$ for the *graph* of ∂f, defined by

$$G(\partial f) := \{(x, x^*) : x \in E, x^* \in \partial f(x)\} \subset E \times E^*.$$

The maximal monotonicity theorem, which was proved by Rockafellar in [**17**], states that if $(q, q^*) \in E \times E^*$ and (q, q^*) is *monotonically related* to $G(\partial f)$, that is to say

$$(z, z^*) \in G(\partial f) \quad \Longrightarrow \quad \langle z - q, z^* - q^* \rangle \geq 0$$

then

$$(q, q^*) \in G(\partial f).$$

In contrapositive form, this result reads: if

$$(q, q^*) \in E \times E^* \setminus G(\partial f)$$

then

$$\text{there exists } (z, z^*) \in G(\partial f) \text{ such that } \langle z - q, z^* - q^* \rangle < 0.$$

1991 *Mathematics Subject Classification.* Primary 49J45, 47H05; Secondary 47N10, 46G05.

This contrapositive form shows two things. Firstly that the result is, in fact, an *existence theorem* and, secondly, that there a phenonemon of *action at a distance* in that the hypotheses assume that (q, q^*) is, in some sense, far from being related to f, while the conclusion is that there a pair (z, z^*) closely related to f that produces the desired inequality.

In this paper, we collect together and discuss various generalizations of the maximal monotonicity theorem. Some of these are new, and some of them have appeared before, usually with different proofs.

We first observe that the maximal monotonicity theorem seems to show a certain symmetry between E and E^*. Of course, this symmetry may be an illusion if E is not reflexive, and so the first main division of the results is into *primal* and *dual*. The primal results use as tools the formula for the subdifferential of a sum (Theorem 1) and the Brøndsted-Rockafellar theorem (Theorem 2). The dual results use conjugate functions — and sometimes, the formula for the conjugate for a sum (Theorem 15) — which necessitates the introduction of the concept of the *episum* (or *inf-convolution*) of convex functions.

What do we mean by "primal" and "dual" results? Formally, the result dual to a given primal result about a convex function f is that which can be obtained by interchanging E and E^*, interchanging f and f^*, but leaving statements of the form "$(z, z^*) \in G(\partial f)$" unchanged. If a primal result is given, the dual result may be true (we give many examples of this), or may not be true (see, for instance, Remark 22) or we may simply just not know (see, for instance Conjecture 18 and Conjecture 43). In fact, the final section of the paper contains a summary of what we know about the truth or otherwise of dual results.

One is tempted to attack dual problems by applying the primal result to f^*. The problem then is that we end up with a statement about E^{**} and f^{**}, and passing back to a statement about E and f (if possible) will frequently involve weak*-separation in E^{**} which, in turn, requires the development of the theory of locally convex spaces, and a further layer of functional analytic complexity. (This technique was introduced by Rockafellar in his original proof of the maximal monotonicity theorem.) The proofs that we shall provide will stay within the context of E and E^*, and consequently use the smallest number of abstract functional analytic tools.

We next discuss the *bootstrapping* procedure. This is the substitution

$$g(x) := f(x + q) - \langle x, q^* \rangle \quad (x \in E).$$

We shall use this substitution so many times that we shall describe it simply as the *usual substitution.* The effect of this is to replace (q, q^*) by $(0, 0)$, which simplifies the algebra considerably. We should emphasize that the bootstrapped results do not involve significant extra work — the additional generality is essentially cosmetic. Our first use of this usual substitution will be in the proof of the maximal monotonicity theorem in Corollary 5.

Why is it that the maximal monotonicity theorem was considered so long to be a "difficult" result? We believe that this can be traced to the fact that the inequality in the statement "$\langle z - q, z^* - q^* \rangle < 0$" is *strict.* Paradoxically, it is easier to prove the harder statement "$\langle z - q, z^* - q^* \rangle \leq \mu$" for some appropriate $\mu < 0$. The problem, of course, it to find the appropriate value of μ. We show one way of achieving this in Theorem 4.

We have used two different approaches for obtaining Theorem 4 and similar results. The first approach, that adopted in [**19**], is through the Brøndsted-Rockafellar theorem (applied to a suitably penalized version of the function under consideration) and the sum formula for subdifferentials. The second approach, that adopted in [**20**], [**21**] and [**22**], is through Ekeland's variational principle (a generalization of the Brøndsted-Rockafellar theorem, again applied to a suitably penalized version of the function under consideration), and various results on the manipulation of sublinear functionals, such as the Mazur-Orlicz theorem. The second approach is more powerful, and has the advantage that it leads to a more or less mechanical procedure for finding constants like the constant μ above. On the other hand, the Brøndsted-Rockafellar theorem and the sum formula for subdifferentials are now part of the folklore of convex analysis, and it is for this reason that we have chosen to use the first approach in this paper. However, we do not know if we can use the first approach on certain other problems on subdifferentials, so in these cases we will point the reader to the papers where these results have been established using the second approach. (Theorems 34, 36, 39, 40, 41 and 42 fall into this category.)

Returning to the results that we present in this paper, suppose that $g : E \to \mathbb{R} \cup \{\infty\}$ is a proper, convex, lower semicontinuous function and $\inf_E g < g(0)$. In Corollaries 6 and 7, we discuss the problem of finding an element (x, x^*) of $G(\partial g)$ for which $\langle x, x^* \rangle < 0$ and x is constrained to lie in a specific set. These results are bootstrapped in Theorem 11 and, as a consequence, we derive in Corollary 12 the result that subdifferentials are maximal monotone locally (see Definition 28).

If g is as above, we will call the negative quantity

$$L := \inf_{y \in \operatorname{dom} g,\ y \neq 0} \frac{g(y) - g(0)}{\|y\|}$$

the *lower conical slope* of g at 0. The reason for this nomenclature is that, in the special case when $g(0) \in \mathbb{R}$, L is the largest number such that the cone with equation $y \to g(0) + L\|y\|$ lies below the graph of g. In Corollary 8, we give a number of other formulae for L in terms of $G(\partial g)$. Corollary 8 is bootstrapped as part of Theorem 13, in which we give two formulae for the quantity

$$\inf_{z \in \operatorname{dom} f,\ z \neq q} \frac{f(z) - f(q) - \langle z - q, q^* \rangle}{\|z - q\|},$$

(in which the quotient is exactly that which appears in the definition of Fréchet derivative) in terms of the extent of the departure of (q, q^*) from being monotonically related to $G(\partial f)$.

The conclusion of Theorem 4 contains the constant K (defined in (4.1)) as both an upper bound and as a lower bound. In Corollary 9, we eliminate K by division and obtain the *almost negative alignment* property of subdifferentials. Corollary 9 is also bootstrapped as part of Theorem 13.

We now discuss our investigation of dual results. We establish in Theorem 17 the dual result of Theorem 4. Theorem 17 is certainly the most technical result in this paper. It leads to Corollary 21, in which we give formulae for the lower conical slope of g^* at 0 in terms of $G(\partial g)$. Corollary 21 gives 80% of the result dual to Corollary 8. We show in Remark 22 that the other 20% fails in a spectacular way. Corollary 21 is bootstrapped as part of Theorem 26.

We next consider the problem of finding an element (x, x^*) of $G(\partial g)$ for which $\langle x, x^* \rangle < 0$ and x^* is constrained to lie in a specific set. We do not know if the dual of Corollary 6 is true (we state this as Conjecture 18), but we prove in Lemma 19 a slightly weaker result leading to Corollary 20, which is the dual of Corollary 7. Corollary 20 is bootstrapped in Theorem 24 and, as a consequence, we derive in Corollary 25 the result that subdifferentials are locally maximal monotone (see Definition 29).

In Corollary 23, we eliminate the constant $\widetilde{K}$ of Theorem 17 by division, and obtain more results on the the almost negative alignment property of subdifferentials. Corollary 23 is also bootstrapped as part of Theorem 26. It is interesting to note that the *hypotheses* of Theorem 26 are identical with those of Theorem 13, but the *conclusions* are very different. This observation leads to the question of the possibility of intermediate results. Remark 27 contains one such question.

In Definitions 28, 29, 30, 31, 33, 35 and 37, we introduce formally seven subclasses of the maximal monotone operators which have been proved (either here or elsewhere) to contain all subdifferentials. Very little is known about these classes of operators other than what appears here in the text.

In Theorem 38, we give the statement of the *dominating subtangent theorem* (proved in [**20**]) that the constant K defined in Theorem 4 as the supremum of the slopes of line segments from $(0,0)$ to the graph of g (with g as above) is also the infimum of the Lipschitz constants of the subtangents to the graph of g that dominate $(0,0)$. In Theorem 39, we give the statement of the more general *sharp separation theorem* (also proved in [**20**]), in which it was shown that $(0,0)$ can be replaced by a suitably positioned nonempty bounded closed subset of $E \times \mathbb{R}$.

As we have already indicated, in the final section of the paper we catalog what we know about dual results.

Primal results

For our primal results, we shall use two standard results from convex analysis. The first of these, Theorem 1, is the "sum formula for subdifferentials", which follows from Rockafellar's generalization of a finite dimensional result of Fenchel:

THEOREM 1. *If $g_0 : E \to \mathbb{R} \cup \{\infty\}$ is proper and convex, and $g_1, \ldots, g_p : E \to \mathbb{R}$ are convex and continuous then*

$$\partial(g_0 + g_1 + \cdots + g_p) = \partial g_0 + \partial g_1 + \cdots + \partial g_p.$$

PROOF. See [**18**], Theorem 20, p. 56. □

The second of these, Theorem 2, is the Brøndsted-Rockafellar theorem on the existence of subgradients:

THEOREM 2. *Let $f : E \to \mathbb{R} \cup \{\infty\}$ be a proper, convex, lower semicontinuous function. Suppose that f is bounded below, $\alpha, \beta > 0$, $y \in E$ and $f(y) \le \inf_E f + \alpha\beta$. Then there exists $(x, x^*) \in G(\partial f)$ such that $\|x - y\| \le \alpha$ and $\|x^*\| \le \beta$.*

PROOF. See [**5**], p. 608. □

Theorem 4 and Lemma 19 are both proved by applying Theorems 1 and 2 to a convex function to which a continuous sublinear functional has been added as a penalty term. The following result will be useful in this connection. We leave the proof of it to the reader.

LEMMA 3. *Let $S : E \to \mathbb{R}$ be continuous and sublinear and $z^* \in \partial S(x)$. Then $z^* \leq S$ on E and $\langle x, z^* \rangle = S(x)$.*

The following result leads rapidly to a proof of the maximal monotonicity theorem. It is also the prototype for other results on the existence of elements of the subdifferential: Theorem 17, which will be crucial in our study of dual results, and Lemma 19, which will be used in our proof that subdifferentials are locally maximal monotone. Graphically, the constant K defined in (4.1) is the supremum of the slopes of the line segments from $(0, \lambda)$ to points on the graph of g that lie below $(0, \lambda)$.

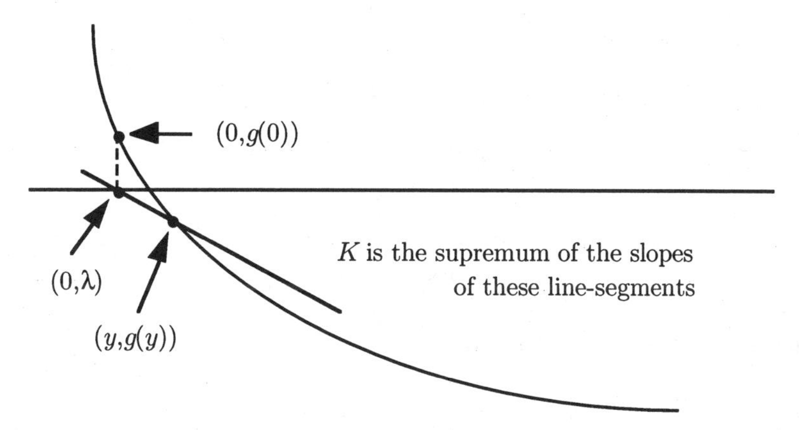

We can also think of K as the infimum of the slopes of the downward cones with vertex $(0, \lambda)$ that miss the graph of g.

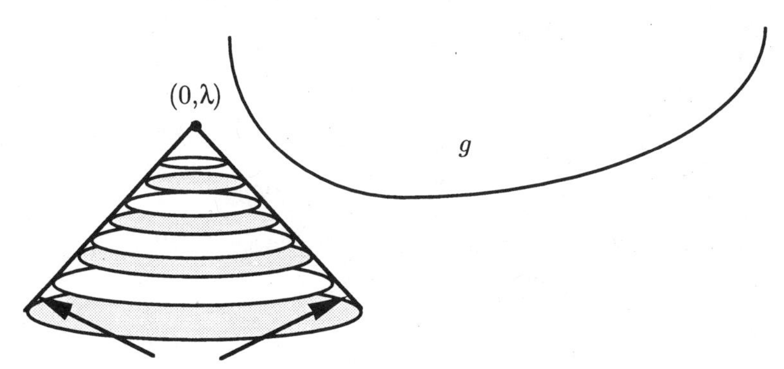

K is the infimum of the slopes of these cones

We will say more about K in the sections titled "Graphical interpretations of K" and "Separating subtangents". Theorem 4 appeared with a different proof in [**22**], Main Theorem, p. 328-331.

THEOREM 4. *Let $g : E \to \mathbb{R} \cup \{\infty\}$ be a proper, convex, lower semicontinuous function and $\inf_E g < \lambda < g(0)$ $(\leq \infty)$. Let $F := \{y : y \in E,\ g(y) < \lambda\} \neq \emptyset$, and write*

$$K := \sup_{y \in F} \frac{\lambda - g(y)}{\|y\|} \tag{4.1}$$

Then: $0 < K < \infty$ *and, for all* $\varepsilon \in (0,1)$*, there exists* $(x, x^*) \in G(\partial g)$ *such that*

$$\|x^*\| \le (1+\varepsilon)K \quad \text{and} \quad -\langle x, x^*\rangle \ge (1-\varepsilon)K\|x\| > 0. \tag{4.2}$$

PROOF. It can be proved in a number of different ways that $0 < K < \infty$. We refer the reader to [**19**], Lemma 2.2(a), pp. 130-131 for a proof using a separation theorem in $E \times \mathbb{R}$, and to [**22**], Main Theorem (a), p. 329-330 for a different proof that does not use a separation theorem. Clearly

$$y \in F \implies \frac{\lambda - g(y)}{\|y\|} \le K \implies \lambda \le g(y) + K\|y\|$$

and

$$y \in E \setminus F \implies g(y) \ge \lambda \implies \lambda \le g(y) + K\|y\|.$$

Combining together these two inequalities:

$$\lambda \le \inf_E (g + K\|\ \|). \tag{4.3}$$

Let $\varepsilon \in (0,1)$. Since $(1-\varepsilon/2)K < K$, from (4.1), there exists $y \in F$ (from which $\|y\| > 0$) such that

$$\frac{\lambda - g(y)}{\|y\|} > \left(1 - \frac{\varepsilon}{2}\right) K.$$

Clearing of fractions and using (4.3),

$$g(y) + K\|y\| \le \lambda + \frac{\varepsilon K\|y\|}{2} \le \inf_E (g + K\|\ \|) + \frac{\|y\|}{2}\varepsilon K.$$

From Theorem 2 with $f := g + K\|\ \|$, $\alpha := \|y\|/2$ and $\beta := \varepsilon K$, and Theorem 1 with $g_0 = g$ and $g_1 = K\|\ \|$, there exist $x \in \operatorname{dom}(g + K\|\ \|) = \operatorname{dom} g$, $x^* \in \partial g(x)$ and $z^* \in \partial K\|\ \|(x)$ such that

$$\|x - y\| \le \frac{\|y\|}{2} \tag{4.4}$$

and

$$\|x^* + z^*\| \le \varepsilon K. \tag{4.5}$$

We first note from (4.4) that $\|x\| > 0$. Next, since $z^* \in \partial K\|\ \|(x)$, from Lemma 3,

$$\|z^*\| \le K, \tag{4.6}$$

and

$$\langle x, z^*\rangle = K\|x\|. \tag{4.7}$$

It follows from (4.6) and (4.5) that

$$\|x^*\| \le \|z^*\| + \|x^* + z^*\| \le (1+\varepsilon)K,$$

which gives the first part of (4.2). Further, from (4.7) and (4.5),

$$-\langle x, x^*\rangle = \langle x, z^*\rangle - \langle x, x^* + z^*\rangle \ge K\|x\| - \varepsilon K\|x\| = (1-\varepsilon)K\|x\|,$$

which gives the second part of (4.2). This completes the proof of Theorem 4. □

Here, for the record, is the contrapositive form of the maximal monotonicity theorem.

COROLLARY 5. *Let $f : E \to \mathbb{R} \cup \{\infty\}$ be a proper, convex, lower semicontinous function and $(q, q^*) \in E \times E^* \setminus G(\partial f)$. Then there exists $(z, z^*) \in G(\partial f)$ such that*

$$\langle z - q, z^* - q^* \rangle < 0.$$

PROOF. Define $g : E \to \mathbb{R} \cup \{\infty\}$ by the usual substitution:

$$g(x) := f(x + q) - \langle x, q^* \rangle \quad (x \in E).$$

Since $q^* \notin \partial f(q)$, $\inf_E g < g(0)$. Let λ be any number such that $\inf_E g < \lambda < g(0)$ and x and x^* be found as in Theorem 4. Let $z := x + q$ and $z^* := x^* + q^*$. Then $z^* \in \partial f(z)$ and

$$\langle z - q, z^* - q^* \rangle = \langle x, x^* \rangle < 0.$$

□

In Theorem 4, we do not obtain any specific information about the location of x other than that $x \neq 0$. In Corollary 6, we give a result in which we do have more specific information about this. In order to achieve this, we replace the hypothesis $\inf_E g < g(0)$ by the more concrete

$$w \in E \quad \text{and} \quad g(w) < g(0),$$

and we consider the question of finding a small set in which we can assert that there is an element (x, x^*) of $G(\partial g)$ such that $\langle x, x^* \rangle < 0$. One possible conjecture is that

- such an element x can always be found on the segment $[0, w]$.

The diagram below shows that this is false even with $E = \mathbb{R}$.

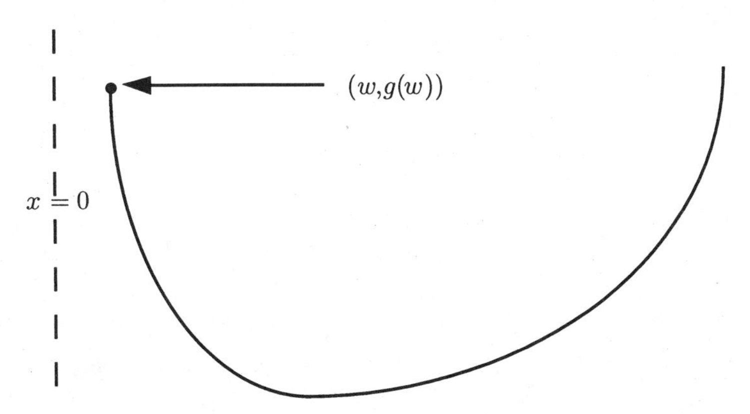

Another possible conjecture is that:

- such an element x can always be found arbitrarily close to w.

The diagram below shows that this is also false with $E = \mathbb{R}$.

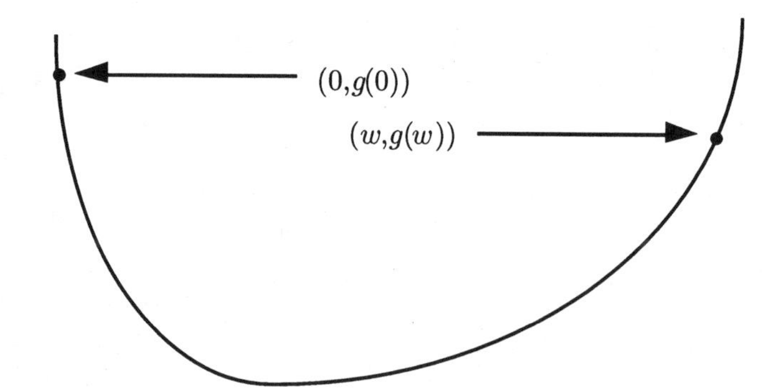

However, we will show in Corollary 6 that a "convex combination" of the above two conjectures always holds. We do not know if it is true that

- either such an element x can always be found on the segment $[0, w]$ or such an element x can always be found arbitrarily close to w.

Corollary 6 was suggested by [**25**], Theorem 3, p. 269.

COROLLARY 6. *Let $g : E \to \mathbb{R} \cup \{\infty\}$ be a proper, convex lower semicontinuous function, $w \in E$ and $g(w) < g(0)$ $(\le \infty)$. Let D be a closed ball centered at w. Then there exists $(x, x^*) \in G(\partial g)$ such that*

$$x \in \operatorname{co}(\{0\} \cup D) \quad \textit{and} \quad \langle x, x^* \rangle < 0.$$

PROOF. Let $A := \operatorname{co}(\{0\} \cup D)$ and h be the indicator function of A. Since A is nonempty, convex and closed, h is proper, convex and lower semicontinuous. Further, since $h(w) = h(0) = 0$, $(g + h)(w) < (g + h)(0)$. Consequently

$$\inf_E (g + h) < (g + h)(0).$$

Even though neither of g and h is necessarily continuous, the formula $\partial(g + h) = \partial g + \partial h$ is still true because there exists a point in $\operatorname{dom} g \cap \operatorname{dom} h$ at which h is continuous. (See [**14**], Theorem 3.16, p. 47.) Thus, from Theorem 4 or Corollary 5, there exist $x \in E$, $x^* \in \partial g(x)$ and $y^* \in \partial h(x)$ such that

$$\langle x, x^* + y^* \rangle < 0. \tag{6.1}$$

Now the statement "$y^* \in \partial h(x)$" is equivalent to:

$$x \in A \quad \text{and} \quad \langle x, y^* \rangle = \max_A y^*.$$

In particular, since $0 \in A$, $\langle x, y^* \rangle \ge 0$. From (6.1), $\langle x, x^* \rangle < 0$. This completes the proof of Corollary 6. □

We now parlay Corollary 6 into a result on open sets.

COROLLARY 7. *Let $g : E \to \mathbb{R} \cup \{\infty\}$ be a proper, convex lower semicontinuous function and $\inf_E g < g(0)$ $(\le \infty)$. Let V be a convex open subset of E such that $V \ni 0$ and $V \cap \operatorname{dom} g \ne \emptyset$. Then there exists $(x, x^*) \in G(\partial g)$ such that*

$$x \in V \quad \textit{and} \quad \langle x, x^* \rangle < 0.$$

PROOF. Fix $y \in V \cap \operatorname{dom} g$ and $z \in E$ such that $g(z) < g(0)$. If $g(0) = \infty$ then $g(y) < g(0)$, and we write $w := y$. If, on the other hand, $g(0) \in \mathbb{R}$ then, for all $\lambda \in (0,1]$, $g(\lambda z) < g(0)$. We choose $\lambda \in (0,1]$ so small that $\lambda z \in V$, and write $w := \lambda z$. In either case, we have found $w \in V$ such that $g(w) < g(0)$. We now let D be any closed ball centered at w such that $D \subset V$. The result now follows from Corollary 6 since $\operatorname{co}(\{0\} \cup D) \subset V$. □

The lower conical slope of g

Corollary 5 uses only a fraction of the information contained in Theorem 4. Corollary 8 shows how Theorem 4 leads to a reasonable interpretation of the lower conical slope of g at 0 in terms of the subgradients of g. Corollary 8 is a strengthening of [**19**], Theorem 2.3, p. 132-133 (for the case when $x = 0$).

COROLLARY 8. *Let* $g : E \to \mathbb{R} \cup \{\infty\}$ *be a proper, convex, lower semicontinuous function and* $\inf_E g < g(0)$ $(\leq \infty)$. *Let*

$$L_1 := -\min_{y^* \in \partial g(0)} \|y^*\|,$$

$$L_2 := \inf_{y \in E,\ g(y) < g(0)} \frac{g(y) - g(0)}{\|y\|},$$

$$L_3 := \inf_{y \in \operatorname{dom} g,\ y \neq 0} \frac{g(y) - g(0)}{\|y\|},$$

$$L_4 := \inf_{(x,x^*) \in G(\partial g),\ x \neq 0} \frac{\langle x, x^* \rangle}{\|x\|},$$

and

$$L_5 := \inf_{(x,x^*) \in G(\partial g),\ g(x) < g(0)} \frac{\langle x, x^* \rangle}{\|x\|}.$$

Then

$$L_1 = L_2 = L_3 = L_5 = L_4 \in [-\infty, 0).$$

PROOF. We shall prove this result in the order $L_1 = L_2 = L_3$, $L_2 = L_3 = L_4$ and $L_4 = L_5$.

We first prove that $L_1 \geq L_2$. For this, we can clearly suppose that $L_2 > -\infty$. Thus, from the definition of L_2,

$$y \in E \text{ and } g(y) < g(0) \quad \Longrightarrow \quad g(y) \geq g(0) + L_2 \|y\|.$$

Since $L_2 < 0$, it follows from this that

$$y \in E \quad \Longrightarrow \quad g(y) \geq g(0) + L_2 \|y\|.$$

Using the Eidelheit separation theorem in $E \times \mathbb{R}$, or the sandwich theorem (see [**11**], Theorem 1.7, p. 112 — see also [**9**]) we can now derive that there exists $y^* \in \partial g(0)$ such that $\|y^*\| \leq -L_2$. Consequently, $L_1 \geq L_2$. Since

$$g(y) < g(0) \quad \Longrightarrow \quad y \in \operatorname{dom} g \quad \text{and} \quad y \neq 0,$$

it follows that $L_2 \geq L_3$. If $y^* \in \partial g(0)$ and $y \neq 0$ then

$$g(0) - \|y\| \|y^*\| \leq g(0) + \langle y, y^* \rangle \leq g(y)$$

from which

$$-\|y^*\| \le \frac{g(y) - g(0)}{\|y\|}.$$

Taking the supremum over y^* and the infimum over y, $L_3 \ge L_1$. This completes the proof that $L_1 = L_2 = L_3$. Next, since

$$(8.1) \qquad (x, x^*) \in G(\partial g) \implies g(x) - g(0) \le \langle x, x^* \rangle,$$

$L_3 \le L_4$. Now, suppose that $y \in E$ and $g(y) < g(0)$. Let λ be an arbitrary number in $(g(y), g(0))$ and write K_λ for the constant K defined in (4.1). For any $\varepsilon \in (0,1)$, we find $(x, x^*) \in G(\partial g)$ as in (4.2). Then

$$\|x\| > 0 \quad \text{and} \quad \frac{\langle x, x^* \rangle}{\|x\|} \le -(1-\varepsilon)K_\lambda,$$

consequently, $L_4 \le -(1-\varepsilon)K_\lambda$. Letting $\varepsilon \to 0$, $L_4 \le -K_\lambda$. From the definition of K_λ,

$$L_4 \le \frac{g(y) - \lambda}{\|y\|}.$$

Letting $\lambda \to g(0)$,

$$L_4 \le \frac{g(y) - g(0)}{\|y\|}.$$

Taking the infimum over y, $L_4 \le L_2$. Since $L_2 = L_3$, this completes the proof that $L_2 = L_3 = L_4$. Finally,

$$L_4 = L_5 \wedge \inf_{(x,x^*) \in G(\partial g),\ x \ne 0,\ g(x) \ge g(0)} \frac{\langle x, x^* \rangle}{\|x\|}.$$

From (8.1),

$$(x, x^*) \in G(\partial g),\ x \ne 0 \text{ and } g(x) \ge g(0) \implies \frac{\langle x, x^* \rangle}{\|x\|} \ge 0.$$

Thus, since $L_4 = L_2 < 0$, $L_4 = L_5$. This completes the proof of Corollary 8. □

Almost negatively aligned elements of ∂g.

Even Corollary 8 does not use the full force of Theorem 4. Specifically, it does not use the inequality $\|x^*\| \le (1+\varepsilon)K$ established in (4.2). In Corollary 9, we use this additional information to show that we can find $(x, x^*) \in G(\partial g)$ to be "almost negatively aligned".

COROLLARY 9. *Let $g : E \to \mathbb{R} \cup \{\infty\}$ be a proper, convex, lower semicontinuous function, $\inf_E g < g(0)$ $(\le \infty)$, and define $L_2 \in [-\infty, 0)$ as in Corollary 8. Then, for all $n \ge 1$, there exists $(x_n, x_n^*) \in G(\partial g)$ such that*

$$x_n \ne 0, \quad \|x_n^*\| \to -L_2 \text{ in } (0, \infty] \quad \text{and} \quad \frac{\langle x_n, x_n^* \rangle}{\|x_n\| \, \|x_n^*\|} \to -1.$$

PROOF. Let $0 < \gamma < 1 < \delta$ and $L_2 < P < 0$. Using the definition of L_2, we first fix $y \in E$ such that $g(y) < g(0) + P\|y\|$. We next find $\lambda \in \mathbb{R}$ such that $\inf_E g < \lambda < g(0)$ and $g(y) < \lambda + P\|y\|$, and we write K_λ for the constant K defined in (4.1). Then:

$$K_\lambda \geq \frac{\lambda - g(y)}{\|y\|} > -P \tag{9.1}$$

Finally, we choose $\varepsilon \in (0,1)$ such that

$$\frac{1-\varepsilon}{1+\varepsilon} \geq \gamma, \quad \frac{\lambda - g(y)}{\|y\|} > \frac{-P}{1-\varepsilon} \quad \text{and} \quad 1+\varepsilon \leq \delta, \tag{9.2}$$

and $(x, x^*) \in G(\partial g)$ as in (4.2). Then $\|x\| > 0$,

$$\langle x, x^* \rangle \leq -(1-\varepsilon)K_\lambda \|x\| \quad \text{and} \quad -(1+\varepsilon)K_\lambda \leq -\|x^*\|. \tag{9.3}$$

Combining with the first inequality in (9.2),

$$-\|x\|\,\|x^*\| \leq \langle x, x^* \rangle \leq -(1-\varepsilon)K_\lambda \|x\| \leq -\gamma(1+\varepsilon)K_\lambda \|x\| \leq -\gamma \|x\|\,\|x^*\|,$$

from which

$$(1-\varepsilon)K_\lambda \leq \|x^*\| \quad \text{and} \quad -1 \leq \frac{\langle x, x^* \rangle}{\|x\|\,\|x^*\|} \leq -\gamma. \tag{9.4}$$

From the second inequality in (9.2), the first inequality in (9.1) and the first inequality in (9.4),

$$-P < (1-\varepsilon)\frac{\lambda - g(y)}{\|y\|} \leq (1-\varepsilon)K_\lambda \leq \|x^*\|. \tag{9.5}$$

It also follows from the definitions of K_λ and L_2 that

$$K_\lambda = \sup_{g(y)<\lambda} \frac{\lambda - g(y)}{\|y\|} \leq \sup_{g(y)<g(0)} \frac{\lambda - g(y)}{\|y\|} \leq \sup_{g(y)<g(0)} \frac{g(0) - g(y)}{\|y\|} = -L_2.$$

Combining this with the second inequality in (9.3) and the third inequality in (9.2),

$$\|x^*\| \leq (1+\varepsilon)K_\lambda \leq -(1+\varepsilon)L_2 \leq -\delta L_2. \tag{9.6}$$

Thus, from the second inequality in (9.4), (9.5) and (9.6), we have

$$-P < \|x^*\| \leq -\delta L_2 \quad \text{and} \quad -1 \leq \frac{\langle x, x^* \rangle}{\|x\|\,\|x^*\|} \leq -\gamma.$$

The result follows by letting $\gamma \to 1-$, $P \to L_2+$ and $\delta \to 1+$. □

REMARK 10. Corollary 9 hides another, somewhat more subtle, extremal property of L_2 than those listed in Corollary 8. Suppose that $(x, x^*) \in G(\partial g)$ and $\langle x, x^* \rangle < 0$. Then, from (8.1),

$$L_2 = L_3 \leq \frac{g(x) - g(0)}{\|x\|} \leq \frac{\langle x, x^* \rangle}{\|x\|}$$

thus,

$$\|x^*\| \leq L_2 \frac{\|x\|\,\|x^*\|}{\langle x, x^* \rangle}.$$

Suppose now that, for all $n \geq 1$,

$$(x_n, x_n^*) \in G(\partial g) \setminus \{(0,0)\} \quad \text{and} \quad \frac{\langle x_n, x_n^* \rangle}{\|x_n\| \, \|x_n^*\|} \to -1.$$

Then it follows from the above that

$$\limsup_n \|x_n^*\| \leq -L_2.$$

So Corollary 9 is saying simply that we can find sequences x_n and x_n^* as above for which this upper bound for $\limsup_n \|x_n^*\|$ is attained.

Bootstrapped primal results

Most of the results that we have already obtained can be given a much more general appearance by using the usual substitution:

$$g(x) := f(x+q) - \langle x, q^* \rangle \quad (x \in E).$$

We should emphasize that we have already done nearly all the work necessary to prove these bootstrapped results — the additional generality is essentially cosmetic.

Theorem 11 follows from Corollary 6 using the usual substitution. It is worth noting that if $(q, q^*) \in E \times E^* \setminus G(\partial f)$ then, from the definition of $\partial f(q)$, there always exists at least one element v of E that satisfies (11.1).

THEOREM 11. *Let $f : E \to \mathbb{R} \cup \{\infty\}$ be a proper convex, lower semicontinuous function, $q \in E$, $q^* \in E^*$, $v \in E$ and*

$$f(v) < f(q) + \langle v - q, q^* \rangle . \tag{11.1}$$

Let B be a closed ball centered at v. Then there exists $(z, z^*) \in G(\partial f)$ such that

$$z \in \operatorname{co}(\{q\} \cup B) \quad \textit{and} \quad \langle z - q, z^* - q^* \rangle < 0.$$

Corollary 12 is equivalent to the result proved in [**8**], Corollary 3.4, p. 66 and [**25**], Theorem 3, p. 269 that subdifferentials are *maximal monotone locally*. It can be deduced either from Corollary 7 using the usual substitution, or from Theorem 11 using the technique of Corollary 7. We will give more details about some of the properties of maximal monotone locally operators in Definition 28.

COROLLARY 12. *Let $f : E \to \mathbb{R} \cup \{\infty\}$ be a proper convex, lower semicontinuous function and $(q, q^*) \in E \times E^* \setminus G(\partial f)$. Let U be a convex open subset of E such that $U \ni q$ and $U \cap \operatorname{dom} f \neq \emptyset$. Then there exists $(z, z^*) \in G(\partial f)$ such that*

$$z \in U \quad \textit{and} \quad \langle z - q, z^* - q^* \rangle < 0.$$

The composite Theorem 13 follows by applying the usual substitution to Corollary 8, Corollary 9 and Remark 10. We note that the first quotient in the statement of Theorem 13 is exactly that which appears in the definition of Fréchet derivative. The second and third quotients measure (in different ways) the extent of the departure of (q, q^*) from being monotonically related to $G(\partial f)$. Theorem 13 is a slight strengthening of [**19**], Theorem 2.4, p. 133 and half of [**23**], Theorem 4.5, p. 367-369.

THEOREM 13. *Let $f : E \to \mathbb{R} \cup \{\infty\}$ be a proper convex, lower semicontinous function and $(q, q^*) \in E \times E^* \setminus G(\partial f)$. Then:*

$$\inf_{z \in \mathrm{dom} f,\ z \neq q} \frac{f(z) - f(q) - \langle z - q, q^* \rangle}{\|z - q\|} = \inf_{(z,z^*) \in G(\partial f),\ z \neq q} \frac{\langle z - q, z^* - q^* \rangle}{\|z - q\|} \in [-\infty, 0)$$

and the common value is $-\mathrm{dist}(q^*, \partial f(q))$, which we interpret as $-\infty$ if $\partial f(q) = \emptyset$. If $(z, z^*) \in G(\partial f)$ and $\langle z - q, z^* - q^* \rangle < 0$ then

$$\|z^* - q^*\| \leq -\mathrm{dist}(q^*, \partial f(q)) \frac{\|z - q\| \, \|z^* - q^*\|}{\langle z - q, z^* - q^* \rangle}.$$

If,

$$\text{for all } n \geq 1, \quad (z_n, z_n^*) \in G(\partial f) \setminus \{(q, q^*)\} \quad \text{and} \quad \frac{\langle z_n - q, z_n^* - q^* \rangle}{\|z_n - q\| \, \|z_n^* - q^*\|} \to -1$$

then

$$\limsup_n \|z_n^* - q^*\| \leq \mathrm{dist}(q^*, \partial f(q)).$$

Finally, for all $n \geq 1$, there exist $(z_n, z_n^*) \in G(\partial f)$ such that

$$z_n \neq q, \quad \|z_n^* - q^*\| \to \mathrm{dist}(q^*, \partial f(q)) \text{ in } (0, \infty] \quad \text{and} \quad \frac{\langle z_n - q, z_n^* - q^* \rangle}{\|z_n - q\| \, \|z_n^* - q^*\|} \to -1.$$

Dual results

In order to explain dual results, we will have to introduce the *conjugate* of a convex function and the *episum* or *inf-convolution* of two convex functions.

NOTATION 14. If $f : E \to \mathbb{R} \cup \{\infty\}$ is proper and convex, we define $f^* : E^* \to \mathbb{R} \cup \{\infty\}$ by $f^*(x^*) := \sup_E (x^* - f)$. f^* is the *conjugate* of f. f^* is convex and $w(E^*, E)$-lower semicontinuous. If f is also lower semicontinuous, then f^* is proper. If $f, g : E \to \mathbb{R} \cup \{\infty\}$, we define $f \underset{e}{+} g$ by $(f \underset{e}{+} g)(x) := \inf_{y \in E} \big(f(x - y) + g(y) \big)$. $f \underset{e}{+} g$ is the *episum* or *inf-convolution* of f and g. We write $f \vee g$ for the pointwise maximum of f and g.

We shall use the following version of the "sum formula for conjugates" which follows, like Theorem 1, from [**18**], Theorem 20, p. 56:

THEOREM 15. *If $g_0 : E \to \mathbb{R} \cup \{\infty\}$ is proper and convex, and $g_1, \ldots, g_p : E \to \mathbb{R}$ are convex and continuous then*

$$(g_0 + g_1 + \cdots + g_p)^* = g_0^* \underset{e}{+} g_1^* \underset{e}{+} \cdots \underset{e}{+} g_p^*.$$

In order to focus the discussion, we will establish in Theorem 17 the result dual to Theorem 4. In order to do this, we will first prove a technical lemma.

LEMMA 16. *Let $\eta \in (0, 1)$, $u^* \in E^*$ with $\|u^*\| = 1$ and the convex continuous functions $g_1, g_2 : E \to \mathbb{R}$ be defined by:*

$$g_1(x) := (\|x\| - 1 - \eta)) \vee 0 \quad \text{and} \quad g_2(x) := (\langle x, u^* \rangle + 1 - \eta) \vee 0.$$

Then

$$\text{for all } w^* \in E^*, \quad (g_1^* \underset{e}{+} g_2^*)(u^* - w^*) \geq \|w^*\| + \eta \, \|u^* - w^*\| - 1$$

PROOF. We can compute directly that,

$$\text{for all } v^* \in E^*, \quad g_1^*(v^*) \geq \sup_{\|x\| \leq 1+\eta} [\langle x, v^* \rangle - g_1(x)] = (1+\eta)\|v^*\|$$

and,

$$\text{for all } x^* \in E^*, \quad g_2^*(x^*) \quad = \quad \begin{cases} (\eta - 1)\|x^*\| & \text{if } x^* \in [0, u^*] \\ \infty & \text{otherwise.} \end{cases}$$

If now $v^* \in E^*$ and $x^* \in [0, u^*]$ then, since $\|x^*\| + \|u^* - x^*\| = \|u^*\| = 1$, we have

$$\begin{aligned} g_1^*(v^*) + g_2^*(x^*) &\geq (1+\eta)\|v^*\| + (\eta - 1)\|x^*\| \\ &= \|v^*\| - \|x^*\| + \eta(\|v^*\| + \|x^*\|) \\ &= \|v^*\| + \|u^* - x^*\| - 1 + \eta(\|v^*\| + \|x^*\|) \\ &\geq \|v^* + x^* - u^*\| + \eta\|v^* + x^*\| - 1. \end{aligned}$$

If, on the other hand, $v^* \in E^*$ and $x^* \in E \setminus [0, u^*]$ then

$$g_1^*(v^*) + g_2^*(x^*) = \infty \geq \|v^* + x^* - u^*\| + \eta\|v^* + x^*\| - 1.$$

Thus we have proved that

$$\text{for all } v^*, x^* \in E^*, \quad g_1^*(v^*) + g_2^*(x^*) \geq \|v^* + x^* - u^*\| + \eta\|v^* + x^*\| - 1.$$

The result follows by taking the infimum over all $v^*, x^* \in E^*$ such that $v^* + x^* = u^* - w^*$. □

As we have already announced, Theorem 17 is the result dual to Theorem 4. As seems to be the general rule, its proof is much harder than the proof of its primal counterpart. Theorem 17 appeared with a much more complicated proof in [**23**], Theorem 4.1(b), p. 364-365.

THEOREM 17. *Let $g : E \to \mathbb{R} \cup \{\infty\}$ be a proper, convex, lower semicontinuous function. Let $\lambda \in \mathbb{R}$ and $\inf_{E^*} g^* < \lambda < g^*(0)$ $(\leq \infty)$. Let $\widetilde{F} := \{y^* : y^* \in E^*,\ g^*(y^*) < \lambda\}$ and write*

$$\widetilde{K} := \sup\left\{\frac{\lambda - g^*(y^*)}{\|y^*\|} : y^* \in \widetilde{F}\right\}.$$

Then $0 < \widetilde{K} < \infty$ and, for all $\varepsilon \in (0,1)$, there exists $(x, x^*) \in G(\partial g)$ such that

$$(17.1) \qquad \|x\| \leq (1+\varepsilon)\widetilde{K} \quad \text{and} \quad -\langle x, x^* \rangle \geq (1-\varepsilon)\widetilde{K}\|x^*\| > 0.$$

PROOF. Let $\eta := \varepsilon/5$. It follows exactly as in Theorem 4 that $0 < \widetilde{K} < \infty$ and there exists $p^* \in E^*$ such that $\|p^*\| > 0$ and

$$(17.2) \qquad g^*(p^*) + \widetilde{K}\|p^*\| < \inf_{E^*}(g^* + \widetilde{K}\|\ \|) + \eta\widetilde{K}\|p^*\|.$$

Since g^* and $\|\ \|$ are $w(E^*, E)$-lower semicontinuous, it follows from $w(E^*, E)$-compactness that we can choose $y^* \in \operatorname{dom} g^*$ to minimize $g^*(y^*) + \widetilde{K}\|y^*\| +$

$\eta\widetilde{K}\,\|y^* - p^*\|$. We then observe from (17.2) that

$$\begin{aligned} g^*(y^*) + \widetilde{K}\,\|y^*\| + \eta\widetilde{K}\,\|y^* - p^*\| &\le g^*(p^*) + \widetilde{K}\,\|p^*\| + \eta\widetilde{K}\,\|p^* - p^*\| \\ &= g^*(p^*) + \widetilde{K}\,\|p^*\| \\ &< g^*(y^*) + \widetilde{K}\,\|y^*\| + \eta\widetilde{K}\,\|p^*\|. \end{aligned}$$

Consequently, $\|y^* - p^*\| < \|p^*\|$, from which $\|y^*\| > 0$. From the minimizing property of y^*,

$$\text{for all } z^* \in E^*, \quad g^*(z^*)+\widetilde{K}\,\|z^*\|+\eta\widetilde{K}\,\|z^* - p^*\| \ge g^*(y^*)+\widetilde{K}\,\|y^*\|+\eta\widetilde{K}\,\|y^* - p^*\|.$$

We now use the triangle inequality to eliminate p^* from the above and obtain:

$$\text{(17.3)} \qquad \text{for all } z^* \in E^*, \quad g^*(z^*) + \widetilde{K}\,\|z^*\| + \eta\widetilde{K}\,\|y^* - z^*\| \ge g^*(y^*) + \widetilde{K}\,\|y^*\|.$$

We now perform some scaling to simplify the arithmetic. We define $u^* \in E^*$ and $h : E \to \mathbb{R} \cup \{\infty\}$ by:

$$u^* := \frac{y^*}{\|y^*\|} \quad \text{and} \quad h(x) := \frac{g(\widetilde{K}x)}{\widetilde{K}\,\|y^*\|}.$$

Then $\|u^*\| = 1$ and, further,

$$\text{for all } z^* \in E^*, \quad g^*(z^*) = \widetilde{K}\,\|y^*\|\,h^*\left(\frac{z^*}{\|y^*\|}\right),$$

and so (17.3) becomes:

$$\text{for all } w^* \in E^*, \quad h^*(w^*) + \|w^*\| + \eta\,\|u^* - w^*\| \ge h^*(u^*) + 1.$$

Let $g_1(x) := (\|x\| - 1 - \eta)) \vee 0$ and $g_2(x) := (\langle x, u^*\rangle + 1 - \eta) \vee 0$. We then obtain from Lemma 16 that,

$$\text{for all } w^* \in E^*, \quad h^*(w^*) + (g_1^* \underset{e}{+} g_2^*)(u^* - w^*) \ge h^*(u^*).$$

It follows by taking the infimum over w^* that $(h^* \underset{e}{+} g_1^* \underset{e}{+} g_2^*)(u^*) \ge h^*(u^*)$ hence, from Theorem 15, that $(h + g_1 + g_2)^*(u^*) \ge h^*(u^*)$. Consequently, there exists $u \in E$ such that

$$\langle u, u^*\rangle - h(u) - g_1(u) - g_2(u) \ge h^*(u^*) - \eta^2.$$

Since $\langle u, u^*\rangle - h(u) \le h^*(u^*)$, $g_1 \ge 0$ and $g_2 \ge 0$, we derive that

$$\text{(17.4)} \qquad \langle u, u^*\rangle - h(u) \ge h^*(u^*) - \eta^2, \quad g_1(u) \le \eta^2 \quad \text{and} \quad g_2(u) \le \eta^2.$$

It follows from the first inequality in (17.4) that

$$\langle u, u^*\rangle \ge h(u) + h^*(u^*) - \eta^2.$$

Applying Theorem 2 to $f := h - u^*$, we obtain $(z, z^*) \in G(\partial h)$ such that

$$\text{(17.5)} \qquad \|z - u\| \le \eta$$

and

$$\text{(17.6)} \qquad \|z^* - u^*\| \le \eta.$$

From the second inequality in (17.4), $\|u\| \le 1 + \eta + \eta^2 < 1 + 2\eta$. Combining this with (17.5),

$$\|z\| \le \|u\| + \|z - u\| \le (1 + 2\eta) + \eta = 1 + 3\eta. \tag{17.7}$$

From the third inequality in (17.4), $\langle u, u^* \rangle \le \eta - 1 + \eta^2 < 2\eta - 1$. Combining this also with (17.5),

$$\langle z, u^* \rangle = \langle u, u^* \rangle + \langle z - u, u^* \rangle \le (2\eta - 1) + \eta = 3\eta - 1. \tag{17.8}$$

Combining (17.8), (17.7) and (17.6),

$$\begin{aligned} \langle z, z^* \rangle + (1-\varepsilon)\|z^*\| &\le \langle z, u^* \rangle + \langle z, z^* - u^* \rangle + (1-\varepsilon)(\|u^*\| + \|z^* - u^*\|) \\ &\le (3\eta - 1) + (1 + 3\eta)\eta + (1 - 5\eta)(1 + \eta) \\ &= -2\eta^2 < 0. \end{aligned}$$

We define $x := \widetilde{K} z$ and $x^* := \|y^*\|\, z^*$. Since $(z, z^*) \in G(\partial h)$, $(x, x^*) \in G(\partial g)$. We will show that (x, x^*) has the required properties. From the above inequality,

$$\langle x, x^* \rangle + (1-\varepsilon)\widetilde{K}\,\|x^*\| = \widetilde{K}\,\|y^*\|\,\big[\,\langle z, z^* \rangle + (1-\varepsilon)\,\|z^*\|\,\big] < 0. \tag{17.9}$$

From (17.7),

$$\|x\| = \widetilde{K}\,\|z\| \le (1 + 3\eta)\widetilde{K} \le (1+\varepsilon)\widetilde{K}.$$

This gives (17.1) — we note that, since the inequality in (17.9) is strict, it follows that $x^* \ne 0$, hence $\|x^*\| > 0$. This completes the proof of Theorem 17. □

We next consider the possibility of a result dual to Corollary 6. Since we do not know it this result is true (in fact, for certain technical reasons, we suspect that it is false), we state it as a conjecture.

CONJECTURE 18. Let $g : E \to \mathbb{R} \cup \{\infty\}$ be a proper, convex lower semicontinuous function, $w^* \in E^*$ and $g^*(w^*) < g^*(0)$ $(\le \infty)$. Let $\widetilde{D}$ be a closed ball in E^* centered at w^*. Then there exists $(x, x^*) \in G(\partial g)$ such that

$$x^* \in \operatorname{co}\Big(\{0\} \cup \widetilde{D}\Big) \quad \text{and} \quad \langle x, x^* \rangle < 0.$$

We now show in Lemma 19 that a slightly weakened version of Conjecture 18 is true. We shall see that this weaker result is still strong enough to have some significant consequences. A weakening of Lemma 19 appears with a more complicated proof in [**21**], Lemma 5, p. 470.

LEMMA 19. *Let $g : E \to \mathbb{R} \cup \{\infty\}$ be a proper, convex lower semicontinuous function, $w^* \in E^*$ and $g^*(w^*) < g^*(0)$ $(\le \infty)$. Let $\widetilde{C}$ be a closed ball in E^* centered at 0. Then there exists $(x, x^*) \in G(\partial g)$ such that*

$$x^* \in \widetilde{C} + [0, w^*] \quad \textit{and} \quad \langle x, x^* \rangle < 0.$$

PROOF. Since $g^*(w^*) < g^*(0)$, $\|w^*\| > 0$. We suppose that $\widetilde{C}$ has radius $\eta > 0$ where, for simplicity, $\eta < \|w^*\|$. We write $\widetilde{A}$ for the weak*-compact set $\widetilde{C} + [0, w^*]$, and define the continuous sublinear functional $S : E \to \mathbb{R}$ by the formula

$$S(y) := \max\left\{\langle y, y^*\rangle : y^* \in -\widetilde{A}\right\} \quad (y \in E).$$

(In fact, $S(y)$ is given explicitly by the formula $\eta\,\|y\| + 0 \vee \langle y, -w^*\rangle$, but we do not need this in what follows.) We also write $M := \|w^*\| + \eta$. Since

$$\|y^*\| \le \eta \quad \Longrightarrow \quad y^* \in -\widetilde{A} \quad \Longrightarrow \quad \|y^*\| \le M,$$

it follows that,

$$\text{for all } y \in E, \quad \eta\,\|y\| \le S(y) \le M\,\|y\|. \tag{19.1}$$

Since $Mw^* = \eta w^* + \|w^*\|\,w^* \in \|w^*\|\,\widetilde{C} + \|w^*\|\,[0, w^*] = \|w^*\|\,\widetilde{A}$, it also follows that,

$$\text{for all } y \in E, \quad -M\,\langle y, w^*\rangle \le \|w^*\|\,S(y). \tag{19.2}$$

Let $\widehat{F} := \{y : y \in E,\ g(y) + g^*(w^*) < 0\}$. We note that

$$\inf_E g = -g^*(0) < -g^*(w^*),$$

and consequently $\widehat{F} \ne \emptyset$. Using (19.2) and the definition of $g^*(w^*)$,

$$\text{for all } y \in \widehat{F}, \quad 0 < -Mg(y) - Mg^*(w^*) \le -M\,\langle y, w^*\rangle \le \|w^*\|\,S(y). \tag{19.3}$$

Define

$$\widehat{K} := \sup_{y \in \widehat{F}} \frac{-g(y) - g^*(w^*)}{S(y)} > 0.$$

It is clear from this definition and (19.3) that

$$\left(1 + \frac{\eta}{\|w^*\|}\right)\widehat{K} = \frac{M}{\|w^*\|}\widehat{K} \le 1. \tag{19.4}$$

It also follows from the definition of $\widehat{K}$ that

$$y \in \widehat{F} \quad \Longrightarrow \quad \frac{-g(y) - g^*(w^*)}{S(y)} \le \widehat{K} \quad \Longrightarrow \quad -g^*(w^*) \le g(y) + \widehat{K}S(y).$$

On the other hand, since $S \ge 0$ on E,

$$y \in E \setminus \widehat{F} \quad \Longrightarrow \quad -g^*(w^*) \le g(y) \quad \Longrightarrow \quad -g^*(w^*) \le g(y) + \widehat{K}S(y).$$

Combining together these two inequalities,

$$-g^*(w^*) \le \inf_E (g + \widehat{K}S). \tag{19.5}$$

Again from the definition of $\widehat{K}$, there exists $y \in \widehat{F}$ (from which $S(y) > 0$) such that

$$\frac{-g(y) - g^*(w^*)}{S(y)} \ge \left(1 - \frac{\eta^2}{2M\,\|w^*\|}\right)\widehat{K}.$$

Clearing of fractions and using (19.5),

$$g(y) + \widehat{K}S(y) \le -g^*(w^*) + \frac{\eta^2 \widehat{K} S(y)}{2M\,\|w^*\|} \le \inf_E (g + \widehat{K}S) + \frac{S(y)}{2M}\frac{\eta^2 \widehat{K}}{\|w^*\|}.$$

From Theorem 2 with $f := g + \widehat{K}S$, $\alpha := S(y)/2M$ and $\beta := \eta^2\widehat{K}/\|w^*\|$, and Theorem 1, with $g_0 = g$ and $g_1 = \widehat{K}S$, there exist $x \in \mathrm{dom}(g + \widehat{K}S) = \mathrm{dom} g$, $x^* \in \partial g(x)$ and $z^* \in \partial\widehat{K}S(x)$ such that

$$\|x - y\| \le \frac{S(y)}{2M} \tag{19.6}$$

and

$$\|x^* + z^*\| \le \frac{\eta^2\widehat{K}}{\|w^*\|} \quad \text{— from which } x^* + z^* \in \frac{\eta\widehat{K}}{\|w^*\|}\widetilde{A}. \tag{19.7}$$

We first note from (19.1) and (19.6) that

$$S(y - x) \le M\|y - x\| = M\|x - y\| \le \frac{S(y)}{2},$$

and consequently

$$S(x) \ge S(y) - S(y - x) \ge \frac{S(y)}{2} > 0. \tag{19.8}$$

Since $z^* \in \partial\widehat{K}S(x)$, we next note from Lemma 3 that

$$\langle x, z^* \rangle = \widehat{K}S(x), \tag{19.9}$$

and $z^* \le \widehat{K}S$ on E. Using the weak* separation theorem in E^* (or alternatively, the explicit formula for S given above), we deduce from this that

$$z^* \in -\widehat{K}\widetilde{A}. \tag{19.10}$$

We now show that x and x^* have the desired properties. On the one hand, by combining (19.9), (19.7), (19.1) and (19.8), we obtain:

$$\begin{aligned}\langle x, x^* \rangle &= -\langle x, z^* \rangle + \langle x, x^* + z^* \rangle \le -\widehat{K}S(x) + \frac{\eta^2\widehat{K}}{\|w^*\|}\|x\| \\ &\le -\widehat{K}S(x) + \frac{\eta\widehat{K}}{\|w^*\|}S(x) = -\left(1 - \frac{\eta}{\|w^*\|}\right)\widehat{K}S(x) < 0.\end{aligned}$$

On the other hand, from (19.10), (19.7), the convexity of $\widetilde{A}$ and (19.4),

$$x^* = -z^* + (x^* + z^*) \in \widehat{K}\widetilde{A} + \frac{\eta\widehat{K}}{\|w^*\|}\widetilde{A} = \left(1 + \frac{\eta}{\|w^*\|}\right)\widehat{K}\widetilde{A} \subset \widetilde{A} = \widetilde{C} + [0, w^*],$$

as required. This completes the proof of Lemma 19. □

We now parlay Lemma 19 into a result on open sets. Despite the fact that Lemma 19 is not as strong as Conjecture 18, Corollary 20 is the exact dual of Corollary 7. We observe that the hypothesis $\inf_{E^*} g^* < g^*(0)$ that appears in Corollaries 20, 21 and 23 is equivalent to the hypothesis $\inf_E g < g(0)$ that appears in Corollaries 7, 8 and 9. This follows easily since, from the Fenchel-Moreau formula, $g^{**}(0) = g(0)$.

COROLLARY 20. *Let $g : E \to \mathbb{R} \cup \{\infty\}$ be a proper, convex lower semicontinuous function and $\inf_{E^*} g^* < g^*(0)$ $(\le \infty)$. Let $\widetilde{V}$ be a convex open subset of E^* such that $\widetilde{V} \ni 0$ and $\widetilde{V} \cap \operatorname{dom} g^* \neq \emptyset$. Then there exists $(x, x^*) \in G(\partial g)$ such that*

$$x^* \in \widetilde{V} \quad \text{and} \quad \langle x, x^* \rangle < 0.$$

PROOF. Arguing exactly as in Corollary 7, we can find $w^* \in \widetilde{V}$ such that $g^*(w^*) < g^*(0)$. Since $\widetilde{V}$ is convex, $[0, w^*] \subset \widetilde{V}$. Since $[0, w^*]$ is compact, there exists a closed ball $\widetilde{C}$ in E^* centered at 0 such that $\widetilde{C} + [0, w^*] \subset \widetilde{V}$. The result now follows from Lemma 19. □

The lower conical slope of g^*

Corollary 21 shows that 80% of the dual analog of Corollary 8 holds. The failure of the other 20%, which we will discuss in Remark 22, gives a very striking illustration of the difference between the primal and the dual situation. Corollary 21 is a slight strengthening of [**23**], Corollary 4.3(b), p. 366-367.

COROLLARY 21. *Let $g : E \to \mathbb{R} \cup \{\infty\}$ be a proper, convex, lower semicontinuous function and $\inf_{E^*} g^* < g^*(0)$ $(\le \infty)$. Let*

$$\widetilde{L}_2 := \inf_{y^* \in E^*,\ g^*(y^*) < g^*(0)} \frac{g^*(y^*) - g^*(0)}{\|y^*\|},$$

$$\widetilde{L}_3 := \inf_{y^* \in \operatorname{dom} g^*,\ y^* \neq 0} \frac{g^*(y^*) - g^*(0)}{\|y^*\|},$$

$$\widetilde{L}_4 := \inf_{(x, x^*) \in G(\partial g),\ x^* \neq 0} \frac{\langle x, x^* \rangle}{\|x^*\|},$$

and

$$\widetilde{L}_5 := \inf_{(x, x^*) \in G(\partial g),\ g^*(x^*) < g^*(0)} \frac{\langle x, x^* \rangle}{\|x^*\|}.$$

Then

$$\widetilde{L}_2 = \widetilde{L}_3 = \widetilde{L}_3 = \widetilde{L}_5 \in [-\infty, 0).$$

PROOF. We first observe that

$$\widetilde{L}_3 = \widetilde{L}_2 \wedge \inf_{y^* \in \operatorname{dom} g^*,\ y^* \neq 0,\ g^*(y^*) \ge g^*(0)} \frac{g^*(y^*) - g^*(0)}{\|y^*\|}.$$

Since $\widetilde{L}_2 < 0$, it follows immediately from this that $\widetilde{L}_2 = \widetilde{L}_3$. The rest of the proof follows exactly as that of Corollary 8, using Theorem 17 instead of Theorem 4, and the relation

$$(x, x^*) \in G(\partial g) \implies g^*(x^*) - g^*(0) \le \langle x, x^* \rangle \tag{21.1}$$

instead of (8.1). □

REMARK 22. The exact dual analog of the equality $L_1 = L_2$ in Corollary 8 would be the relation,

$$-\widetilde{L}_2 = \min\{\|y\| : y \in E,\ \partial g(y) \ni 0\}.$$

Of course, it would be unreasonable to expect this minimum to be attained if E is not reflexive. It would, however, be reasonable to expect that

$$-\widetilde{L}_2 < \infty \quad \Longrightarrow \quad -\widetilde{L}_2 = \inf\{\|y\| : y \in E,\ \partial g(y) \ni 0\}.$$

However, even this weaker assertion is *never* true if E is not reflexive. In this case, we use James's theorem to find $t^* \in E^*$ such that t^* does not attain its minimum on the unit ball of E. Let f be the indicator function of the unit ball of E and define $g : E \to \mathbb{R} \cup \{\infty\}$ by $g := f + t^*$. Then, from Theorem 15, for all $y^* \in E^*$, $g^*(y^*) = \|y^* - t^*\|$, the conditions of Corollary 21 are satisfied, and $\widetilde{L}_2 = -1$. On the other hand, g does not attain its minimum on E, so $\{y : y \in E,\ \partial g(y) \ni 0\} = \emptyset$.

More on almost negatively aligned elements of ∂g.

Corollary 23 is the dual analog of Corollary 9 and Remark 10. The proof of this is exactly analogous to that of Corollary 9 and Remark 10, only using Theorem 17 instead of Theorem 4, and (21.1) instead of (8.1).

COROLLARY 23. *Let $g : E \to \mathbb{R} \cup \{\infty\}$ be a proper, convex, lower semicontinuous function and $\inf_{E^*} g^* < g^*(0)$ $(\le \infty)$, and define $\widetilde{L}_2$ as in Corollary 21. If $(x, x^*) \in G(\partial g)$ and $\langle x, x^*\rangle < 0$ then*

$$\|x\| \le \widetilde{L}_2 \frac{\|x\|\,\|x^*\|}{\langle x, x^*\rangle}.$$

If, for all $n \ge 1$,

$$(x_n, x_n^*) \in G(\partial g) \setminus \{(0,0)\} \quad \text{and} \quad \frac{\langle x_n, x_n^*\rangle}{\|x_n\|\,\|x_n^*\|} \to -1.$$

Then

$$\limsup_n \|x_n\| \le -\widetilde{L}_2.$$

Finally, for all $n \ge 1$, there exist $(x_n, x_n^*) \in G(\partial g)$ such that

$$x_n^* \ne 0, \quad \|x_n\| \to -\widetilde{L}_2 \text{ in } (0, \infty] \quad \text{and} \quad \frac{\langle x_n, x_n^*\rangle}{\|x_n\|\,\|x_n^*\|} \to -1.$$

Bootstrapped dual results

Theorem 24 follows from Lemma 19 using the usual substitution. It is worth noting that if $(q, q^*) \in E \times E^* \setminus G(\partial f)$ then there always exists at least one element v^* of E^* that satisfies (24.1). This is seen most easily by using the Fenchel-Moreau formula that $f^{**} = f$ on E.

THEOREM 24. *Let $f : E \to \mathbb{R} \cup \{\infty\}$ be a proper convex, lower semicontinuous function, $q \in E$, $q^* \in E^*$, $v^* \in E^*$ and*

$$f^*(v^*) < f^*(q^*) + \langle q, v^* - q^* \rangle . \tag{24.1}$$

Let $\widetilde{C}$ be a closed ball in E^* centered at 0. Then there exists $(z, z^*) \in G(\partial f)$ such that

$$z^* \in \widetilde{C} + [q^*, v^*] \quad \text{and} \quad \langle z - q, z^* - q^* \rangle < 0.$$

Corollary 25 is equivalent to the result proved in [**21**], Main Theorem, p. 466 and [**25**], Theorem 3, p. 269 that subdifferentials *are locally maximal monotone.* It can be deduced either from Corollary 20 using the usual substitution, or from Theorem 24 using the technique of Corollary 20. We will give more details about some of the properties of locally maximal monotone operators in Definition 29.

COROLLARY 25. *Let $f : E \to \mathbb{R} \cup \{\infty\}$ be a proper convex, lower semicontinuous function and $(q, q^*) \in E \times E^* \setminus G(\partial f)$. Let $\widetilde{U}$ be a convex open subset of E^* such that $\widetilde{U} \ni q^*$ and $\widetilde{U} \cap \operatorname{dom} f^* \neq \emptyset$. Then there exists $(z, z^*) \in G(\partial f)$ such that*

$$z^* \in \widetilde{U} \quad \text{and} \quad \langle z - q, z^* - q^* \rangle < 0.$$

The composite Theorem 26 follows by applying the usual substitution to Corollaries 21 and 23. Theorem 26 is a generalization of half of [**23**], Corollary 4.4, p. 367 and half of [**23**], Theorem 4.5, p. 367-369.

THEOREM 26. *Let $f : E \to \mathbb{R} \cup \{\infty\}$ be a proper convex, lower semicontinous function and $(q, q^*) \in E \times E^* \setminus G(\partial f)$. Then:*

$$\inf_{z^* \in \operatorname{dom} f^*,\ z^* \neq q^*} \frac{f^*(z^*) - f^*(q^*) - \langle q, z^* - q^* \rangle}{\|z^* - q^*\|} = \inf_{(z, z^*) \in G(\partial f),\ z^* \neq q^*} \frac{\langle z - q, z^* - q^* \rangle}{\|z^* - q^*\|}.$$

We write $\widetilde{M} \in [-\infty, 0)$ for the common value above. If $(z, z^*) \in G(\partial f)$ and

$$\langle z - q, z^* - q^* \rangle < 0$$

then

$$\|z - q\| \leq \widetilde{M} \frac{\|z - q\| \, \|z^* - q^*\|}{\langle z - q, z^* - q^* \rangle}.$$

If, for all $n \geq 1$,

$$(z_n, z_n^*) \in G(\partial f) \setminus \{(q, q^*)\} \quad \text{and} \quad \frac{\langle z_n - q, z_n^* - q^* \rangle}{\|z_n - q\| \, \|z_n^* - q^*\|} \to -1.$$

then

$$\limsup_n \|z_n - q\| \leq -\widetilde{M}.$$

Finally, for all $n \geq 1$, there exist $(z_n, z_n^*) \in G(\partial f)$ such that

$$z_n^* \neq q^*, \quad \|z_n - q\| \to -\widetilde{M} \text{ in } (0, \infty] \quad \text{and} \quad \frac{\langle z_n - q, z_n^* - q^* \rangle}{\|z_n - q\| \, \|z_n^* - q^*\|} \to -1.$$

REMARK 27. Is there a result that unifies the "almost negative alignment" parts of Theorem 13 and Theorem 26? More specifically, if $(q,q^*) \in E\times E^*\setminus G(\partial f)$, write $\mathcal{N}$ for the family of all subsets of $(\mathbb{R}\setminus\{0\})\times(\mathbb{R}\setminus\{0\})$ with the property that, for all $n\geq 1$, there exists $(z_n,z_n^*)\in G(\partial f)$ such that

$$(\|z_n - q\|, \|z_n^* - q^*\|) \in \mathcal{N} \quad \text{and} \quad \frac{\langle z_n - q, z_n^* - q^*\rangle}{\|z_n - q\|\,\|z_n^* - q^*\|} \to -1.$$

It follows from Theorem 13 and Theorem 26 that if U is a neighborhood of the extended real number $\operatorname{dist}(q^*,\partial f(q))$ in $(0,\infty]$ and $\delta > 1$ then

$$(0,-\delta\widetilde{M}]\times(U\setminus\{\infty\})\in\mathcal{N},$$

and if U is a neighborhood of $-\widetilde{M}$ in $(0,\infty]$ and $\delta>1$ then

$$(U\setminus\{\infty\})\times(0,\delta\operatorname{dist}(q^*,\partial f(q))]\in\mathcal{N}.$$

Are there members of $\mathcal{N}$ in which both $\|z_n - q\|$ and $\|z_n^* - q^*\|$ are significantly constrained from below?

Multifunction properties

We suppose througthout this section that $S : E \to 2^{E^*}$ is a *monotone* multifunction, that is to say,

$$(x,x^*)\in G(S) \text{ and } (y,y^*)\in G(S) \quad\Longrightarrow\quad \langle x-y, x^*-y^*\rangle \geq 0,$$

where $G(S) := \{(x,x^*) : x\in E,\ x^*\in Sx\} \subset E\times E^*$. S is said to be *maximal monotone* if, whenever $(q,q^*)\in E\times E^*$ and

$$(z,z^*)\in G(S) \quad\Longrightarrow\quad \langle z-q, z^*-q^*\rangle \geq 0$$

then

$$(q,q^*)\in G(S).$$

We now introduce formally seven subclasses of the class of maximal monotone operators. It follows from Corollary 12, Corollary 25, Theorem 32, Theorem 34, Theorem 36, and Theorems 13 and 26 that subdifferentials belong to all five classes.

DEFINITION 28. We say that S is *maximal monotone locally* if, whenever U is a convex open subset of E such that $U\cap D(S)\neq\emptyset$, $(q,q^*)\in U\times E^*$ and

$$(z,z^*)\in G(S) \text{ and } z\in U \quad\Longrightarrow\quad \langle z-q, z^*-q^*\rangle \geq 0$$

then

$$(q,q^*)\in G(S),$$

where $D(S) := \{x : x\in E,\ Sx\neq\emptyset\}$.

Maximal monotone locally operators were introduced by Fitzpatrick and Phelps in [**8**], p. 65 and Verona and Verona in [**25**], p. 268. It follows from [**8**], Theorem 3.10, p. 68 that

- if E is reflexive then every maximal monotone operator is maximal monotone locally,

and

- if S is maximal monotone and $D(S) = E$ then S is maximal monotone locally.

It seems to be unknown whether there exists a maximal monotone operator that is not maximal monotone locally.

DEFINITION 29. We say that S is *locally maximal monotone* if, whenever $\widetilde{U}$ is a convex open subset of E^* such that $\widetilde{U} \cap R(S) \neq \emptyset$, $(q, q^*) \in E \times \widetilde{U}$ and

$$(z, z^*) \in G(S) \text{ and } z^* \in \widetilde{U} \quad \Longrightarrow \quad \langle z - q, z^* - q^* \rangle \geq 0$$

then

$$(q, q^*) \in G(S),$$

where $R(S) := \bigcup_{x \in E} Sx$.

Locally maximal monotone operators were introduced by Fitzpatrick and Phelps in [**7**], and shown to have important approximation properties. Fitzpatrick and Phelps continued their study in [**8**]. It follows from [**7**], Proposition 3.3, p. 585 and Theorem 3.5, p. 585 that

- if E is reflexive then every maximal monotone operator is locally maximal monotone,

and

- if S is locally maximal monotone then $\overline{R(S)}$ is convex,

and, from [**8**], Theorem 3.7, p. 67 that

- if S is maximal monotone and $R(S) = E^*$, or S is coercive and $\overline{R(S)} = E^*$ then S is locally maximal monotone.

Further, Bauschke and Borwein proved in [**4**], Theorem 3.3 that

- if $S : E \to E^*$ is *continuous and linear* then S is locally maximal monotone if, and only if, S^* is monotone.

There is an example of a maximal monotone operator that is not locally maximal monotone in [**8**], Example 3.2, p. 63.

We now discuss two classes of operators that use the bidual E^{**} of E. If $y \in E$, we write $\widehat{y}$ for the canonical image of y in E^{**}. We define $\overline{S} : E^{**} \to 2^{E^*}$ (see [**10**], Lemme 2.1, p. 375) by

$$x^* \in \overline{S}x^{**} \quad \Longleftrightarrow \quad \inf_{(y,y^*) \in G(S)} \langle y^* - x^*, \widehat{y} - x^{**} \rangle \geq 0.$$

DEFINITION 30. S is said to be *maximal monotone of type (D)* (this is essentially in [**10**]: see [**15**], Section 3 for an exposition) if S is maximal monotone and, for all $(x^{**}, x^*) \in G(\overline{S})$, there exists a bounded net (y_α, y_α^*) in $G(S)$ such that $\widehat{y_\alpha} \to x^{**}$ in $w(E^{**}, E^*)$ and $\|y_\alpha^* - x^*\| \to 0$.

DEFINITION 31. S is *maximal monotone of type (WD)* (see [**24**], Definition 14, p. 187) if S is maximal monotone and, for all $x^* \in R(\overline{S})$, there exists a bounded net (y_α, y_α^*) in $G(S)$ such that $\|y_\alpha^* - x^*\| \to 0$. ("W" stands for "weak".)

The following was proved by Gossez:

THEOREM 32. *If $f : E \to \mathbb{R} \cup \{\infty\}$ is a proper convex, lower semicontinous function then ∂f is maximal monotone of type (D).*

PROOF. See [**10**], Théorème 3.1, p. 376. □

It is easy to see that

- if E is reflexive then every maximal monotone operator is of type (D) and

- maximal monotone operators of type (D) are of type (WD).

It was proved in [**24**], Corollary 16, p. 188 that

- If S is maximal monotone of type (WD) then $\overline{R(S)}$ is convex.

Further, Bauschke and Borwein proved in [**4**], Theorem 3.3 that if S is *continuous, linear and maximal monotone* then

$$S \text{ is of type (D)} \iff S \text{ is of type (WD)} \iff S^* \text{ is monotone.}$$

From what we have already said, these three conditions are also equivalent to:

$$S \text{ is locally maximal monotone.}$$

However, in the nonlinear case, the relationships between the locally maximal operators and the operators of type (D) and (WD) is still uncharted.

DEFINITION 33. We say that S is *primally strongly maximal monotone* if, whenever Q is a nonempty weakly compact convex subset of E and $q^* \in E^*$ and,

$$\text{for all } (z, z^*) \in G(S), \quad \text{there exists } q \in Q \text{ such that } \langle z - q, z^* - q^* \rangle \geq 0$$

then

$$q^* \in S(Q).$$

It seems that nothing is known about these operators, other than the following result:

THEOREM 34. *If $f : E \to \mathbb{R} \cup \{\infty\}$ is a proper convex, lower semicontinous function then ∂f is primally strongly maximal monotone.*

PROOF. See [**20**], Theorem 6.1, p. 1386. □

DEFINITION 35. We say that S is *dually strongly maximal monotone* if, whenever $\widetilde{Q}$ is a nonempty weak*-compact convex subset of E^* and $q \in E$ and,

$$\text{for all } (z, z^*) \in G(S), \quad \text{there exists } q^* \in \widetilde{Q} \text{ such that } \langle z - q, z^* - q^* \rangle \geq 0$$

then

$$Sq \cap \widetilde{Q} \neq \emptyset.$$

It seems that nothing is known about these operators, other than the following result:

THEOREM 36. *If $f : E \to \mathbb{R} \cup \{\infty\}$ is a proper convex, lower semicontinous function then ∂f is dually strongly maximal monotone.*

PROOF. See [**20**], Theorems 6.2, p. 1386. □

One is tempted by Theorems 34 and 36 to hypothesize that if Q is a nonempty weakly compact convex subset of E and $\widetilde{Q}$ is a nonempty weak* compact convex subset of E^* and,

$$\text{for all } (z, z^*) \in G(\partial f), \quad \text{there exists } (q, q^*) \in Q \times \widetilde{Q} \text{ such that } \langle z - q, z^* - q^* \rangle \geq 0$$

then

$$(Q \times \widetilde{Q}) \cap G(\partial f) \neq \emptyset.$$

It was shown in [**12**] and [**13**] that this is true if $E = \mathbb{R}$. It was also shown in [**13**] that this fails if $E = \mathbb{R}^2$ with the Euclidean norm, even when $\widetilde{Q}$ is the unit ball of E^*.

DEFINITION 37. We say that S is *almost negative alignment maximal monotone* if, whenever $(q, q^*) \in E \times E^* \setminus G(S)$ then, for all $n \geq 1$, there exist $(z_n, z_n^*) \in G(S)$ such that

$$\frac{\langle z_n - q, z_n^* - q^* \rangle}{\|z_n - q\| \, \|z_n^* - q^*\|} \to -1.$$

It seems that nothing is known about these operators, other than what was proved in Theorems 13 and 26.

We note that the first six subclasses of maximal monotone operators introduced above are invariant under the renorming of E. We do not know what the situation is with the almost negative alignment maximal monotone operators.

We conclude this section with some brief comments about the nature of $D(\partial f)$, $D(S)$, $R(\partial f)$ and $R(S)$ when $f : E \to \mathbb{R} \cup \{\infty\}$ is a proper convex, lower semicontinuous function and $S : E \to 2^{E^*}$ is maximal monotone.

There is an easy example (see, for instance, [**3**], Section 4, p. 70) with $E := \mathbb{R}^2$ in which $D(\partial f)$ is not convex. On the other hand, from the Brøndsted-Rockafellar theorem, $\overline{D(\partial f)} = \overline{D(\operatorname{dom} f)}$ and, using the fact that f is continuous on $\operatorname{int}(\operatorname{dom} f)$, $\operatorname{int}(D(\partial f)) = \operatorname{int}(\operatorname{dom} f)$. Thus both $\overline{D(\partial f)}$ and $\operatorname{int}(D(\partial f))$ are convex.

Most of these results generalize to the $D(S)$ situation, since Rockafellar proved in [**16**], Theorem 1, p. 398 (see also [**15**], Theorem 1.9, p. 6) that $\operatorname{int} D(S)$ is convex and, if $\operatorname{int}(\operatorname{co} D(S)) \neq \emptyset$, then $\overline{D(S)}$ is convex. Indeed, more is true: one can define a proper convex, lower semicontinuous function $\chi_S : E \to \mathbb{R} \cup \{\infty\}$ such that $\operatorname{int} D(S) = \operatorname{int}(\operatorname{dom} \chi_S)$ and, if $\operatorname{int}(\operatorname{co} D(S)) \neq \emptyset$, then $\overline{D(S)} = \overline{\operatorname{dom} \chi_S}$. (See [**6**] for the definition and more of the properties of χ_S.) It does not seem to be known whether $\overline{D(S)}$ is necessarily convex if $\operatorname{int}(\operatorname{co} D(S)) = \emptyset$.

The situations for $R(\partial f)$ is different from that for $D(\partial f)$. It follows (for instance, from the results on locally maximal monotone operators or operators of type (D) and (WD) mentioned above) that $\overline{R(\partial f)}$ is convex. However, Borwein, Fitzpatrick and Vanderwerff proved in [**3**], Theorem 3.1, p. 68 that if E is not reflexive then there is always a continuous convex function $f : E \to \mathbb{R}$ such that $\operatorname{int}(R(\partial f))$ is not convex.

The situation for $R(S)$ is also different from that for $D(S)$. There is an example of a coercive maximal monotone operator S such that $\overline{R(S)}$ is not convex in [**8**], Example 3.2, p. 63.

Graphical interpretations of K

We now give a graphical interpretation of Theorem 4 in terms of *subtangents* to the graph of g. We suppose for simplicity that $\lambda = 0$, so $g : E \to \mathbb{R} \cup \{\infty\}$ is a proper, convex, lower semicontinuous function, $\inf_E g < 0 < g(0)$ and

$$K = \sup_{g(y) < 0} \frac{-g(y)}{\|y\|}.$$

If $(x, x^*) \in G(\partial g)$, we write $\sigma g(x, x^*)$ for the subtangent to g at $(x, g(x))$ with slope x^*, that is to say the real function on E given by

$$\sigma g(x, x^*)(y) := g(x) + \langle y - x, x^* \rangle.$$

Theorem 4 then says that there exist a point P on the graph of g and a subtangent f to g at P such that both the Lipschitz constant of f and the slope of the line going from P to $(0, f(0))$ in the graph of f are arbitrarily close to K. Hence this line is arbitrarily close to being a line of steepest ascent in f. (Specifically, $P = (x, g(x))$ and $f = \sigma g(x, x^*)$.) In all the remaining figures, the dashed line represents the slope K, so we cannot assume that there is a point of contact between the dashed line and the graph of g, or between the dashed line and the set B.

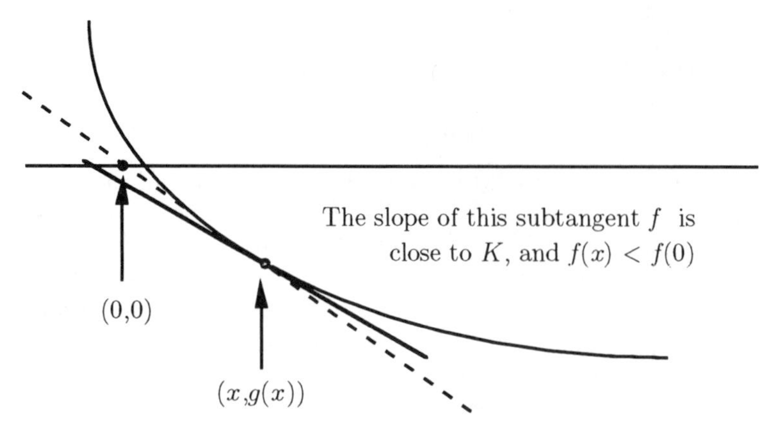

Let $(x, x^*) \in G(\partial g)$. We say that $\sigma g(x, x^*)$ *dominates* $(0, 0)$ if

$$\sigma g(x, x^*)(0) = g(x) - \langle x, x^* \rangle > 0.$$

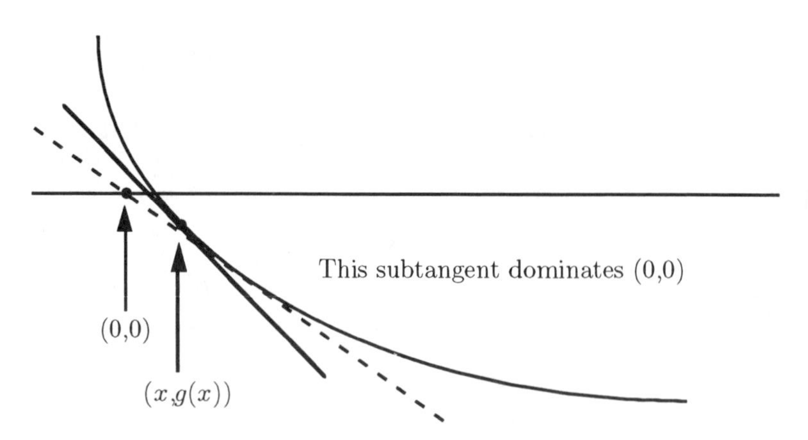

Suppose now that $\sigma g(x, x^*)$ dominates $(0, 0)$. Then, for all $y \in E$,

$$\|y\|\,\|x^*\| \geq \langle -y, x^* \rangle = \langle x - y, x^* \rangle - \langle x, x^* \rangle \geq [g(x) - g(y)] - g(x) = -g(y).$$

Dividing by $\|y\|$ and taking the supremum over y, we see that $\|x^*\| \geq K$. This statement cannot be materially improved, as can be seen from the *dominating subtangent theorem*, which follows from [**20**], Theorem 5.4, p. 1384:

THEOREM 38. *Let $g : E \to \mathbb{R} \cup \{\infty\}$ be a proper, convex, lower semicontinuous function, $\inf_E g < 0 < g(0)$ and $\varepsilon > 0$. Then there exists $(x, x^*) \in G(\partial g)$ such that*

$$\|x^*\| \leq (1 + \varepsilon)K \quad \textit{and} \quad \sigma g(x, x^*) \textit{ dominates } (0, 0).$$

Thus we can give an alternative characterization of K as the infimum of the slopes of the subtangents to the graph of g that dominate $(0,0)$. We should point out that Theorem 38 does *not* lead to a proof of the maximal monotonicity theorem, since it may well happen that $\langle x, x^*\rangle \geq 0$.

Here is an interesting open question on the existence of subtangents: we note from Theorem 4 that there always exists $(x, x^*) \in G(\partial g)$ such that $\langle x, x^*\rangle < 0$ and, from Theorem 38, that there always exists $(x, x^*) \in G(\partial g)$ such that $\langle x, x^*\rangle < g(x)$. We do not know if there always exists $(x, x^*) \in G(\partial g)$ such that $\langle x, x^*\rangle < g(x)$ and $\langle x, x^*\rangle < 0$. In the language of subtangents, we can rephrase this question as: does there always exist a subtangent f to the graph of g at a point $(x, g(x))$ such that $f(0) > 0$ and $f(0) > f(x)$?

Separating subtangents

We now extend the definition of K to a situation more general than that considered in (4.1). Specifically, let B be a nonempty bounded closed convex subset of $E \times \mathrm{IR}$, $\mathrm{gap}(B, \mathrm{epi}g) > 0$ and $\sup_B pr_2 > \inf_E g$. ($\mathrm{gap}(B, \mathrm{epi}g)$ stands for the distance between B and the epigraph of g measured by any norm that gives the product topology on $E \times \mathrm{IR}$, and pr_2 stands for the projection map from $E \times \mathrm{IR}$ to IR.) Write

$$K := \sup_{(x,\lambda)\in B,\ y\in E,\ \lambda > g(y)} \frac{\lambda - g(y)}{\|x-y\|}.$$

The figure below explains the definition of K.

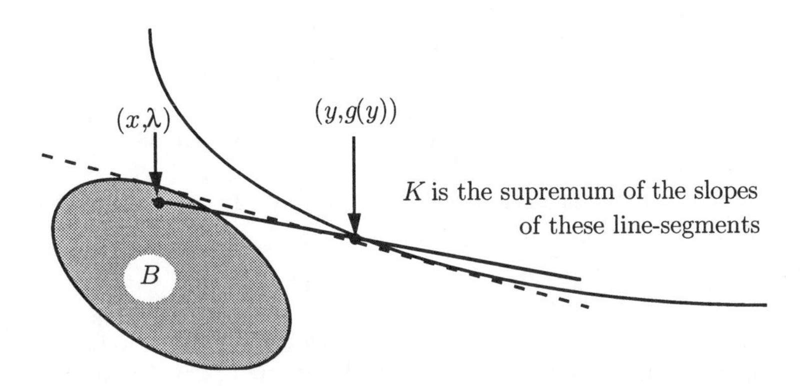

Just as in the case when $B = \{(0,0)\}$, one can prove that $K \in (0, \infty)$.

If $(z, z^*) \in G(\partial g)$, we shall say that $\sigma g(z, z^*)$ *dominates* B if, for all $(x, \lambda) \in B$, $\sigma g(z, z^*)(x) > \lambda$. Then one can prove just as in the case when $B = \{(0,0)\}$ that: if $(z, z^*) \in G(\partial g)$ and $\sigma g(z, z^*)$ dominates B then $\|z^*\| \geq K$. Again, this statement cannot be materially improved, as can be seen from the following *sharp separation theorem* generalizing Theorem 38, which was proved in [**20**], Theorem 5.4, p. 1384:

THEOREM 39. *Let $g : E \to \mathrm{IR}\cup\{\infty\}$ be a proper, convex, lower semicontinuous function, B and K be as above and $\varepsilon > 0$. Then there exists $(z, z^*) \in G(\partial g)$ such that*

$$\|z^*\| \leq (1+\varepsilon)K \quad \textit{and} \quad \sigma g(z, z^*) \textit{ dominates } B.$$

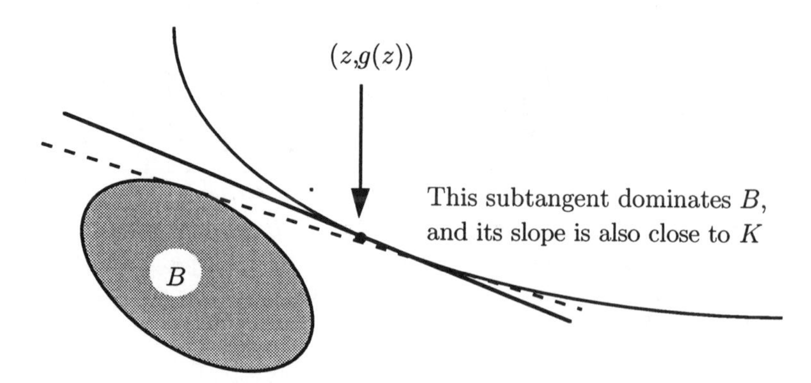

Beer established in [**2**], Lemma 4.10, p. 286 the existence of a subtangent to g lying above B. The significant feature of Theorem 39 is the sharp bound on the possible slopes of such subtangents.

More on dual results

In this final section, we summarize what we have proved about dual pairs of result, as well as discussing further examples of the phenomenon. The situation is most satisfactory in the following cases:

- Theorem 17 is the dual of Theorem 4.
- Corollary 20 is the dual of Corollary 7.
- Corollary 23 is the dual of Corollary 9 and Remark 10.
- Corollary 25 is the dual of Corollary 12.
- Theorem 36 is the dual of Theorem 34.

The situation is less satisfactory in the following cases:

- Corollary 21 is 80% of the dual of Corollary 8, but Remark 22 shows that the other 20% fails.
- Theorem 26 is the dual of Theorem 13, except that $\widetilde{M}$ is not defined by a distance.

As we indicated in the remarks preceding Conjecture 18, we suspect that

- the result dual to Corollary 6 is false.

Of course,

- Corollary 5 is self-dual.

Now let $f : E \to \mathbb{R} \cup \{\infty\}$ be a proper, convex, lower semicontinuous function. The following result on the approximation of elements of $\mathrm{dom} f$ was proved in [**20**], Theorem 3.2, p. 1379:

THEOREM 40. *If $x \in \mathrm{dom} f$ and $\varepsilon > 0$ then there exists $(z, z^*) \in G(\partial f)$ such that*

$$\|z - x\| < \varepsilon, \quad \langle z, z^* \rangle \leq \langle x, z^* \rangle \quad \textit{and} \quad f(z) \leq f(x).$$

The dual result on the approximation of elements of $\mathrm{dom} f^*$ is also true. It was proved in [**20**], Corollary 4.2, p. 1382, and generalizes [**1**], Proposition 4.3, p. 398:

THEOREM 41. *If $x^* \in \mathrm{dom} f^*$ and $\varepsilon > 0$ then there exists $(z, z^*) \in G(\partial f)$ such that*

$$\|z^* - x^*\| < \varepsilon, \quad \langle z, z^* \rangle \leq \langle z, x^* \rangle \quad \textit{and} \quad f^*(z^*) \leq f^*(x^*).$$

The following result follows from [**20**], Theorem 3.1, p. 1379:

THEOREM 42. *If* $\inf_E(f + \| \|) > -\infty$ *and* $\varepsilon > 0$ *then there exists* $(z, z^*) \in G(\partial f)$ *such that*

$$\|z^*\| \leq 1 + \varepsilon \quad \textit{and} \quad -\langle z, z^*\rangle \geq (1-\varepsilon)\|z\|.$$

We do not know if the corresponding dual result holds. That would be:

CONJECTURE 43. If $\inf_{E^*}(f^* + \| \|) > -\infty$ and $\varepsilon > 0$ then there exists $(z, z^*) \in G(\partial f)$ such that

$$\|z\| \leq 1 + \varepsilon \quad \text{and} \quad -\langle z, z^*\rangle \geq (1-\varepsilon)\|z^*\|.$$

References

[1] H. Attouch and G. Beer, *On the convergence of subdifferentials of convex functions*, Arch. Math. **60** (1993), 389–400.

[2] G. Beer, *The slice topology: A viable alternative to Mosco convergence in nonreflexive spaces*, Nonlinear Analysis **19** (1992), 271–290.

[3] J. Borwein, S. Fitzpatrick and J. Vanderwerff, *Examples of convex functions and classification of normed spaces*, J. Convex Analysis **1** (1994), 61–73.

[4] Heinz H. Bauschke and Jonathan M. Borwein, *Continuous Linear Monotone Operators on Banach Spaces*, Preprint.

[5] A. Brøndsted and R.T. Rockafellar, *On the Subdifferentiability of Convex Functions*, Proc. Amer. Math. Soc. **16** (1965), 605–611.

[6] M. Coodey and S. Simons, *The convex function determined by a multifunction*, Bull. Austral. Math. Soc. **54** (1996), 87–97.

[7] S. Fitzpatrick and R. R. Phelps, *Bounded approximants to monotone operators on Banach Spaces*, Ann. Inst. Henri Poincaré, Analyse non linéaire **9** (1992), 573–595.

[8] S. P. Fitzpatrick and R. R. Phelps, *Some properties of maximal monotone operators on nonreflexive Banach spaces*, Set-Valued Analysis **3** (1995), 51–69.

[9] D. Gale, *A Geometric Duality Theorem with Economic Applications*, Review of Economic Studies **34** (1967), 19–24.

[10] J.- P. Gossez, *Opérateurs monotones non linéaires dans les espaces de Banach non réflexifs*, J. Math. Anal. Appl. **34** (1971), 371–395.

[11] H. König, *Some basic theorems in convex analysis*, Optimization and operations research, Edited by B. Korte, North-Holland, Amsterdam, Holland, 1982, pp. 107–144.

[12] S. Kum, *Maximal monotone operators in the one dimensional case*, Preprint.

[13] D. T. Luc, *A resolution of Simons' maximal monotonicity conjecture*, Preprint.

[14] R. R. Phelps, *Convex functions, monotone operators and differentiablity*, 2nd ed., Springer-Verlag, Berlin, Germany, 1993.

[15] R. R. Phelps paper Lectures on Maximal Monotone Operators, 2nd Summer School on Banach Spaces, Related Areas and Applications, Prague and Paseky, August 15–28, 1993. (Preprint, 30 pages.) TeX file: Banach space bulletin board archive: <ftp://ftp:@math.okstate.edu/pub/banach/phelpsmaxmonop.tex>. Posted Nov. 1993.

[16] R. T. Rockafellar, *Local boundedness of Nonlinear, Monotone Operators*, Michigan Math. J. **16** (1969), 397–407.

[17] R. T. Rockafellar, *On the Maximal Monotonicity of Subdifferential Mappings*, Pacific Journal of Mathematics **33** (1970), 209–216.

[18] R. T. Rockafellar, *Conjugate duality and optimization*, Conference Board of the Mathematical Sciences 16, SIAM publications, 1974.

[19] S. Simons, *The least slope of a convex function and the maximal monotonicity of its subdifferential*, Journal of Optimization Theory **71** (1991), 127–136.

[20] S. Simons, *Subtangents with controlled slope*, Nonlinear Analysis **22** (1994), 1373–1389.

[21] S. Simons, *Subdifferentials are locally maximal monotone*, Bull. Australian Math. Soc. **47** (1993), 465–471.

[22] S. Simons, *Swimming below icebergs*, Set-Valued Analysis **2** (1994), 327–337.

[23] S. Simons, *Simultaneous almost minimization of convex functions and duals of results related to maximal monotonicity*, Journal of Convex Analysis **2** (1995), 359–373.

[24] S. Simons, *The range of a monotone operator*, J. Math. Anal. Appl. **199** (1996), 176–201.
[25] A. and M. E. Verona, *Remarks on subgradients and ε-subgradients*, Set-Valued Analysis **1** (1993), 261-272.

Department of Mathematics, University of California, Santa Barbara, California 93106-3080
E-mail address: simons@math.ucsb.edu

Contemporary Mathematics
Volume **204**, 1997

Existence and Structure of Optimal Solutions of Variational Problems

Alexander J. Zaslavski

ABSTRACT. In this paper we consider the existence and the structure of extremals of autonomous variational problems. Given an $x_0 \in R^n$ we study the infinite horizon problem of minimizing the expression $\int_0^T f(x(t), x'(t))\, dt$ as T grows to infinity where the function x: $[0, \infty) \to R^n$ is locally absolutely continuous and satisfies the initial condition $x(0) = x_0$ and the integrand $f = f(x, u)$ belongs to a complete metric space of functions $\mathfrak{M}$. We discuss the results establishing the existence of a set $F \subset \mathfrak{M}$ which is a countable intersection of open everywhere dense sets in $\mathfrak{M}$ and such that the following properties hold: (i) for each $f \in F$ and each $z \in R^n$ there exists an (f)-weakly optimal function x: $[0, \infty) \to R^n$ satisfying $x(0) = z$; (ii) for each $f \in F$ there exists a compact set $H(f) \subset R^n$ such that for any finite horizon the turnpike property holds with the set $H(f)$ being the attractor.

1. Introduction

The study of variational and optimal control problems defined on infinite intervals has recently been a rapidly growing area of research. These problems arise in engineering (see Anderson and Moore [1], Artstein and Leizarowitz [2]), in models of economic growth (see Rockafellar [20], Carlson [6]), in infinite discrete models of solid-state physics related to dislocations in one-dimensional crystals which are under discussion in Aubry and Le Daeron [3], Zaslavski [23] and in the theory of thermodynamical equilibrium for materials (see Leizarowitz and Mizel [15], Coleman, Marcus and Mizel [9], Marcus [17], Zaslavski [26,27]).

In this paper we discuss the existence and the structure of optimal solutions of autonomous variational problems. Given an $x_0 \in R^n$, we study the infinite horizon problem of minimizing the expression $\int_0^T f(x(t), x'(t))\, dt$ as T grows to infinity where the function x: $[0, \infty) \to R^n$ is locally absolutely continuous (a.c.) and satisfies the initial condition $x(0) = x_0$, and $f = f(x, u)$ is an appropriate integrand.

The following notion known as *the overtaking optimality criterion* was introduced in the economic literature by Gale [10] and von Weizsacker [22] and has been used

1991 *Mathematics Subject Classification.* 49J99, 58F99.

in control theory by Artstein and Leizarowitz [2], Brock and Haurie [5], Carlson [6], Carlson, Haurie and Jabrane [7], Carlson, Haurie and Leizarowitz [8].

An a.c. function x: $[0, \infty) \to R^n$ is called *(f)-overtaking optimal* if for any a.c. function y: $[0, \infty) \to R^n$ satisfying $y(0) = x(0)$

$$\limsup_{T \to \infty} \int_0^T [f(x(t), x'(t)) - f(y(t), y'(t))] \, dt \leq 0.$$

In this paper we employ the following weakened version of this criterion.

An a.c. function $x:$ $[0, \infty) \to R^n$ is called *(f)-weakly optimal* if for any a.c. function y: $[0, \infty) \to R^n$ satisfying $y(0) = x(0)$

$$\liminf_{T \to \infty} \int_0^T [f(x(t), x'(t)) - f(y(t), y'(t))] \, dt \leq 0.$$

Another type of optimality criterion for infinite horizon problems was introduced by Aubry and Le Daeron [3] in their study of the discrete Frenkel-Kontorova model related to dislocations in one-dimensional crystals. More recently this notion was used by Moser [18,19], Bangert [4], Leizarowitz and Mizel [15] and Zaslavski [23].

Let I be either $[0, +\infty)$ or $(-\infty, +\infty)$. An a.c. function x: $I \to R^n$ is called an *(f)-minimal solution* if for each numbers $T_1 \in I, T_2 > T_1$ and each a.c. function y: $[T_1, T_2] \to R^n$ which satisfies $y(T_i) = x(T_i), i = 1, 2$ the following relation holds

$$\int_{T_1}^{T_2} [f(x(t), x'(t)) - f(y(t), y'(t))] \, dt \leq 0.$$

Clearly every (f)-weakly optimal function is an (f)-minimal solution. We can establish the existence of (f)-minimal solutions using the following scheme.

Let $x_0 \in R^n$, $N > 0$ be an integer and let x_N: $[0, N] \to R^n$ be an optimal solution of the following variational problem

$$\int_0^N f(x(t), x'(t))dt \to \min, \quad x(0) = x_0.$$

When the integrand f satisfies certain assumptions we can show that there exist a subsequence $\{x_{N_i}\}_{i=1}^{\infty}$ and an a.c. function x: $[0, \infty) \to R^n$ such that for any natural number k

$$x_{N_i}(t) \underset{i \to \infty}{\longrightarrow} x(t) \text{ uniformly in } [0, k], \quad x'_{N_i} \underset{i \to \infty}{\longrightarrow} x' \text{ weakly in } L^1([0, k]; R^n)$$

and x is an (f)-minimal solution. Various existence results of overtaking optimal and weakly optimal functions are nicely collected in Carlson, Haurie and Leizarowitz [8]. The most typical infinite horizon optimization problem for which the existence of overtaking optimal function has been established is an autonomous variational problem with a convex integrand $\int_0^T f(x(t), x'(t))dt$ studied by Rockafellar [20], Brock and Haurie [5] and Leizarowitz [13]. When the existence of optimal solutions over an infinite horizon was established for nonconvex problems a convexity assumption was usually replaced by another assumptions which were restrictive and did not hold in general.

Our first goal is to show that the existence of weakly optimal solutions is a general phenomenon which holds for various classes of variational problems.

An optimal solution v: $[0, T] \to R^n$ of the variational problem

$$(P) \qquad \int_0^T f(z(t), z'(t))dt \to \min, \quad z(0) = x, \quad z(T) = y,$$

$$z\colon [0, T] \to R^n \text{ is an a.c. function}$$

always depends on the integrand f and on x, y, T. We say that the integrand f has the *turnpike property* if for large enough T the dependence on x, y, T is not essential. Namely, for any $\epsilon > 0$ there exist constants $L_1, L_2 > 0$ which depend only on $|x|$, $|y|$ and ϵ and a measurable set $E \subset [0, T]$ such that the Lebesgue measure of E does not exceed L_1 and for each $\tau \in [0, T - L_2] \setminus E$ the set $\{v(t) \colon t \in [\tau, \tau + L_2]\}$ is equal to a set $H(f)$ up to ϵ in the Hausdorff metric where $H(f) \subset R^n$ is a compact set depending only on the integrand f. Moreover, the set E is a finite union of intervals and their number does not exceed a constant which depends only on $|x|$, $|y|$ and ϵ. If for each optimal solution of the variational problem (P) with the integrand f the set $E = [0, L_1] \cup [T - L_1, T]$, then we say that the integrand f has the strong turnpike property.

More formally, an integrand $f = f(x, u) \in C(R^{2n})$ has the turnpike property if there exists a compact set $H(f) \subset R^n$ such that:

for each bounded set $K \subset R^n$ and each $\epsilon > 0$ there exist a number $l > 0$ and integers $L, Q \geq 1$ such that for each $T \geq L + lQ$, each $x, y \in K$ and an optimal solution v: $[0, T] \to R^n$ for the variational problem (P) there exist sequences of numbers $\{b_j\}_{j=1}^q, \{c_j\}_{j=1}^q \subset [0, T]$ for which $q \leq Q$, $0 \leq c_j - b_j \leq l$, $j = 1, ...q$, and

$$\mathrm{dist}(H(f), \{v(t) \colon t \in [\tau, \tau + L]\}) \leq \epsilon \text{ for each } \tau \in [0, T - L] \setminus \bigcup_{j=1}^{q} [b_j, c_j].$$

(Here $\mathrm{dist}(\cdot, \cdot)$ is the Hausdorff metric).

We say that an integrand $f = f(x, u) \in C(R^{2n})$ has the strong turnpike property if there exists a compact set $H(f) \subset R^n$ such that:

for each bounded set $K \subset R^n$ and each $\epsilon > 0$ there exist numbers $L_1 > L_2 > 0$ such that for each $T \geq 2L_1$, each $x, y \in K$ and an optimal solution v: $[0, T] \to R^n$ for the variational problem (P), the relation

$$\mathrm{dist}(H(f), \{v(t) \colon t \in [\tau, \tau + L_2]\}) \leq \epsilon$$

for each $\tau \in [L_1, T - L_1]$.

The turnpike property is well known in mathematical economics. It was studied by many authors for optimal trajectories of a von Neumann-Gale model determined by a superlinear set-valued mapping (see Makarov and Rubinov [16] and the survey [21]) and for optimal trajectories of convex autonomous systems (see Carlson, Haurie and Leizarowitz [8, Ch. 4,6]). In control theory the turnpike property was established by Artstein and Leizarowitz [2] for a tracking periodic problem. In all these cases we have an optimal control problem with a convex cost function and a convex set of trajectories.

Our second goal is to show that the turnpike property is a general phenomenon which holds for various classes of variational problems. We consider a complete metric space of integrands or cost functions $\mathfrak{A}$ and establish the existence of a set $F \subset \mathfrak{A}$ which is a countable intersection of open everywhere dense sets in $\mathfrak{A}$ and such that for each $f \in F$ the following properties hold:

for each $z \in R^n$ there exists an (f)-weakly optimal function $Z\colon [0, \infty) \to R^n$ satisfying $Z(0) = z$;

the integrand f has the turnpike property.

Moreover we show that the turnpike property holds for approximate solutions of variational problems with a generic integrand f and that the turnpike phenomenon is stable under small pertubations of a generic integrand f.

The paper is organized as follows. In section 2 we discuss the existence and structure of optimal solutions of discrete-time control systems. In section 3 we consider a class of one-dimensional variational problems arising in continuum mechanics and in Section 4 we discuss the dynamic properties of extremals of variational problems with vector-valued functions.

2. Discrete-time control systems

2.1. Optimal programs on infinite horizon. In this section we consider the infinite horizon problem of minimizing the expression $\sum_{i=0}^{N-1} v(x_i, x_{i+1})$ as N grows to infinity where $\{x_i\}_{i=0}^{\infty}$ is a sequence in a compact metric space K and v is a continuous function defined on $K \times K$. This provides a convenient setting for the study of various optimization problems, e.g., continuous time control systems which are represented by ordinary differential equations whose cost integrand contains a discounting factor [12], the infinite-horizon deterministic control problem of minimizing $\int_0^T L(z(t), z'(t))dt$ as $T \to \infty$ [14], the analysis of a long slender bar of a polymeric material under tension [15], and the analysis of an infinite discrete model for crystals which undergo phase transitions [3,23].

Let K be a compact metric space, R^n the Euclidean n-dimensional space, $C(K \times K)$ the space of all continuous functions $v\colon K \times K \to R^1$ with the topology of the uniform convergence ($||v|| = \sup\{|v(x,y)|\colon x, y \in K\}$). Let $C(K)$ be the space of all continuous functions $v\colon K \to R^1$ with the topology of uniform convergence ($||v|| = \sup\{|v(x)|\colon x \in K\}$), and $B(K \times K)$ the set of all bounded and lower semicontinuous functions $v\colon K \times K \to R^1$.

Consider any $v \in B(K \times K)$. We are interested in the limit behavior as $N \to \infty$ of the expression $\sum_{i=0}^{N-1} v(x_i, x_{i+1})$ where $\{x_i\}_{i=0}^{\infty}$ is an infinite sequence in K which we call a program (or a configuration) (see [3,12,23]), and which occasionally will be denoted by a bold face letter $\mathbf{x}$ (similarly $\{y_i\}_{i=0}^{\infty}$ will be denoted by $\mathbf{y}$, etc.) A finite sequence $\{x_i\}_{i=0}^{N} \subset K$ $(N = 0, 1, ...)$ will be also called a program.

A program $\{x_i\}_{i=0}^{\infty}$ is an (v)-overtaking optimal program if for every program

$\{z_i\}_{i=0}^{\infty}$ satisfying $z_0 = x_0$ the following inequality holds:

$$\limsup_{N\to\infty} \sum_{i=0}^{N-1} [v(x_i, x_{i+1}) - v(z_i, z_{i+1})] \le 0.$$

A program $\{x_i\}_{i=0}^{\infty}$ is (v)-weakly optimal [12] if for every program $\{z_i\}_{i=0}^{\infty}$ satisfying $z_0 = x_0$ the following inequality holds:

$$\liminf_{N\to\infty} \sum_{i=0}^{N-1} [v(x_i, x_{i+1}) - v(z_i, z_{i+1})] \le 0.$$

Of special interest is the minimal long-run average cost growth rate

$$\mu(v) = \inf\{\liminf_{N\to\infty} N^{-1} \sum_{i=0}^{N-1} v(z_i, z_{i+1}) : \{z_i\}_{i=0}^{\infty} \text{ is a program}\}.$$

A program $\{z_i\}_{i=0}^{\infty}$ is called a (v)-good program [12] if the sequence

$$\left\{\sum_{i=0}^{N-1} [v(z_i, z_{i+1}) - \mu(v)]\right\}_{N=1}^{\infty}$$

is bounded. It was proved in [12] that for every program $\{z_i\}_{i=0}^{\infty}$ the sequence $\{\sum_{i=0}^{N-1} [v(z_i, z_{i+1}) - \mu(v)]\}_{N=1}^{\infty}$ is either bounded or diverges to infinity and that for every initial value z there is a (v)-good program $\{z_i\}_{i=0}^{\infty}$ satisfying $z_0 = z$.

In [24,25] we investigated the structure of (v)-good programs and established for a generic $v \in C(K \times K)$, and for every given $x \in K$ the existence of (v)-weakly optimal program $\{x_i\}_{i=0}^{\infty}$ satisfying $x_0 = x$.

2.2. Definitions and theorems. Let K be a compact metric space, and let $v \in B(K \times K)$. We define

$$a(v) = \sup\{v(x, y) \colon x, y \in K\}, \; b(v) = \inf\{v(x, y) \colon x, y \in K\},$$

$$\mu(v) = \inf\left\{\liminf_{N\to\infty} N^{-1} \sum_{i=0}^{N-1} v(z_i, z_{i+1}) \colon \{z_i\}_{i=0}^{\infty} \text{ is a program}\right\}.$$

The following two results established in [12] are very useful in the study of infinite horizon control problems.

THEOREM 2.1. *1. For every program* $\{z_i\}_{i=0}^{\infty}$

$$\sum_{i=0}^{N-1} [v(z_i, z_{i+1}) - \mu(v)] \ge b(v) - a(v) \; (N = 1, 2, \dots);$$

2. For every program $\{z_i\}_{i=0}^{\infty}$ *the sequence* $\left\{\sum_{i=0}^{N-1} [v(z_i, z_{i+1}) - \mu(v)]\right\}_{N=1}^{\infty}$ *is either bounded or it diverges to infinity;*

3. For every initial value z_0 *there is a program* $\{z_i\}_{i=0}^{\infty}$ *which satisfies*

$$\left|\sum_{i=0}^{N} [v(z_i, z_{i+1}) - \mu(v)]\right| \le 4|a(v) - b(v)| \; (N = 1, 2, \dots).$$

THEOREM 2.2. *Let $v \in C(K \times K)$ and define*

$$\pi^v(x) = \inf\left\{\liminf_{N\to\infty}\sum_{i=0}^{N-1}[v(z_i, z_{i+1}) - \mu(v)] \colon \mathbf{z} \in K, z_0 = x\right\},$$

$$\theta^v(x,y) = v(x,y) - \mu(v) + \pi^v(y) - \pi^v(x) \text{ for } x, y \in K.$$

Then π^v, θ^v are continuous functions, θ^v is nonnegative and $E(x) = \{y \in K\colon \theta^v(x,y) = 0\}$ is nonempty for every $x \in K$.

In [12] these theorems were established when K was a compact subset of R^n, but their proofs remain in force also when K is a compact metric space.

For a program $\mathbf{x}$ we denote by $\omega(\mathbf{x})$ the set of all points $z \in K$ such that some subsequence $\{x_{i_k}\}_{k=1}^{\infty}$ converge to z, and denote by $\Omega(\mathbf{x})$ the set of all points $(z_1, z_2) \in K \times K$ such that some subsequence $\{(x_{i_k}, x_{i_k+1})\}_{k=1}^{\infty}$ converge to (z_1, z_2). Denote by $d(x,y)$ $(x, y \in K)$ the metric in K and define the metric d_1 on $K \times K$ by

$$d_1((x_1,x_2),(y_1,y_2)) = d(x_1,y_1) + d(x_2,y_2) \quad (x_1,x_2,y_1,y_2 \in K).$$

We denote $d(x,B) = \inf\{d(x,y)\colon y \in B\}$ for $x \in K, B \subset K$ and

$$d_1((x_1,x_2),A) = \inf\{d_1(x_1,x_2),(y_1,y_2))\colon (y_1,y_2) \in A\}$$

for $(x_1, x_2) \in K \times K$ and $A \subset K \times K$.

Denote by $\text{dist}(A, B)$ the Hausdorff metric for two sets $A \subset K$ and $B \subset K$ and denote by $\text{Card}(A)$ the cardinality of a set A.

A sequence $\{x_i\}_{i=-\infty}^{\infty} \subset K$ is called almost periodic if for every $\epsilon > 0$ there exists an integer $m \geq 1$ such that the relation $d(x_i, x_{i+mp}) \leq \epsilon$ holds for any i and p. A program $\{x_i\}_{i=0}^{\infty}$ is called asymptotically almost periodic if for every $\epsilon > 0$ there exist integers $k \geq 1, m \geq 1$ such that $d(x_i, x_{i+mj}) \leq \epsilon$ for any $i \geq k$ and any $j \geq 1$.

In [24,25] we proved the existence of a set $F \subset C(K \times K)$ which is a countable intersection of open everywhere dense sets in $C(K \times K)$ and for which the following theorems are valid.

THEOREM 2.3 [24]. *1. For every $u \in F$ there are closed sets $H(u) \subset K \times K$, $H_0(u) \subset K$ such that for every (u)-good program $\mathbf{x}$ we have*

$$H(u) = \Omega(\mathbf{x}), \; H_0(u) = \omega(\mathbf{x}).$$

2. Let $u \in F$. Then every (u)-good program $\mathbf{x}$ is asymptotically almost periodic.

3. Let $u \in F$ and δ be a positive number. Then there is a neighborhood $W(u)$ of u in $C(K \times K)$ such that for every $w \in W(u)$, and for every (w)-good program $\mathbf{x}$ we have $\text{dist}(H(u), \Omega(\mathbf{x})) \leq \delta$.

THEOREM 2.4 [24]. *Let $u \in F$, and let $\{x_i\}_{i=0}^{\infty}$ be a program such that*

$$\theta^u(x_i, x_{i+1}) = 0, i = 0, 1, \ldots$$

Then $\{x_i\}_{i=0}^{\infty}$ is a (u)-weakly optimal program. Moreover, there exists a strictly increasing sequence of natural numbers $\{i_k\}_{k=1}^{\infty}$ such that for every program $\{y_i\}_{i=0}^{\infty}$ satisfying $y_0 = x_0$ the inequality

$$\liminf_{k\to\infty} \sum_{j=0}^{i_k-1} [u(y_j, y_{j+1}) - u(x_j, x_{j+1})] \geq 0$$

holds, and if for some program $\{y_i\}_{i=0}^{\infty}$ satisfying $y_0 = x_0$,

$$\liminf_{k\to\infty} \sum_{j=0}^{i_k-1} [u(y_j, y_{j+1}) - u(x_j, x_{j+1})] = 0,$$

then $\theta^u(y_j, y_{j+1}) = 0$ $(j = 0, 1, \dots)$.

For every $u \in C(K \times K)$, every number $\Delta > 0$, and every integer $N \geq 1$ we denote by $A(u, N, \Delta)$ the set of all sequences $\{y_i\}_{i=0}^{N} \subset K$ such that for every sequence $\{z_i\}_{i=0}^{N} \subset K$ satisfying $z_0 = y_0, z_N = y_N$ the following inequality holds:

$$\sum_{i=0}^{N-1} [u(y_i, y_{i+1}) - u(z_i, z_{i+1})] \leq \Delta.$$

We define L: $C(K \times K) \to R^1 \times C(K) \times C(K \times K)$ by

$$L(v) = (\mu(v), \pi^v, \theta^v), \; v \in C(K \times K).$$

The second assertion of Theorem 2.5 and Theorem 2.6 establish the turnpike property for every $u \in F$.

THEOREM 2.5 [25].

1. The set of continuity points of the operator L contains F.

2. Let $u \in F$ and δ be a positive number. Then there are a neighborhood $W(u)$ of u in $C(K\times K)$, and positive numbers Q_1, Q_2 such that for every $w \in W(u)$, for every integer $N \geq 1$, for every integer $M > 0$, and every program $\{y_i\}_{i=0}^{N} \in A(w, N, M)$ the following relation holds

$$Card\{i \in \{0, \dots, N-1\}: d_1((y_i, y_{i+1}), H(u)) > \delta\} \leq Q_1 + MQ_2.$$

THEOREM 2.6 [25].

1. Let $u \in F$, ϵ be a positive number. Then there exist a neighborhood $W(u)$ of u in $C(K \times K)$ and $\delta > 0$ such that for every $w \in W(u)$, and for every program $\{x_i\}_{i=0}^{\infty}$ satisfying $\theta^w(x_i, x_{i+1}) = 0$, $i = 0, 1, \dots$, and $d(x_0, H_0(u)) \leq \delta$, the relation $d_1((x_i, x_{i+1}), H(u)) \leq \epsilon$ holds for $i = 0, 1, \dots$

2. Let $u \in F$, ϵ be a positive number. Then there exist a neighborhood $W(u)$ of u in $C(K \times K)$ and an integer $N \geq 1$ such that for every $w \in W(u)$, and for every program $\{x_i\}_{i=0}^{\infty}$ satisfying $\theta^w(x_i, x_{i+1}) = 0$, $i = 0, 1, \dots$, the following relation holds: $d_1((x_i, x_{i+1}), H(u)) \leq \epsilon$ (i is an integer, $i \geq N$).

COROLLARY 2.1. *Let $u \in F$, $\{x_i\}_{i=-\infty}^{\infty}$ be a program such that $\theta^u(x_i, x_{i+1}) = 0$, $i = 0, \pm 1, \ldots$. Then $(x_i, x_{i+1}) \in H(u)$, $i = 0, \pm 1, \ldots$.*

COROLLARY 2.2. *Let $u \in F$, $\epsilon > 0$. Then there exists a neighborhood $W(u)$ of u in $C(K \times K)$ such that for every $w \in W(u)$, and for every program $\{x_i\}_{i=-\infty}^{\infty}$ satisfying $\theta^w(x_i, x_{i+1}) = 0$, $i = 0, \pm 1, \ldots$ the relation $d_1((x_i, x_{i+1}), H(u)) \leq \epsilon$ holds for $i = 0, \pm 1, \ldots$.*

Theorem 2.7 implies that for each $w \in F$ every sequence $\{y_i\}_{i=-\infty}^{\infty}$ satisfying $\theta^w(y_i, y_{i+1}) = 0$, $i = 0, \pm 1, \ldots$ is almost periodic and shows that given any ϵ we can choose m (see the definition of an almost periodic program) uniformly for all w belonging to some small neighborhood of $u \in F$.

THEOREM 2.7 [25]. *Let $u \in F$. Then every sequence $\{y_i\}_{i=-\infty}^{\infty}$ which satisfies $\theta^u(y_i, y_{i+1}) = 0$, $i = 0, \pm 1, \ldots$ is almost periodic. Moreover, for every $\epsilon > 0$ there exist a neighborhood $W(u)$ of u in $C(K \times K)$, and an integer $m \geq 1$ such that for every $w \in W(u)$, and for every program $\{y_i\}_{i=-\infty}^{\infty}$ satisfying $\theta^w(y_i, y_{i+1}) = 0$, $i = 0, \pm 1, \ldots$, the relation $d(y_i, y_{i+pm}) \leq \epsilon$ holds for any integers i and p.*

THEOREM 2.8 [25]. *1. Let $u \in F$, and let $\{x_i\}_{i=0}^{\infty}$ be a (u)-good program. Then there exists a program $\{y_i\}_{i=-\infty}^{\infty}$ such that*

$$\theta^u(y_i, y_{i+1}) = 0,\ i = 0, \pm 1, \ldots, \qquad \text{and } \lim_{i \to \infty} d(x_i, y_i) = 0.$$

2. Let $u \in F$, and let $\{x_i\}_{i=-\infty}^{\infty}$ be a (u)-minimal energy configuration. Then there exist programs $\{y_i\}_{i=-\infty}^{\infty}$, $\{z_i\}_{i=-\infty}^{\infty}$ such that

$$\theta^u(y_i, y_{i+1}) = 0,\ \theta^u(z_i, z_{i+1}) = 0,\ i = 0, \pm 1, \ldots,$$

$$\lim_{i \to \infty} d(x_i, y_i) = 0, \quad \lim_{i \to -\infty} d(x_i, z_i) = 0.$$

We define $C_0(K \times K) = \{v \in C(K \times K) \colon \ \mu(v) = \max\{v(x,x) \colon x \in K\}\}$. It is easy to see that $C_0(K \times K)$ is a closed subspace of $C(K \times K)$. The space $C_0(K \times K)$ also has the topology of uniform convergence. In [25] we reenforced the previous theorems for $u \in C_0(K \times K)$ and proved the existence of a set $F_0 \subset F \cap C_0(K \times K)$ which is a countable intersection of open everywhere dense in $C_0(K \times K)$ subsets of $C_0(K \times K)$ and for which Theorem 2.9 holds. This theorem shows that for every $u \in F_0$, and every $x \in K$ there is an (u)-overtaking optimal program $\{x_i\}_{i=0}^{\infty}$ satisfying $x_0 = x$ and establishes the strong turnpike theorem for $u \in F_0$.

THEOREM 2.9 [25].

1. Card$(H_0(u)) = 1$ $(u \in F_0)$.

2. Let $u \in F_0$, and let δ be a positive number. Then there exists a neighborhood $W(u)$ of u in $C(K \times K)$ such that for every $w \in W(u)$, and for every (w)-good program $\mathbf{x}$ the relation dist$(\Omega(\mathbf{x}), (H_0(u) \times H_0(u))) \leq \delta$ holds.

3. Let $u \in F_0$, $\{x_i\}_{i=0}^{\infty}$ be a program for which $\theta^u(x_i, x_{i+1}) = 0$, $i = 0, 1, \ldots$. Then $\{x_i\}_{i=0}^{\infty}$ is an (u)-overtaking optimal program and, moreover, if $\{y_i\}_{i=0}^{\infty}$ is a program such that $y_0 = x_0$, and

$$\liminf_{N \to \infty} \sum_{i=0}^{N-1} [u(y_i, y_{i+1}) - u(x_i, x_{i+1})] = 0,$$

then $\theta^u(y_i, y_{i+1}) = 0$, $i = 0, 1, \dots$

4. *Let* $u \in F_0$, $\epsilon > 0$. *Then there exist a neighborhood* $W(u)$ *of* u *in* $C(K \times K)$, *an integer* $Q \geq 1$ *and* $\epsilon_0 \in (0, \epsilon)$ *such that for every* $w \in W(u)$, *for every integer* $N \geq 2Q$ *and every program* $\{y_i\}_{i=0}^N \in A(w, N, \epsilon_0)$ *the following relation holds:* $d(y_i, H_0(u)) \leq \epsilon$, $i = Q, \dots N - Q$ *and if* $d(y_0, H_0(u)) \leq \epsilon_0$, *then* $d(y_i, H_0(u)) \leq \epsilon$, $i = 0, \dots N - Q$.

Two examples were given in [24] - one concerning the nonexistence of (v)-overtaking optimal program for each v belonging to some open set $D \subset C([0, 1] \times [0, 1])$, and the other showing the existence of a (v)-weakly optimal program $\{x_i\}_{i=0}^\infty$ for which the relation $\theta^v(x_i, x_{i+1}) = 0$ $(i = 0, 1, \dots)$ does not hold.

3. Existence and structure of extremals for a class of second order variational problems

3.1. One-dimensional infinite horizon variational problems arising in continuum mechanics. In this section we discuss the existence and the structure of extremals of an infinite horizon variational problem for real valued functions defined on an infinite semiaxis of the line. Given $x \in R^2$ we study the infinite horizon problem of minimizing the expression $\int_0^T f(w(t), w'(t), w''(t))dt$ as T grows to infinity where

$$w \in A_x = \{v \in W_{loc}^{2,1}[0, \infty)\colon (v(0), v'(0)) = x\}.$$

Here $W_{loc}^{2,1}[0, \infty) \subset C^1$ denotes the Sobolev space of functions possessing a locally integrable second derivative and f belongs to a space of functions to be described below.

A function $u \in A_x$ will be called a (f)-weakly optimal solution of the variational problem

$$(P_\infty) \qquad \text{Minimize} \int_0^T f(w(t), w'(t), w''(t))dt \text{ as } T \to \infty$$

if $\liminf_{T\to\infty}[\int_0^T f(u(t), u'(t), u''(t))dt - \int_0^T f(w(t), w'(t), w''(t))dt] \leq 0$ for any $w \in A_x$.

The interest in variational problems of the form (P_∞) stems from the theory of thermodynamical equilibrium for second-order materials developed by Coleman, Marcus and Mizel [9], Leizarowitz and Mizel [15] and Marcus [17].

Denote by $\mathfrak{A}$ the set of all continuous functions $f\colon R^3 \to R$ such that for each $N > 0$ the function $|f(x, y, z)| \to \infty$ as $|z| \to \infty$ uniformly on the set $\{(x, y) \in R^2\colon |x|, |y| \leq N\}$. For the set $\mathfrak{A}$ we consider the uniformity which is determined by the following base

$$E(N, \epsilon, \Gamma) = \{(f, g) \in \mathfrak{A} \times \mathfrak{A} :$$

$$|f(x_1, x_2, x_3) - g(x_1, x_2, x_3)| \leq \epsilon \ (x_i \in R, |x_i| \leq N, \ i = 1, 2, 3),$$

$$(|f(x_1, x_2, x_3)| + 1)(|g(x_1, x_2, x_3)| + 1)^{-1} \in [\Gamma^{-1}, \Gamma]$$

$$(x_1, x_2, x_3) \in R^3, \ |x_1|, |x_2| \leq N)\},$$

where $N > 0$, $\epsilon > 0$, $\Gamma > 1$ [11]. Clearly, the uniform space $\mathfrak{A}$ is Hausdorff and has a countable base. Therefore $\mathfrak{A}$ is metrizable. It is easy to verify that the uniform space $\mathfrak{A}$ is complete.

Let $a = (a_1, a_2, a_3, a_4) \in R^4$, $a_i > 0$ $(i = 1, 2, 3, 4)$ and let α, β, γ be positive numbers such that $1 \leq \beta < \alpha$, $\beta \leq \gamma$, $\gamma > 1$. Denote by $\mathfrak{M}(\alpha, \beta, \gamma, a)$ the set of functions $f \in \mathfrak{A}$ such that:

$$f(w,p,r) \geq a_1|w|^\alpha - a_2|p|^\beta + a_3|r|^\gamma - a_4, \ (w,p,r) \in R^3; \tag{3.1}$$

$$f, \ \partial f/\partial p \in C^2, \ \partial f/\partial r \in C^3, \ \partial^2 f/\partial r^2(w,p,r) > 0 \text{ for all } (w,p,r) \in R^3;$$

there is a monotone increasing function $M_f\colon [0,\infty) \to [0,\infty)$ such that for every $(w,p,r) \in R^3$

$$\sup\{f(w,p,r), \ |\partial f/\partial w(w,p,r)|, \ |\partial f/\partial p(w,p,r)|, \ |\partial f/\partial r(w,p,r)|\} \leq$$

$$M_f(|w| + |p|)(1 + |r|^\gamma).$$

Denote by $\overline{\mathfrak{M}}(\alpha, \beta, \gamma, a)$ the closure of $\mathfrak{M}(\alpha, \beta, \gamma, a)$ in $\mathfrak{A}$ and consider any $f \in \overline{\mathfrak{M}}(\alpha, \beta, \gamma, a)$. Of special interest is the minimal long-run average cost growth rate

$$\mu(f) = \inf\{\liminf_{T\to+\infty} T^{-1} \int_0^T f(w(t), w'(t), w''(t))dt\colon w \in A_x\}. \tag{3.2}$$

It is easy to verify that $\mu(f)$ is well defined and is independent of the initial vector x. A function $w \in W^{2,1}_{loc}[0,\infty)$ is called an (f)-good configuration if the function $\phi^f_w\colon T \to \int_0^T [f(w(t), w'(t), w''(t)) - \mu(f)]dt$, $T \in (0,\infty)$ is bounded. For every $w \in W^{2,1}_{loc}[0,\infty)$ the function ϕ^f_w is either bounded or diverges to $+\infty$ as $T \to +\infty$ and moreover, if ϕ^f_w is a bounded function, then $\sup\{|(w(t), w'(t))|\colon t \in [0,\infty)\} < \infty$.

Leizarowitz and Mizel [15] established that for every $f \in \mathfrak{M}(\alpha, \beta, \gamma, a)$ satisfying $\mu(f) < \inf\{f(w, 0, s)\colon (w,s) \in R^2\}$ there exists a periodic (f)-good configuration. In Zaslavski [26] we generalized their result and proved the following assertion.

THEOREM 3.1. *Let $f \in \mathfrak{M}(\alpha, \beta, \gamma, a)$. Then there exists an (f)-good configuration v and a number $T > 0$ such that $v(t) = v(t + T)$ for all $t \geq 0$ and $\int_0^T f(v(t), v'(t), v''(t))dt = T\mu(f)$.*

In [15], analyzing the problem (P_∞) Leizarowitz and Mizel considered for each $T > 0$ the function $U^f_T\colon R^2 \times R^2 \to R$ which is defined as follows:

$$U^f_T(x,y) = \inf\left\{\int_0^T f(w(t), w'(t), w''(t)dt\colon w \in A^T_{x,y}\right\}, \tag{3.3}$$

where $A^T_{x,y} = \{v \in W^{2,1}[0,T]\colon (v(0), v'(0)) = x, \ (v(T), v'(T)) = y\}$, and established the following representation formula

$$U^f_T(x,y) = T\mu(f) + \pi^f(x) - \pi^f(y) + \theta^f_T(x,y), \ x, y \in R^2, \ T > 0, \tag{3.4}$$

where π^f: $R^2 \to R$ and $(T, x, y) \to \theta_T^f(x, y)$, $x, y \in R^2$, $T > 0$ are continuous functions,

$$(3.5) \quad \pi^f(x) = \inf\{\liminf_{T\to\infty} \int_0^T [f(w(t), w'(t), w''(t)) - \mu(f)]dt: w \in A_x\}, \; x \in R^2,$$

$\theta_T^f(x, y) \geq 0$ for each $T > 0$, and each $x, y \in R^2$, and for every $T > 0$, and every $x \in R^2$ there is $y \in R^2$ satisfying $\theta_T^f(x, y) = 0$.

Leizarowitz and Mizel established the representation formula for any integrand $f \in \mathfrak{M}(\alpha, \beta, \gamma, a)$, but their result also holds for every $f \in \bar{\mathfrak{M}}(\alpha, \beta, \gamma, a)$ without change in the proofs.

In [27] we investigated the structure of (f)-good configurations and established for a generic $f \in \bar{\mathfrak{M}}(\alpha, \beta, \gamma, a)$ and for every given $x \in R^2$ the existence of a (f)-weakly optimal solution $v \in A_x$. Most studies which are concerned with the existence of optimal solutions on an infinite horizon assume convex integrands f. One new contribution of [27] is in establishing optimal solutions without such convexity assumptions.

3.2. Definitions and main results. We denote by $|\cdot|$ the Euclidean norm in R^n. For $\tau > 0$ and $v \in W^{2,1}[0, \tau]$ we define X_v: $[0, \tau] \to R^2$ as follows:

$$X_v(t) = (v(t), v'(t)), \; t \in [0, \tau].$$

We also use this definition for $v \in W_{loc}^{2,1}[0, \infty)$.

Fix $a = (a_1, a_2, a_3, a_4) \in R^4$ and positive numbers α, β, γ such that $a_i > 0$, $i = 1, 2, 3, 4$, $1 \leq \beta < \alpha$, $\beta \leq \gamma$, $\gamma > 1$. We denote be $\mathfrak{M}_0(\alpha, \beta, \gamma, a)$ the set of continuous functions $f = f(w, p, r)$: $R^3 \to R$ satisfying (3.1) for any $(w, p, r) \in R^3$.

Denote by $\mathfrak{M}_1(\alpha, \beta, \gamma, a)$ the set of all functions $f \in \mathfrak{M}_0(\alpha, \beta, \gamma, a)$ such that:
the function $f(w, p, r)$ is convex in r for all $(w, p) \in R^2$;
the function $\partial f / \partial r$: $R^3 \to R$ is continuous;
there exists a monotone increasing function M_f: $[0, \infty) \to [0, \infty)$ such that

$$f(w, p, r) \leq M_f(|w| + |p|)(1 + |r|^\gamma) \text{ for all } (w, p, r) \in R^3.$$

We consider the topological subspaces $\bar{\mathfrak{M}}(\alpha, \beta, \gamma, a)$, $\bar{\mathfrak{M}}_1(\alpha, \beta, \gamma, a) \subset \mathfrak{A}$ which have the relative topology. The set $\bar{\mathfrak{M}}_1(\alpha, \beta, \gamma, a)$ is the closure of $\mathfrak{M}_1(\alpha, \beta, \gamma, a)$ in $\mathfrak{A}$ and the set $\bar{\mathfrak{M}}(\alpha, \beta, \gamma, a)$ was defined in 3.1.

We consider functionals of the form

$$(3.6) \qquad I^f(T_1, T_2, w) = \int_{T_1}^{T_2} f(w(t), w'(t), w''(t))dt$$

where $-\infty < T_1 < T_2 < +\infty$, $w \in W^{2,1}[T_1, T_2]$ and $f \in \mathfrak{M}_0(\alpha, \beta, \gamma, a)$.

Let $f \in \bar{\mathfrak{M}}_1(\alpha, \beta, \gamma, a)$. A function $v \in W_{loc}^{2,1}[0, \infty)$ is (f)-weakly optimal if

$$\liminf_{T\to\infty}[I^f(0, T, v) - I^f(0, T, w)] \leq 0$$

for all $w \in W^{2,1}_{loc}[0,\infty)$ satisfying $X_w(0) = X_v(0)$.

Of special interest is the minimal long-run average cost growth rate $\mu(f)$ defined by (3.2). By a result of Leizarowitz and Mizel [15, p. 164] $\mu(f) \in (-\infty, f(0,0,0)]$.

A function $v \in W^{2,1}_{loc}[0,\infty)$ is called an (f)-good configuration if $\sup\{|I^f(0,\tau,w) - \tau\mu(f)|: \tau \in (0,\infty)\} < \infty$.

For $f \in \mathfrak{M}_0(\alpha,\beta,\gamma,a)$ and $T > 0$ we consider the function $U^f_T: R^2 \times R^2 \to R$ defined by (3.3). In [15] Leizarowitz and Mizel established the representation formula (3.4) where $\pi^f: R^2 \to R$ is a continuous function defined by (3.5) and $(T,x,y) \to \theta^f_T(x,y)$, $T > 0$, $x, y \in R^2$, is a nonnegative continuous function such that for every $T > 0$ and every $x \in R^2$ there is $y \in R^2$ satisfying $\theta^f_T(x,y) = 0$.

Leizarowitz and Mizel established the representation formula for any integrand $f \in \mathfrak{M}(\alpha,\beta,\gamma,a)$, but their result also holds for every $f \in \bar{\mathfrak{M}}_1(\alpha,\beta,\gamma,a)$ without changes in the proofs.

In [27] we investigated the structure of (f)-good configurations and established for a generic $f \in \bar{\mathfrak{M}}(\alpha,\beta,\gamma,a)$ and every $x \in R^2$ the existence of a (f)-weakly optimal function $v \in A_x$.

For a function $w \in W^{2,1}_{loc}[0,\infty)$ we denote by $\Omega(w)$ the set of all points $z \in R^2$ such that $X_w(t_j) \to z$ as $j \to \infty$ for some sequence of numbers $t_j \to \infty$.

We denote $d(x,B) = \inf\{|x-y|: y \in B\}$ for $x \in R^n$, $B \subset R^n$ and denote by $\mathrm{dist}(A,B)$ the Hausdorff metric for two sets $A \subset R^n$ and $B \subset R^n$.

A function $w \in W^{2,1}_{loc}(-\infty,+\infty)$ is called almost subperiodic if for every $\epsilon > 0$ there exists a number $T_\epsilon > 0$ such that for each $\tau_1, \tau_2 \in R$ there is $T \in [0,T_\epsilon)$ which satisfies the conditions

$$|X_w(\tau_1 + t) - X_w(\tau_2 + t + T)| \le \epsilon, \; t \in [0, T_\epsilon - T] \tag{3.7}$$

$$|X_w(\tau_1 + t + T_\epsilon - T) - X_w(\tau_2 + t)| \le \epsilon, \; t \in [0,T].$$

A function $w \in W^{2,1}_{loc}[0,\infty)$ is called asymptotically almost subperiodic if for any $\epsilon > 0$ there exist numbers $T_\epsilon > 0$ and $t_\epsilon > 0$ such that for every $\tau_1 \ge t_\epsilon$ and every $\tau_2 \ge t_\epsilon$ there is $T \in [0,T_\epsilon)$ which satisfies (3.7).

In [27] we proved the existence of a set $F \subset \bar{\mathfrak{M}}(\alpha,\beta,\gamma,a)$ which is a countable intersection of open everywhere dense sets in $\bar{\mathfrak{M}}(\alpha,\beta,\gamma,a)$ and for which the following theorems are valid.

THEOREM 3.2.
1. Let $f \in F$. Then there is a compact set $H(f) \subset R^2$ such that $\Omega(w) = H(f)$ for any (f)-good configuration w. Moreover, for every (f)-good configuration w and every positive number ϵ there exist $T_\epsilon > 0$ and $t_\epsilon > 0$ such that

$$dist(\{(w(t),w'(t)): t \in [\tau, \tau + T_\epsilon]\}, H(f)) \le \epsilon \text{ for any } t \ge t_\epsilon.$$

2. For every $f \in F$ and every $\epsilon > 0$ there exist a number $T > 0$ and a function $v \in C^2[0,\infty)$ such that $v(t+T) = v(t)$ for all $t \ge 0$, and

$$dist(\{(v(t),v'(t)): t \in [0,T]\}, H(f)) \le \epsilon.$$

3. Let $f \in F$. Then every (f)-good configuration is asymptotically almost subperiodic.

4. Let $f \in F$ and let ϵ be a positive number. Then there exist a neighborhood $\mathbf{U}$ of f in $\bar{\mathfrak{M}}_1(\alpha, \beta, \gamma, a)$ and a number $T > 0$ such that for every $g \in \mathbf{U}$ and every (g)-good configuration w

$$dist(\{(w(t), w'(t)) \colon t \in [\tau, \tau + T]\}, H(f)) \leq \epsilon \text{ for all large enough } \tau.$$

For $f \in F$ we may consider $H(f)$ as an analog of a turnpike set [8,16, 21]. Assertion 1 of Theorem 3.2 establishes that for $f \in F$ all (f)-good configurations converge to the turnpike set $H(f)$. Assertion 2 shows that for $f \in F$ the set $H(f)$ is approximated by periodic curves in R^2 and Assertion 4 of Theorem 3.2 shows that for every g belonging to a small neighborhood of f and every (g)-good configuration w, the set $\Omega(w)$ is close enough to $H(f)$ in the Hausdorff metric. If we think of $H(f)$ as an analog of a turnpike set Assertion 4 yields the stability of the turnpike phenomenon.

THEOREM 3.3. *Let $f \in F$, $x \in R^2$. Then there exists $v \in A_x$ such that*

$$I^f(T_1, T_2, v) = (T_2 - T_1)\mu(f) + \pi^f((v(T_1), v'(T_1)) - \pi^f((v(T_2), v'(T_2)) \tag{3.8}$$

for each T_1, T_2 satisfying $0 \leq T_1 < T_2$. Moreover, for every $v \in A_x$ which satisfies (3.8) for each $T_1 \geq 0$, $T_2 > T_1$ there exists a sequence of numbers $t_j \to \infty$ as $j \to \infty$ such that:

$$\limsup_{j \to \infty} [I^f(0, t_j, v) - I^f(0, t_j, w)] \leq 0 \text{ for all } w \in A_x;$$

$$\text{if } w \in A_x \text{ and } \limsup_{j \to \infty} [I^f(0, t_j, v) - I^f(0, t_j, w)] = 0, \text{ then}$$

$$I^f(T_1, T_2, w) = (T_2 - T_1)\mu(f) + \pi^f((w(T_1), w'(T_1)) - \pi^f((w(T_2), w'(T_2))$$

for each T_1, T_2 satisfying $0 \leq T_1 < T_2$.

Theorem 3.3 shows that for every $f \in F$ and every initial value $x \in R^2$ there exists an (f)-weakly optimal solution $v \in A_x$ satisfying (3.8) for each T_1, T_2 such that $0 \leq T_1 < T_2$. This theorem also implies that if $f \in F$ and $v \in W^{2,1}_{loc}[0, \infty)$ is an (f)-overtaking optimal function, then (3.8) holds for each T_1, T_2 satisfying $0 \leq T_1 < T_2$.

THEOREM 3.4. *Let $C(R^n)$ be the space of all continuous functions $g \colon R^n \to R$ with the topology of uniform convergence on bounded subsets of R^n. We define $L \colon \bar{\mathfrak{M}}(\alpha, \beta, \gamma, a) \to R \times C(R^2)$ by*

$$L(f) = (\mu(f), \pi^f), \ f \in \bar{\mathfrak{M}}(\alpha, \beta, \gamma, a).$$

Then the set of continuity points of the operator L contains F.

For $f \in \bar{\mathfrak{M}}(\alpha, \beta, \gamma, a)$ and $x \in R^2$ we set

$$A(f, x) = \{v \in A_x \colon \text{(3.8) holds for each } T_1, T_2 \text{ satisfying } 0 \leq T_1 < T_2\}.$$

Let $f \in F$. Theorem 3.5 establishes that for each g belonging to some small neighborhood of f in $\overline{\mathfrak{M}}_1(\alpha,\beta,\gamma,a)$, each x belonging to some small neighborhood of $H(f)$ and each $v \in A(g,x)$, the point $(v(t),v'(t))$ is contained in a small neighborhood of $H(f)$ for all $t \geq 0$.

Let $f \in F$ and $K > 0$. By Theorem 3.6 for every g belonging to some small neighborhood of f in $\overline{\mathfrak{M}}_1(\alpha,\beta,\gamma,a)$, every $x \in R^2$ satisfying $|x| \leq K$, and every $v \in A(g,x)$, the point $(v(t),v'(t))$ is contained in a small neighborhood of $H(f)$ for all $t \geq Q$, where Q is a constant which depends on K and the neighborhoods, but does not depend on g and x.

THEOREM 3.5. *Let $f \in F$, $\epsilon > 0$. Then there exist a neighborhood $\mathbf{U}$ of f in $\overline{\mathfrak{M}}_1(\alpha,\beta,\gamma,a)$ and numbers $l > 0$, $\delta \in (0,\epsilon)$ such that for every $g \in \mathbf{U}$, each $x \in R^2$ satisfying $d(x,H(f)) \leq \delta$, every $v \in A(g,x)$, and every $T \geq 0$,*

$$dist(\{(v(t+T),v'(t+T))\colon t \in [0,l]\},H(f)) \leq \epsilon. \tag{3.9}$$

THEOREM 3.6. *Let $f \in F$, $\epsilon > 0$, $K > 0$. Then there exist a neighborhood $\mathbf{U}$ of f in $\overline{\mathfrak{M}}_1(\alpha,\beta,\gamma,a)$ and numbers $l > 0$, $Q > 0$ such that for every $g \in \mathbf{U}$, every $x \in R^2$ satisfying $|x| \leq K$ and every $v \in A(g,x)$, Eq. (3.9) holds for all $T \geq Q$.*

COROLLARY 3.1. *Let $f \in F$, $v \in W^{2,1}_{loc}(-\infty,+\infty)$. Assume that (3.8) holds for each T_1, T_2 satisfying $-\infty < T_1 < T_2 < +\infty$ and $\liminf_{t\to-\infty} |(v(t),v'(t))| < \infty$. Then $(v(t),v'(t)) \in H(f)$ for all $t \in R$.*

THEOREM 3.7.

1. Let $f \in F, x \in H(f)$. Then there exists $v \in W^{2,1}_{loc}(-\infty,+\infty)$ such that $(v(t),v'(t)) \in H(f)$ for all $t \in R$, $(v(0),v'(0)) = x$ and (3.8) holds for each T_1, T_2 satisfying $-\infty < T_1 < T_2 < +\infty$.

2. Let $f \in F$. Then each function $v \in W^{2,1}_{loc}(-\infty,+\infty)$ such that (3.8) holds for each T_1, T_2 satisfying $-\infty < T_1 < T_2 < +\infty$ and $\liminf_{t\to-\infty} |(v(t),v'(t))| < \infty$, is almost subperiodic.

EXAMPLE. Set $a_i = 1$, $i = 1,2,3,4$, $\alpha = 4$, $\beta,\gamma = 2$. Consider the space of functions $\overline{\mathfrak{M}}(\alpha,\beta,\gamma,a)$ and let F be as assumed in this section. Consider an integrand

$$f(w,p,r) = 8w^2(w-1)^2 + p^2 + r^2 + b, \text{ where } b > 0.$$

It is easy to see that $f \in \mathfrak{M}(\alpha,\beta,\gamma,a)$ for all b large enough, $\mu(f) = b$ and the functions $w_1(t) = 0$, $t \in [0,\infty)$ and $w_2(t) = 1$, $t \in [0,\infty)$ are (f)-good configurations. Therefore $f \in \overline{\mathfrak{M}}(\alpha,\beta,\gamma,a) \setminus F$ for all b large enough.

3.3. The turnpike property. Fix $a = (a_1,a_2,a_3,a_4) \in R^4$ and positive numbers α,β,γ such that $a_i > 0$, $i = 1,2,3,4$, $1 \leq \beta < \alpha$, $\beta \leq \gamma$, $\gamma > 1$. We consider the topological subspaces $\overline{\mathfrak{M}}(\alpha,\beta,\gamma,a)$, $\overline{\mathfrak{M}}_1(\alpha,\beta,\gamma,a) \subset \mathfrak{A}$ defined in sections 3.1 and 3.2.

For every $f \in \mathfrak{M}_0(\alpha,\beta,\gamma,a)$, every $x \in R^2$, and every $T > 0$, we set

$$\sigma(f,x,T) = \inf\{U^f_T(x,y)\colon y \in R^2\}.$$

For a function $w \in W^{2,1}_{loc}[0,\infty)$ we denote by $\Omega(w)$ the set of all points $z \in R^2$ such that $X_w(t_j) \to z$ as $j \to \infty$ for some sequence of numbers $t_j \to \infty$.

We denote $d(x,B) = \inf\{|x-y|: y \in B\}$ for $x \in R^n$, $B \subset R^n$, and denote by $\mathrm{dist}(A,B)$ the Hausdorff metric for two sets $A \subset R^n$ and $B \subset R^n$.

In [29] we established the existence of a set $F \subset \overline{\mathfrak{M}}(\alpha,\beta,\gamma,a)$ which is a countable intersection of open everywhere dense sets in $\overline{\mathfrak{M}}(\alpha,\beta,\gamma,a)$ and for which the following theorems are valid.

THEOREM 3.8. *Let $f \in F$. Then there exists a compact set $H(f) \subset R^2$ such that $\Omega(w) = H(f)$ for any (f)-good configuration w.*

Theorem 3.8 describes the limit behavior of (f)-good configurations for a generic $f \in \overline{\mathfrak{M}}(\alpha,\beta,\gamma,a)$. The following results show that for a generic $f \in \overline{\mathfrak{M}}(\alpha,\beta,\gamma,a)$ the strong turnpike property holds with the set $H(f)$ being the attractor.

THEOREM 3.9. *Let $f \in F$ and $\epsilon, K > 0$. Then there exist a neighborhood $\mathbf{U}$ of f in $\overline{\mathfrak{M}}_1(\alpha,\beta,\gamma,a)$ and numbers $l_0 > l > 0$, $K_* > K$, $\delta > 0$ such that for each $g \in \mathbf{U}$, each $\tau \geq 2l_0$ and each $v \in W^{2,1}[0,\tau]$ which satisfies*

$$|(v(0),v'(0))|,\ |(v(\tau),v'(\tau))| \leq K, \text{ and}$$

$$I^g(0,\tau,v) \leq U^g_\tau((v(0),v'(0)),(v(\tau),v'(\tau))) + \delta,$$

the relation $|(v(t),v'(t))| \leq K_*$ holds for all $t \in [0,\tau]$, and

$$\mathrm{dist}(H(f),\{(v(t),v'(t)): t \in [T,T+l]\}) \leq \epsilon \tag{3.10}$$

for each $T \in [l_0, \tau - l_0]$.

THEOREM 3.10. *Let $f \in F$ and $\epsilon, K > 0$. Then there exist a neighborhood $\mathbf{U}$ of f in $\overline{\mathfrak{M}}_1(\alpha,\beta,\gamma,a)$ and numbers $l_0 > l > 0$, $K_* > K$, $\delta > 0$ such that for each $g \in \mathbf{U}$, each $\tau \geq 2l_0$ and each $v \in W^{2,1}[0,\tau]$ which satisfies*

$$|(v(0),v'(0))| \leq K,\ I^g(0,\tau,v) \leq \sigma(g,(v(0),v'(0)),\tau) + \delta,$$

the relation $|(v(0),v'(0))| \leq K_*$ holds for all $t \in [0,\tau]$ and (3.10) holds for each $T \in [l_0, \tau - l_0]$.

3.4. Spaces of smooth integrands. Fix $a = (a_1,a_2,a_3,a_4) \in R^4$ and positive numbers α,β,γ such that $a_i > 0$, $i = 1,2,3,4$ and $1 \leq \beta < \alpha$, $\beta \leq \gamma$, $\gamma > 1$. Let $k \geq 2$ be an integer. Denote by $\mathfrak{M}^0_k(\alpha,\beta,\gamma,a)$ the set of all integrands $f = f(w,p,r) \in C^k(R^3)$ such that:

$$f(w,p,r) \geq a_1|w|^\alpha - a_2|p|^\beta + a_3|r|^\gamma - a_4,\ (w,p,r) \in R^3;$$

there is an increasing function $M_f\colon [0,\infty) \to [0,\infty)$ such that for every $(w,p,r) \in R^3$

$$\sup\{f(w,p,r),\ |\partial f/\partial w(w,p,r)|,\ |\partial f/\partial p(w,p,r)|,\ |\partial f/\partial r(w,p,r)|\} \leq$$

$$M_f(|w|+|p|)(1+|r|^\gamma); \quad \partial f/\partial p \in C^2, \ \partial f/\partial r \in C^3.$$

For $q=(q_1,q_2,q_3)\in\{0,\dots k\}^3$ such that $q_1+q_2+q_3\le k$ and $f\in\mathfrak{M}_k^0(\alpha,\beta,\gamma,a)$ we set

$$|q|=q_1+q_2+q_3, \ D^q f=\partial^{|q|}f/\partial w^{q_1}\partial p^{q_2}\partial r^{q_3}. \text{ (Here } D^0 f=f).$$

For the set $\mathfrak{M}_k^0(\alpha,\beta,\gamma,a)$ we consider the uniformity which is determined by the following base

$$E(N,\epsilon,\Gamma)=\{(f,g)\in\mathfrak{M}_k^0(\alpha,\beta,\gamma,a)\times\mathfrak{M}_k^0(\alpha,\beta,\gamma,a):$$

$$|D^q f(x_1,x_2,x_3)-D^q g(x_1,x_2,x_3)|\le\epsilon \ (x_i\in R, \ |x_i|\le N, \ i=1,2,3),$$

$$\text{for each } q\in\{0,\dots k\}^3 \text{ satisfying } |q|\le k, \ \text{for each } q=(q_1,q_2,q_3)\in\{0,1,2,3\}^3$$

$$\text{such that } q_1\ge 1, \ |q|=3 \text{ and for each } q=(q_1,q_2,q_3)\in$$

$$\{0,1,2,3,4\}^3 \text{ such that } q_3\ge 1, \ |q|\in\{3,4\},$$

$$(|D^q f(x_1,x_2,x_3)|+1)(|D^q g(x_1,x_2,x_3)|+1)^{-1}\in[\Gamma^{-1},\Gamma]$$

$$((x_1,x_2,x_3)\in R^3, |x_1|,|x_2|\le N), \ q\in\{0,1\}^3, \ |q|\le 1\},$$

where $N,\epsilon>0$, $\Gamma>1$. Clearly, the uniform space $\mathfrak{M}_k^0(\alpha,\beta,\gamma,a)$ is Hausdorff and has a countable base. Therefore $\mathfrak{M}_k^0(\alpha,\beta,\gamma,a)$ is metrizable [11]. It is easy to verify that the uniform space $\mathfrak{M}_k^0(\alpha,\beta,\gamma,a)$ is complete.

Let $\rho(\cdot,\cdot)$: $\bar{\mathfrak{M}}_1(\alpha,\beta,\gamma,a)\times\bar{\mathfrak{M}}_1(\alpha,\beta,\gamma,a)\to R$ be a metric which generates the uniformity for $\bar{\mathfrak{M}}_1(\alpha,\beta,\gamma,a)$. Set

$$\mathfrak{M}_k(\alpha,\beta,\gamma,a)=\{f\in\mathfrak{M}_k^0(\alpha,\beta,\gamma,a)\text{: } \partial^2 f/\partial r^2(w,p,r)>0 \text{ for all } (w,p,r)\in R^3\}$$

and denote by $\bar{\mathfrak{M}}_k(\alpha,\beta,\gamma,a)$ the closure of $\mathfrak{M}_k(\alpha,\beta,\gamma,a)$ in $\mathfrak{M}_k^0(\alpha,\beta,\gamma,a)$. Clearly

$$\mathfrak{M}_k(\alpha,\beta,\gamma,a)=\mathfrak{M}_k^0(\alpha,\beta,\gamma,a)\cap\mathfrak{M}(\alpha,\beta,\gamma,a), \ \bar{\mathfrak{M}}_k(\alpha,\beta,\gamma,a)\subset\bar{\mathfrak{M}}(\alpha,\beta,\gamma,a),$$

and $\mathfrak{M}_k(\alpha,\beta,\gamma,a)$ is a countable intersection of open everywhere dense sets in $\bar{\mathfrak{M}}_k(\alpha,\beta,\gamma,a)$.

In [29] we considered the topological subspace $\bar{\mathfrak{M}}_k(\alpha,\beta,\gamma,a)\subset\mathfrak{M}_k^0(\alpha,\beta,\gamma,a)$ with the relative topology and established the existence of a set $F_k\subset\mathfrak{M}_k(\alpha,\beta,\gamma,a)$ which is a countable intersection of open everywhere dense sets in $\bar{\mathfrak{M}}_k(\alpha,\beta,\gamma,a)$ and for which the following theorems are valid.

THEOREM 3.11. *Let $f \in F_k$. Then there exist a function $v_f \in C^5(R) \cap C^{k+1}(R)$ and a number $T_f > 0$ such that the following assertions hold:*

1. $v_f(t+T_f) = v_f(t)$ for all $t \in R$ and $I^f(0, T_f, v_f) = T_f \mu(f)$.

2. If $\mu(f) < \inf\{f(z,0,0) \colon z \in R\}$ then $(v_f(t_1), v_f'(t_1)) \neq (v_f(t_2), v_f'(t_2))$ for each t_1, t_2 satisfying $0 \leq t_1 < t_2 < T_f$. Otherwise $v_f(t) = v_f(0)$ for all $t \in R$.

3. For every periodic (f)-good configuration w there exists a number τ such that $w(t) = v_f(t+\tau)$ for all $t \in [0, \infty)$.

4. For every $\epsilon > 0$ there exists a neighborhood $\mathbf{U}$ of f in $\bar{\mathfrak{M}}_1(\alpha, \beta, \gamma, a)$ such that for every $g \in \mathbf{U}$, every (g)-good configuration w and every large enough τ there exists $h \geq 0$ for which

$$\sup\{|(w(t), w'(t)) - (v_f(t+h), (v_f'(t+h))| \colon t \in [\tau, \tau + T_f]\} \leq \epsilon. \tag{3.11}$$

THEOREM 3.12. *Let $f \in F_k$, and let $\epsilon, K > 0$. Then there exist a neighborhood $\mathbf{U}$ of f in $\bar{\mathfrak{M}}_1(\alpha, \beta, \gamma, a)$ and numbers $Q > 0$, $\delta \in (0, \epsilon)$ such that:*

1. For every $g \in \mathbf{U}$, every $x \in R^2$ satisfying $|x| \leq K$, every $w \in A(g, x)$ and every $\tau \geq Q$, Eq. (3.11) holds with some $h \geq 0$.

2. For every $g \in \mathbf{U}$, every $x \in R^2$ satisfying $d(x, \{(v_f(t), v_f'(t)) \colon t \in [0, T_f]\}) \leq \delta$, every $w \in A(g, x)$ and every $\tau \geq 0$, Eq. (3.11) holds with some $h \geq 0$.

COROLLARY 3.2. *Let $f \in F_k$ and $w \in W^{2,1}_{loc}(R)$. Suppose that*

$$I^f(t_1, t_2, w) = \mu(f)(t_2 - t_1) + \pi^f((w(t_1), w'(t_1)) - \pi^f((w(t_2), w'(t_2))$$

for each t_1, t_2 satisfying $-\infty < t_1 < t_2 < +\infty$ and $\liminf_{t \to -\infty} |(w(t), w'(t))| < \infty$. Then there exists a number h such that $w(t) = v_f(t+h)$ for all $t \in R$.

THEOREM 3.13. *Let $f \in F_k$ and $\epsilon, K > 0$. Then there exist a neighborhood $\mathbf{U}$ of f in $\bar{\mathfrak{M}}_1(\alpha, \beta, \gamma, a)$ and numbers $l > T_f$, $K_* > K$, $\delta > 0$ such that:*

1. For each $g \in \mathbf{U}$, each $T \geq 2l$ and each $w \in W^{2,1}[0, T]$ which satisfies

$$|(w(0), w'(0))|, \ |(w(T), w'(T))| \leq K,$$

$$I^g(0, T, w) \leq U_T^g((w(0), w'(0)), (w(T), w'(T))) + \delta$$

the relation $|(w(t), w'(t))| \leq K_*$ holds for all $t \in [0, T]$, and for each $\tau \in [l, T - l]$ Eq. (3.11) holds with some $h \geq 0$.

2. For each $g \in \mathbf{U}$, each $T \geq 2l$ and each $w \in W^{2,1}[0, T]$ which satisfies

$$|(w(0), w'(0))| \leq K, \ I^g(0, T, w) \leq \sigma(g, (w(0), w'(0)), T) + \delta,$$

the relation $|(w(t), w'(t))| \leq K_*$ holds for all $t \in [0, T]$ and for each $\tau \in [l, T - l]$ Eq. (3.11) holds with some $h \geq 0$.

Analogs of Theorems 3.3 and 3.4 hold for every $f \in F_k$. By the analog of Theorem 3.3 for every $f \in F_k$ and every $x \in R^2$ there exists an (f)-weakly optimal solution $v \in A_x$ and the analog of Theorem 3.4 establishes that every $f \in F_k$ is a continuity point of the operator

$$g \to (\mu(g), \pi^g), \ g \in \bar{\mathfrak{M}}(\alpha, \beta, \gamma, a).$$

3.5. Nonexistence of overtaking optimal functions. Set $\alpha = 4, \beta, \gamma = 2$. We will construct a function $\bar{g} \in \mathfrak{M}(\alpha, \beta, \gamma, a)$ where $a = (a_1, a_2, a_3, a_4), a_i > 0, i = 1, 2, 3, 4$, which has the following properties:

for each $x \in R^2$ there exists a $(\bar{g})$-weakly optimal function $v \in W^{2,1}_{loc}[0, \infty)$ satisfying $X_v(0) = x$;

there exists a neighborhood $\mathbf{U}$ of $\bar{g}$ in $\bar{\mathfrak{M}}_1(\alpha, \beta, \gamma, a)$ such that for each $f \in \mathbf{U}$ there are no (f)-overtaking optimal functions.

For any $b = (b_i)_{i=1}^4$ such that $b_i > 0, i = 1, 2, 3, 4$ we define $g_b \in \mathfrak{M}(\alpha, \beta, \gamma, b)$ as

$$g_b(w, p, r) = b_1 w^4 - b_2 p^2 + b_3 r^2 - b_4, (w, p, r) \in R^3. \tag{3.12}$$

Clearly there exists $b^* = (b_1^*, b_2^*, b_3^*, b_4^*) \in R^4$ such that $b_i^* > 0, i = 1, 2, 3, 4$ and

$$\mu(g_{b^*}) < \inf\{g_{b^*}(x, 0, 0) : x \in R^1\} = 0. \tag{3.13}$$

There exists $w_* \in W^{2,1}_{loc}(R^1), T_* > 0$ such that

$$w_*(t + T_*) = w_*(t), t \in R^1, \quad I^{g_{b^*}}(0, T_*, w_*) = T_* \mu(g_{b^*}). \tag{3.14}$$

By (3.13) there is $t \in R^1$ for which $w_*(t) \neq w_*(0)$. Set

$$g = g_{b^*}. \tag{3.15}$$

By Proposition 4.3 in [26] we may assume that $X_{w_*}(t_1) \neq X_{w_*}(t_2)$ for each $t_1, t_2 \in [0, T_*)$ satisfying $t_1 < t_2$. There exists $z_0 \in R^2$ such that

$$z_0 \in \{X_{w_*}(t) : t \in R^1\}, \quad \pi^g(z_0) = \sup\{\pi^g(X_{w_*}(t)) : t \in R^1\}. \tag{3.16}$$

We may assume that

$$X_{w_*}(0) = z_0. \tag{3.17}$$

By Proposition 4.1 in [27] there exists a bounded nonnegative function $\phi \in C^\infty(R^2)$ such that

$$\{x \in R^2 : \phi(x) = 0\} = \{X_{w_*}(t) : t \in R^1\} \tag{3.18}$$

and $\partial^{p+q}\phi / \partial x_1^q \partial x_2^p : R^2 \to R^1$ is a bounded function for any nonnegative integers p, q. We define

$$g_0(w, p, r) = g(w, p, r) + \phi(w, p), \quad (w, p, r) \in R^3. \tag{3.19}$$

Clearly

$$g_0 \in \mathfrak{M}(\alpha, \beta, \gamma, b^*), \quad \mu(g_0) = \mu(g). \tag{3.20}$$

Choose numbers γ_0, d_0, d_1 such that

$$\gamma_0 \in (0,T_*), w_*(0) \neq w_*(T_* - \gamma_0), \quad d_0 > 4 + \sup\{|\pi^g(z)| : z \in \{X_{w_*}(t) : t \in R^1\}\} + \sup\{|X_{w_*}(t)| : t \in R^1\}, \tag{3.21}$$
$$d_1 > \sup\{|U^{g_0}_{\gamma_0}(x,y)| : x, y \in R^2, |x|, |y| \leq 2d_0 + 8\}.$$

There exists a bounded nonnegative function $\psi \in C^\infty(R^1)$ such that

$$\{x \in R^1 : \psi(x) = 0\} = \{w_*(0)\}, \tag{3.22}$$

and for each integer $p \geq 1$ the function $d^p\psi/dx^p : R^1 \to R^1$ is bounded. We may assume without loss of generality that

$$\psi(w_*(T_* - \gamma_0)) > 6(1 + d_0 + d_1). \tag{3.23}$$

We define

$$\bar{g}(w,p,r) = g_0(w,p,r) - \psi'(w)p, \quad (w,p,r) \in R^3. \tag{3.24}$$

There exists $a = (a_1, a_2, a_3, a_4) \in R^4$ such that

$$a_i > b_i^*, i = 1,2,3,4, \quad \bar{g} \in \mathfrak{M}(\alpha, \beta, \gamma, a). \tag{3.25}$$

Clearly for each $T > 0$ and each $v \in W^{2,1}[0,T]$

$$I^{\bar{g}}(0,T,v) = I^{g_0}(0,T,v) - \int_0^T \psi'(v(t))v'(t)dt = \tag{3.26}$$
$$I^{g_0}(0,T,v) - \psi(v(T)) + \psi(v(0)), \quad \mu(\bar{g}) = \mu(g_0) = \mu(g).$$

It follows from Proposition 4.2 in [27] and the definition of g_0 (see (3.18), (3.19)) that the following properties hold:

$\pi^g(x) \to +\infty$ as $|x| \to \infty$;

$U^{g_0}_t(x,y) \geq t\mu(g_0) + \pi^g(x) - \pi^g(y)$ for each $t > 0$ and each $x, y \in R^2$;

for every $u \in W^{2,1}[0,T_*]$ the relation

$$I^{g_0}(0,T_*,u) = T_*\mu(g_0) + \pi^g(X_u(0)) - \pi^g(X_u(T_*))$$

holds if and only if there is $\tau \in [0,T_*)$ such that $u(t) = w_*(t+\tau)$ for all $t \in [0,T_*]$.

It follows from these properties and (3.26) that for each $t > 0$ and $x = (x_1, x_2)$, $y = (y_1, y_2) \in R^2$,

$$U^{\bar{g}}_t(x,y) = U^{g_0}_t(x,y) - \psi(y_1) + \psi(x_1) \tag{3.27}$$
$$\geq \psi(x_1) - \psi(y_1) + \pi^g(x) - \pi^g(y) + t\mu(\bar{g}),$$

and for every $u \in W^{2,1}[0,T_*]$ the relation

$$I^{\bar{g}}(0,T_*,u) = T_*\mu(\bar{g}) + \pi^g(X_u(0)) - \tag{3.28}$$
$$\pi^g(X_u(T_*)) + \psi(u(0)) - \psi(u(T_*))$$

holds if and only if there is $\tau \in [0,T_*)$ such that $u(t) = w_*(t+\tau)$ for all $t \in [0,T_*]$.

This implies that Propositions 4.4-4.6 and Lemmas 4.1-4.19 in [27] hold with $f = \bar{g}, w = w_*, T = T_*, \quad \pi(x_1,x_2) = \pi^g(x_1,x_2) + \psi(x_1), (x_1, x_2 \in R^2)$, and that for each $x \in R^2$ there is a $(\bar{g})$-weakly optimal solution $v \in A_x$.

We will establish the following result.

THEOREM 3.14. *There exists a neighborhood* $\mathbf{U}$ *of* $\bar{g}$ *in* $\bar{\mathfrak{M}}_1(\alpha,\beta,\gamma,a)$ *such that for each* $f \in \mathbf{U}$ *there are no* (f)*-overtaking optimal functions.*

Proof. Choose a number $\epsilon_0 \in (0,1)$ such that for each $x_1, x_2, y_1, y_2 \in R^2$ satisfying $|x_i|, |y_i| \le d_0, |x_i - y_i| \le \epsilon_0, i = 1, 2$, the following relation holds:

$$|U_{\gamma_0}^{\bar{g}}(x_1,x_2) - U_{\gamma_0}^{\bar{g}}(y_1,y_2)| \le 1. \tag{3.29}$$

By Lemma 4.10 in [27] there exists a neighborhood $\mathbf{U}_1$ of $\bar{g}$ in $\bar{\mathfrak{M}}_1(\alpha,\beta,\gamma,a)$ such that for each $f \in \mathbf{U}_1$ and each (f)-good configuration v there is $\tau_0 > 0$ satisfying the following condition: for any $\tau \ge \tau_0$ there is $h \in [0, T_*)$ such that

$$|X_v(t+\tau) - X_{w_*}(t+h)| \le \epsilon_0, \quad t \in [0, 6T_*]. \tag{3.30}$$

By Proposition 3.3 in [27] there exists a neighborhood $\mathbf{U}$ of $\bar{g}$ in $\bar{\mathfrak{M}}_1(\alpha,\beta,\gamma,a)$ such that $\mathbf{U} \subset \mathbf{U}_1$ and for each $f \in \mathbf{U}$, each $\tau \in [8^{-1}\gamma_0, 8T_*]$ and each $x, y \in R^2$ satisfying $|x|, |y| \le d_0 + 2$ the following relation holds:

$$|U_\tau^f(x,y) - U_\tau^{\bar{g}}(x,y)| \le 1. \tag{3.31}$$

Assume that $f \in \mathbf{U}$ and $v \in W_{loc}^{2,1}[0,\infty)$. We will show that there exists $u \in W_{loc}^{2,1}[0,\infty)$ such that

$$X_u(0) = X_v(0), \quad \limsup_{T\to\infty}[I^f(0,T,v) - I^f(0,T,u)] \ge 1. \tag{3.32}$$

We may assume without loss of generality that v is a $(\bar{g})$-good configuration. By the definition of $\mathbf{U}_1$ there is $\tau_0 > 0$ such that for any $\tau \ge \tau_0$ there is $h \in [0, T_*)$ such that (3.30) holds. We may assume that

$$\tau_0 = 0. \tag{3.33}$$

There exists $u \in W_{loc}^{2,1}[0,\infty)$ such that

$$X_u(0) = X_v(0), u(t) = v(t-\gamma_0), t \in [\gamma_0, \infty), \tag{3.34}$$

$$I^f(0,\gamma_0,u) = U_{\gamma_0}^f(X_u(0), X_u(\gamma_0)).$$

Let $\tau \ge \tau_0 + 2\gamma_0 + 2$. It follows from the definition of τ_0 that there is $h \in [0, T_*)$ such that (3.30) holds. This implies that

$$|X_v(\tau+2T_*-h-\gamma_0) - X_{w_*}(T_*-\gamma_0)| \le \epsilon_0, \quad |X_v(\tau+2T_*-h) - X_{w_*}(0)| \le \epsilon_0. \tag{3.35}$$

By (3.34)

$$I^f(0,\tau+2T_*-h,v) - I^f(0,\tau+2T_*-h,u) \ge I^f(\tau+2T_*-h-\gamma_0,\tau+2T_*-h,v) \tag{3.36}$$

$$-U^f_{\gamma_0}(X_u(0), X_u(\gamma_0)) \geq U^f_{\gamma_0}(X_v(\tau + 2T_* - h - \gamma_0), X_v(\tau + 2T_* - h))$$
$$-U^f_{\gamma_0}(X_u(0), X_u(\gamma_0)).$$

We will estimate

$$U^f_{\gamma_0}(X_v(\tau + 2T_* - h - \gamma_0), X_v(\tau + 2T_* - h)) \text{ and } U^f_{\gamma_0}(X_u(0), X_u(\gamma_0)).$$

It follows from (3.34),the definition of τ_0, (3.21),(3.33) that

$$X_u(0), X_u(\gamma_0) = X_v(0), |X_v(0)| < d_0, \quad |U^{g_0}_{\gamma_0}(X_u(0), X_u(\gamma_0))| < d_1. \tag{3.37}$$

By these relations and (3.26), $|U^{\bar{g}}_{\gamma_0}(X_u(0), X_u(\gamma_0))| < d_1$. It follows from this inequality and the definition of $\mathbf{U}$ (see (3.31)) that

$$|U^f_{\gamma_0}(X_u(0), X_u(\gamma_0))| \leq d_1 + 1. \tag{3.38}$$

By (3.35), the definition of ϵ_0 (see (3.29)),(3.21), (3.22),(3.23), and (3.26)

$$U^{\bar{g}}_{\gamma_0}(X_v(\tau + 2T_* - h - \gamma_0), X_v(\tau + 2T_* - h)) \geq U^{\bar{g}}_{\gamma_0}(X_{w_*}(T_* - \gamma_0), X_{w_*}(0)) - 1$$
$$\geq U^{g_0}_{\gamma_0}(X_{w_*}(T_* - \gamma_0), X_{w_*}(0)) - \psi(w_*(0)) + \psi(w_*(T_* - \gamma_0)) - 1 \geq -d_1 - 1$$
$$-\psi(w_*(0)) + \psi(w_*(T_* - \gamma_0)) \geq 5(d_0 + d_1 + 1).$$

It follows from this relation and the definition of $\mathbf{U}$, (3.35), and (3.21) that

$$U^f_{\gamma_0}(X_v(\tau + 2T_* - h - \gamma_0), X_v(\tau + 2T_* - h)) \geq 4(d_0 + d_1 + 1). \tag{3.39}$$

Combining (3.36),(3.38),(3.39) we obtain that

$$I^f(0, \tau + 2T_* - h, v) - I^f(0, \tau + 2T_* - h, u) \geq 1.$$

This implies (3.32). The theorem is proved.

3.6. The structure of periodic good configurations.
Let $f \in \mathfrak{M}(\alpha, \beta, \gamma, a)$. We will establish the following results.

THEOREM 3.15. *Assume that $w \in W^{2,1}_{loc}(R^1), T > 0$,*

$$w(t + T) = w(t), t \in R^1, I^f(0, T, w) = T\mu(f),$$

and $w'(t) \neq 0$ for some $t \in R^1$. Then there exists $\tau > 0$ such that

$$w(t + \tau) = w(t), t \in R^1, X_w(T_1) \neq X_w(T_2)$$

for each $T_1 \in R^1$ and each $T_2 \in (T_1, T_1 + \tau)$.

THEOREM 3.16. *Assume that* $w \in W^{2,1}_{loc}(R^1), \tau > 0,$

$$w(t+\tau) = w(t), t \in R^1, I^f(0,\tau,w) = \tau\mu(f), \tag{3.40}$$

$$w(0) = \inf\{w(t) : t \in R^1\}, \text{ and } w'(t) \neq 0 \text{ for some } t \in R^1.$$

Then there exist $\tau_1 > 0, \tau_2 > \tau_1$ *such that the function* w *is strictly increasing in* $[0,\tau_1]$, w *is strictly decreasing in* $[\tau_1, \tau_2]$, *and*

$$w(\tau_1) = \sup\{w(t) : t \in R^1\}, w(t+\tau_2) = w(t), t \in R^1.$$

To prove Theorems 3.15 and 3.16 we need the following results established in [26,27].

PROPOSITION 3.1 [27]. *Let* $g \in \mathfrak{M}(\alpha,\beta,\gamma,a)$. *Then there exist a neighborhood* U *of* g *in* $\mathfrak{M}(\alpha,\beta,\gamma,a)$ *and a number* $S > 0$ *such that for each* $f \in U$ *and each* (f)*-good configuration* $v \in W^{2,1}_{loc}[0,\infty)$

$$|X_v(t)| \leq S \text{ for all large enough } t.$$

PROPOSITION 3.2 [26]. *Let* $f \in \mathfrak{M}(\alpha,\beta,\gamma,a)$, $x_i, y_i \in R^2$, $T_i > 0$, $i = 1,2$, *and let*

$$w_i \in A^{T_i}_{x_i,y_i}, U^f_{T_i}(x_i,y_i) = I^f(0,T_i,w_i), i = 1,2, \tau \in (0, \min\{T_1,T_2\}),$$

and

$$(w_1(\tau), w_1'(\tau), w_1''(\tau), w_1^{(3)}(\tau)) = (w_2(\tau), w_2'(\tau), w_2''(\tau), w_2^{(3)}(\tau)).$$

Then $w_1(t) = w_2(t)$ *for all* $t \in [0, \min\{T_1,T_2\}]$.

Proposition 3.2 implies the following

PROPOSITION 3.3. *Let* $f \in \mathfrak{M}(\alpha,\beta,\gamma,a)$, $T_1, T_2 \in R^1$, $T_1 < T_2$, *and let* $w_i \in W^{2,1}[T_1,T_2]$,

$$I^f(T_1,T_2,w_i) = (T_2 - T_1)\mu(f) + \pi^f(X_{w_i}(T_1)) - \pi^f(X_{w_i}(T_2)), i = 1,2.$$

Assume that $X_{w_1}(\tau) = X_{w_2}(\tau)$ *with some* $\tau \in (T_1,T_2)$. *Then* $w_1(t) = w_2(t), t \in [T_1,T_2]$.

PROPOSITION 3.4 [26]. *Let* $f \in \mathfrak{M}(\alpha,\beta,\gamma,a)$, $x,y \in R^2$, $T > 0$, *and let*

$$w \in A^T_{x,y}, U^f_T(x,y) = I^f(0,T,w),$$

$$T_0 \in (0,T), \{t_i\}_{i=1}^{\infty} \subset [0,T] \setminus \{T_0\},$$

$$t_i \to T_0 \text{ as } i \to \infty, w'(t_i) = 0, i = 1,2,\dots$$

Then $w(t) = w(0)$ *for all* $t \in [0,T]$.

Proof of Theorem 3.15. Define

$$E = \{s \in (0,\infty) : w(t+s) = w(t), t \in R^1\}, \tag{3.41}$$

$$\tau = \inf\{s : s \in E\}.$$

We will show that $\tau > 0$. Assume the contrary. Then there exists a sequence $\{s_i\}_{i=1}^{\infty} \subset E$ such that $\lim_{i\to\infty} s_i = 0$. By Proposition 3.1 there exists a number $\sigma > 0$ such that

$$|X_w(t)| \le \sigma, t \in [0, \infty).$$

For each integer $i \ge 1$

$$|w(t) - w(0)| \le \sigma s_i, t \in [0, s_i],$$

$$\{w(t) : t \in R^1\} \subset [w(0) - \sigma s_i, w(0) + \sigma s_i].$$

This implies that $w(t) = w(0), t \in R^1$. The obtained contradiction proves that $\tau > 0$.

We will show that $\tau \in E$. There exists a sequence $\{s_i\}_{i=1}^{\infty} \subset E$ such that $s_i \to \tau$ as $i \to \infty$. We have

$$X_w(0) = X_w(s_i) \to X_w(\tau) \text{ as } i \to \infty.$$

It follows from this relation and Proposition 3.3 that $w(t+\tau) = w(t)$, $t \in R^1$. This completes the proof of the theorem.

Proof of Theorem 3.16. By Theorem 3.15 we may assume without loss of generality that

$$X_w(T_1) \ne X_w(T_2) \text{ for each } T_1 \in R^1, \text{ each } T_2 \in (T_1, T_1 + \tau). \tag{3.42}$$

Denote by E the set of nonnegative numbers T for which there exists a sequence of numbers $\{s_i\}_{i=0}^{\infty}$ such that $s_i \to T$ as $i \to \infty$ and

$$w'(s_{2i}) > 0, w'(s_{2i+1}) < 0, i = 0, 1, \dots \tag{3.43}$$

By (3.40) and Proposition 3.4

$$\{i\tau : i = 0, 1, \dots\} \subset E. \tag{3.44}$$

Proposition 3.4 and (3.40) imply that for every integer $N \ge 1$ the set $E \cap [0, N]$ is finite. Consequently there exists a strictly increasing sequence of nonnegative numbers $\{t_i\}_{i=0}^{\infty}$ such that $\{t_i : i = 0, 1, \dots\} = E$. It is easy to see that for each integer $i \ge 0$ the function

$$w \text{ is strictly increasing in } [t_{2i}, t_{2i+1}] \tag{3.45}$$

$$\text{and strictly decreasing in } [t_{2i+1}, t_{2i+2}].$$

By (3.44), (3.45), (3.40) there exists an integer $j \ge 1$ for which

$$\tau = t_{2j}. \tag{3.46}$$

To prove the assertion of the theorem it is sufficient to show that $j = 1$.

Assume the contrary. Then by (3.40), (3.42)

$$w(t_2) \in (w(t_0), w(t_1)) \text{ and } \quad j \geq 2. \tag{3.47}$$

It follows from (3.45) that for each integer $i \geq 0$ there exists a continuous function

$$g_i : [\inf\{w(t_i), w(t_{i+1})\}, \sup\{w(t_i), w(t_{i+1})\}] \to [t_i, t_{i+1}]$$

such that $g_i(w(t)) = t$ for all $t \in [t_i.t_{i+1}]$.

Assume that an integer $i \in [0, j)$ and

$$w(t_{2i+2}) > w(t_{2i}). \tag{3.48}$$

(3.40), (3.44), (3.42), (3.46) imply that

$$w(t_{2i}) > w(0). \tag{3.49}$$

By (3.46), (3.40), (3.48), (3.49)

$$i + 1 < j. \tag{3.50}$$

We will show that

$$w(t_{2i+3}) < w(t_{2i+1}), \tag{3.51}$$

$$w(t_{2i+4}) > w(t_{2i+2}). \tag{3.52}$$

Assume that (3.51) does not hold. Together with (3.50), (3.42), (3.46) this implies that

$$w(t_{2i+3}) > w(t_{2i+1}).$$

By (3.42), (3.50), (3.46)

$$w'(g_{2i+2}(w(t_{2i+2}))) - w'(g_{2i}(w(t_{2i+2}))) < 0,$$

$$w'(g_{2i+2}(w(t_{2i+1}))) - w'(g_{2i}(w(t_{2i+1}))) > 0.$$

Consequently there exists $y \in (w(t_{2i+2}), w(t_{2i+1}))$ such that

$$w'(g_{2i+2}(y)) = w'(g_{2i}(y)).$$

Set

$$T_0 = g_{2i}(y) \in [t_{2i}, t_{2i+1}], T_1 = g_{2i+2}(y) \in [t_{2i+2}, t_{2i+3}].$$

It follows from the definition of y, T_0, T_1 that

$$(w(T_0), w'(T_0)) = (y, w'(g_{2i}(y))) = (w(T_1), w'(T_1)). \tag{3.53}$$

Since $0 \leq T_0 < T_1 < \tau$ (see (3.50), (3.46)) relation (3.53) is contradictory to (3.42). The obtained contradiction proves (3.51).

Assume that (3.52) does not hold. Together with (3.42), (3.46) and (3.50) this implies that

$$w(t_{2i+4}) < w(t_{2i+2}). \tag{3.54}$$

By (3.42), (3.50) and (3.46)

$$w'(g_{2i+1}(w(t_{2i+2}))) - w'(g_{2i+3}(w(t_{2i+2}))) > 0,$$

$$w'(g_{2i+1}(w(t_{2i+3}))) - w'(g_{2i+3}(w(t_{2i+3}))) < 0.$$

Consequently there exists $y \in (w(t_{2i+2}), w(t_{2i+3}))$ such that

$$w'(g_{2i+1}(y)) = w'(g_{2i+3}(y)).$$

Set

$$T_0 = g_{2i+1}(y) \in [t_{2i+1}, t_{2i+2}], T_1 = g_{2i+3}(y) \in [t_{2i+3}, t_{2i+4}].$$

It follows from the definition of y, T_0, T_1 that

$$(w(T_0), w'(T_0)) = (y, w'(g_{2i+1}(y))) = (w(T_1), w'(T_1)). \tag{3.55}$$

Since $0 < T_0 < T_1 \le \tau$ (see (3.50), (3.46)) relation (3.55) is contradictory to (3.42). The obtained contradiction proves (3.52).

Therefore we have shown that if an integer $i \in [0, j)$ satisfies (3.48) then (3.52) and (3.50) hold. Together with (3.47) this implies that the sequence $\{w(t_{2i})\}_{i=0}^{\infty}$ is strictly increasing. This is contradictory to (3.46). The obtained contradiction proves that $j = 1$. This completes the proof of the theorem.

4. Dynamic properties of extremals of variational problems with vector-valued functions

4.1. The class of variational problems. Denote by $|\cdot|$ the Euclidean norm in R^n and denote by $\mathfrak{A}$ the set of continuous functions $f\colon R^n \times R^n \to R^1$ which satisfies the following assumptions:

(A) (i) for each $x \in R^n$ the function $f(x, \cdot)\colon R^n \to R^1$ is convex;

(ii) the function f is bounded on any bounded subset of $R^n \times R^n$;

(iii) $f(x, u) \ge \sup\{\psi(|x|), \psi(|u|)|u|\} - a$ for each $(x, u) \in R^n \times R^n$ where $a > 0$ is a constant and $\psi\colon [0, \infty) \to [0, \infty)$ is an increasing function such that $\psi(t) \to +\infty$ as $t \to \infty$;

(iv) for each M, ϵ there exist $\Gamma, \delta > 0$ such that

$$|f(x_1, u_1) - f(x_2, u_2)| \le \epsilon \sup\{f(x_1, x_2), f(x_2, u_2)\}$$

for each $u_1, u_2, x_1, x_2 \in R^n$ which satisfy

$$|x_i| \le M, \ |u_i| \ge \Gamma, \ i = 1, 2, \ \sup\{|x_1 - x_2|, |u_1 - u_2|\} \le \delta.$$

It is an elementary exercise to show that an integrand $f = f(x, u) \in C^1(R^{2n})$ belongs to $\mathfrak{A}$ if f satisfies assumptions (Ai), (Aiii) with a constant $a > 0$ and a function ψ: $[0, \infty) \to [0, \infty)$, and there exists an increasing function ψ_0: $[0, \infty) \to [0, \infty)$ such that for each $x, u \in R^n$

$$\sup\{|\partial f/\partial x(x,u)|, \ |\partial f/\partial u(x,u)|\} \le \psi_0(|x|)(1 + \psi(|u|)|u|).$$

For the set $\mathfrak{A}$ we consider the uniformity which is determined by the following base

$$E(N, \epsilon, \lambda) = \{(f,g) \in \mathfrak{A} \times \mathfrak{A}\colon |f(x,u) - g(x,u)| \le \epsilon, \ u, x \in R^n, \ |x|, |u| \le N,$$

$$(|f(x,u)| + 1)(|g(x,u)| + 1)^{-1} \in [\lambda^{-1}, \lambda], \ u, x \in R^n, \ |x| \le N\}$$

where $N, \epsilon > 0$, $\lambda > 1$ [11].

Clearly, the uniform space $\mathfrak{A}$ is Hausdorff and has a countable base. Therefore $\mathfrak{A}$ is metrizable [11]. We can show (see [28]) that the uniform space $\mathfrak{A}$ is complete.

We consider functionals of the form

$$I^f(T_1, T_2, x) = \int_{T_1}^{T_2} f(x(t), x'(t))dt, \tag{4.1}$$

where $f \in \mathfrak{A}$, $0 \le T_1 < T_2 < +\infty$ and x: $[T_1, T_2] \to R^n$ is an absolutely continuous function.

For $f \in \mathfrak{A}$, $y, z \in R^n$ and numbers T_1, T_2 satisfying $0 \le T_1 < T_2$ we set

$$U^f(T_1, T_2, y, z) = \inf\{I^f(T_1, T_2, x)\colon x\colon [T_1, T_2] \to R^n \tag{4.2}$$

$$\text{is an a.c. function satisfying } x(T_1) = y, \ x(T_2) = z\},$$

$$\sigma^f(T_1, T_2, y) = \inf\{U^f(T_1, T_2, y, u)\colon u \in R^n\}. \tag{4.3}$$

It is easy to see that $-\infty < U^f(T_1, T_2, y, z) < +\infty$ for each $f \in \mathfrak{A}$, each $y, z \in R^n$ and all T_1, T_2 satisfying $0 \le T_1 < T_2$. Here we follow Leizarowitz [12] in defining "good functions" for the variational problem.

Let $f \in \mathfrak{A}$. An a.c. function x: $[0, \infty) \to R^n$ is called an (f)-good function if for any a.c. function y: $[0, \infty) \to R^n$ there is a number M_y such that

$$I^f(0, T, y) \ge M_y + I^f(0, T, x) \text{ for each } T \in (0, \infty).$$

We have the following result.

THEOREM 4.1 [28, THEOREM 1.1]. *For each $f \in \mathfrak{A}$ and each $z \in R^n$ there exists an (f)-good function Z^f: $[0, \infty) \to R^n$ satisfying $Z^f(0) = z$ and such that:*

1. For each $f \in \mathfrak{A}$, each $z \in R^n$ and each a.c. function y: $[0, \infty) \to R^n$ one of the following properties holds:

(i) $I^f(0, T, y) - I^f(0, T, Z^f) \to +\infty$ as $T \to \infty$;

(ii) $\sup\{|I^f(0, T, y) - I^f(0, T, Z^f)|\colon T \in (0, \infty)\} < \infty$, $\sup\{|y(t)|\colon t \in [0, \infty)\} < \infty$.

2. *For each $f \in \mathfrak{A}$ and each number $M > 0$ there exist a neighborhood $\mathbf{U}$ of f in $\mathfrak{A}$ and a number $Q > 0$ such that $\sup\{|Z^g(t)|: t \in [0, \infty)\} \leq Q$ for each $g \in \mathbf{U}$ and each $z \in R^n$ satisfying $|z| \leq M$.*

3. *For each $f \in \mathfrak{A}$ and each number $M > 0$ there exist a neighborhood $\mathbf{U}$ of f in $\mathfrak{A}$ and a number $Q > 0$ such that for each $g \in \mathbf{U}$, each $z \in R^n$ satisfying $|z| \leq M$, each $T_1 \geq 0$, $T_2 > T_1$ and each a.c. function y: $[T_1, T_2] \to R^n$ satisfying $|y(T_1)| \leq M$, the relation $I^g(T_1, T_2, Z^g) \leq I^g(T_1, T_2, y) + Q$ holds.*

4. *For each $f \in \mathfrak{A}$, each $z \in R^n$, and each $T_1 \geq 0$, $T_2 > T_1$,*

$$U^f(T_1, T_2, Z^f(T_1), Z^f(T_2)) = I^f(T_1, T_2, Z^f).$$

Let $f \in \mathfrak{A}$. For any a.c. function x: $[0, \infty) \to R^n$ we set

$$J(x) = \liminf_{T \to \infty} T^{-1} I^f(0, T, x). \tag{4.4}$$

Of special interest is the minimal long-run average cost growth rate

$$\mu(f) = \inf\{J(x): x: [0, \infty) \to R^n \text{ is an a.c. function}\}. \tag{4.5}$$

Clearly $-\infty < \mu(f) < +\infty$ and for every (f)-good function x: $[0, \infty) \to R^n$

$$\mu(f) = J(x). \tag{4.6}$$

In [28] we established the following result.

PROPOSITION 4.1. *For any a.c. function x: $[0, \infty) \to R^n$ either*

$$I^f(0, T, x) - T\mu(f) \to +\infty \text{ as } T \to \infty \text{ or}$$

$$\sup\{|I^f(0, T, x) - T\mu(f)|: T \in (0, \infty)\} < \infty. \tag{4.7}$$

Moreover (4.7) holds if and only if x is an (f)-good function.

4.2. Definitions and theorems. Recall that we denote $d(x, B) = \inf\{|x - y|: y \in B\}$ for $x \in R^n$, $B \subset R^n$ and denote by $\mathrm{dist}(A, B)$ the Hausdorff metric for two sets $A \subset R^n$, $B \subset R^n$.

For every bounded a.c. function x: $[0, \infty) \to R^n$ define

$$\Omega(x) = \{y \in R^n: \text{ there exists a sequence } \{t_i\}_{i=0}^{\infty} \subset (0, \infty)$$
$$\text{for which } t_i \to \infty,\ x(t_i) \to y \text{ as } i \to \infty\}.$$

In [28] we established the following results.

THEOREM 4.2. *There exists a set $F \subset \mathfrak{A}$ which is a countable intersection of open everywhere dense subsets of $\mathfrak{A}$ and such that each $f \in F$ has the following property: (B) $\Omega(v_2) = \Omega(v_1)$ for each (f)-good functions v_i: $[0, \infty) \to R^n$, $i = 1, 2$.*

Theorem 4.2 describes the limit behavior of (f)-good functions for a generic $f \in \mathfrak{A}$. The following result establishes the existence of an (f)-weakly optimal function for each $f \in \mathfrak{A}$ which has property B and each initial state $x \in R^n$.

THEOREM 4.3. *Assume that $f \in \mathfrak{A}$ and there exists a compact set $H(f) \subset R^n$ such that $\Omega(v) = H(f)$ for each (f)-good function v: $[0, \infty) \to R^n$. Then for each $x \in R^n$ there exists an (f)-weakly optimal function X: $[0, \infty) \to R^n$ satisfying $X(0) = x$.*

It follows from Theorems 4.2 and 4.3 that for a generic $f \in \mathfrak{A}$ and every $x \in R^n$ there exists an (f)-weakly optimal function X: $[0, \infty) \to R^n$ satisfying $X(0) = x$.

Theorems 4.4 and 4.5 establish the turnpike property for each $f \in \mathfrak{A}$ which has property B.

THEOREM 4.4. *Assume that $f \in \mathfrak{A}$ and there exists a compact set $H(f) \subset R^n$ such that $\Omega(v) = H(f)$ for each (f)-good function v: $[0, \infty) \to R^n$. Let ϵ be a positive number. Then there exist an integer $L \geq 1$ and a neighborhood $\mathbf{U}$ of f in $\mathfrak{A}$ such that for each $g \in \mathbf{U}$ and each (g)-good function v: $[0, \infty) \to R^n$*

$$dist(H(f), \{v(t): t \in [T, T + L]\}) \leq \epsilon \text{ for all large } T.$$

Theorems 4.2 and 4.4 show that for a generic $f \in \mathfrak{A}$ there exists a compact set $H(f) \subset R^n$ such that for any (f)-good function the turnpike property holds with the set $H(f)$ being the attractor.

THEOREM 4.5. *Assume that $f \in \mathfrak{A}$ and there exists a compact set $H(f) \subset R^n$ such that $\Omega(v) = H(f)$ for each (f)-good function v: $[0, \infty) \to R^n$. Let $M_0, M_1, \epsilon > 0$. Then there exist a neighborhood $\mathbf{U}$ of f in $\mathfrak{A}$, numbers $l, S > 0$ and integers $L, Q_* \geq 1$ such that for each $g \in \mathbf{U}$, each $T_1 \in [0, \infty)$, $T_2 \in [T_1 + L + lQ_*, \infty)$ and each a.c. function v: $[T_1, T_2] \to R^n$ which satisfies*

$$|v(T_i)| \leq M_1, \ i = 1, 2, \ I^g(T_1, T_2, v) \leq U^g(T_1, T_2, v(T_1), v(T_2)) + M_0,$$

the relation $|v(t)| \leq S$ holds for all $t \in [T_1, T_2]$ and there exist sequences of numbers $\{b_i\}_{i=1}^Q$, $\{c_i\}_{i=1}^Q \subset [T_1, T_2]$ such that

$$Q \leq Q_*, \ 0 \leq c_i - b_i \leq l, \ i = 1, \ldots Q, \text{ and}$$

$$dist(H(f), \{v(t): t \in [T, T + L]\}) \leq \epsilon \text{ for each } T \in [T_1, T_2 - L] \setminus \bigcup_{i=1}^{Q} [b_i, c_i].$$

Theorems 4.2 and 4.5 imply that for a generic $f \in \mathfrak{A}$ there exists a compact set $H(f) \subset R^n$ such that for any finite horizon problem the turnpike property holds with the set $H(f)$ being the attractor.

Let $k \geq 1$ be an integer. Denote by $\mathfrak{A}_k$ the space of all integrands $f \in \mathfrak{A} \cap C^k(R^n)$. For $p = (p_1, \ldots p_{2n}) \in \{0, \ldots k\}^{2n}$ and $f \in C^k(R^{2n})$ we set

$$|p| = \sum_{i=1}^{2n} p_i, \ D^p f = \partial^{|p|} f / \partial y_1^{p_1} \ldots \partial y_{2n}^{p_{2n}}.$$

For the set $\mathfrak{A}_k$ we consider the uniformity which is determined by the following base

$$E(N, \epsilon, \lambda) = \{(f, g) \in \mathfrak{A}_k \times \mathfrak{A}_k: |D^p f(x, u) - D^p g(x, u)| \leq \epsilon$$

$$(u.x \in R^n,\ |x|,|u| \le N,\ p \in \{0,\dots k\}^{2n},\ |p| \le k),$$
$$|f(x,u) - g(x,u)| \le \epsilon\ (u,x \in R^n,\ |x|,|u| \le N),$$
$$(|f(x,u)|+1)(|g(x,u)|+1)^{-1} \in [\lambda^{-1},\lambda]\ (x,u \in R^n,\ |x| \le N)\}$$

where $N > 0,\ \epsilon > 0,\ \lambda > 1$ [11].

Clearly the uniform space $\mathfrak{A}_k$ is Hausdorff and has a countable base. Therefore $\mathfrak{A}_k$ is metrizable [11]. It is easy to verify that the uniform space $\mathfrak{A}_k$ is complete. In [28] we established the following result.

THEOREM 4.6. *Let $k \ge 1$ be an integer. There exist a set $F_k \subset \mathfrak{A}_k$ which is a countable intersection of open everywhere dense subsets of $\mathfrak{A}_k$ and such that for each $f \in F_k$ and each (f)-good functions v_i: $[0,\infty) \to R^n,\ i = 1,2,$*

$$\Omega(v_2) = \Omega(v_1).$$

4.3. The strong turnpike property. For each integrand $f \in \mathfrak{A}$ which has property B there exists a compact set $H(f) \subset R^n$ such that $\Omega(v) = H(f)$ for each (f)-good function v: $[0,\infty) \to R^n$.

Denote by $\mathfrak{M}$ the set of all functions $f \in C^1(R^{2n})$ satisfying the following assumptions:

$$\partial f/\partial u_i \in C^1(R^{2n}) \text{ for } i = 1,\dots n;$$

for all $(x,u) \in R^{2n}$ the matrix $(\partial^2 f/\partial u_i \partial u_j)(x,u),\ i,j = 1,\dots n$ is positive definite;

$$f(x,u) \ge \sup\{\psi(|x|),\ \psi(|u|)|u|\} - a \text{ for all } (x,u) \in R^n \times R^n;$$

there exist a number $c_0 > 1$ and increasing functions $\phi_i : [0,\infty) \to [0,\infty),\ i = 0,1,2$ such that

$$\phi_0(t)t^{-1} \to +\infty \text{ as } t \to +\infty,\ f(x,u) \ge \phi_0(c_0|u|) - \phi_1(|x|),\ x,u \in R^n;$$

$$\sup\{|\partial f/\partial x_i(x,u)|,\ |\partial f/\partial u_i(x,u)|\} \le \phi_2(|x|)(1+\phi_0(|u|)),$$
$$x,u \in R^n,\ i = 1,\dots n.$$

It is easy to see that $\mathfrak{M} \subset \mathfrak{A}$. In [30] we established the following result.

THEOREM 4.7. *Assume that an integrand $f \in \mathfrak{M}$ has property B and $\epsilon, K > 0$. Then there exists a neighborhood* **U** *of f in $\mathfrak{A}$ and numbers $M > K,\ l_0 > l > 0,\ \delta > 0$ such that for each $g \in \mathbf{U}$, each $T \ge 2l_0$, and each a.c. function v: $[0,T] \to R^n$ which satisfies*

$$|v(0)|,|v(T)| \le K,\ I^g(0,T,v) \le U^g(0,T,v(0),v(T)) + \delta,$$

the relation $|v(t)| \le M$ holds for all $t \in [0,T]$, and

$$dist(H(f), \{v(t) : t \in [\tau,\tau + l]\}) \le \epsilon \qquad (4.8)$$

for each $\tau \in [l_0, T - l_0]$. *Moreover if* $d(v(0), H(f)) \le \delta$, *then (4.8) holds for each* $\tau \in [0, T - l_0]$, *and if* $d(v(T), H(f)) \le \delta$, *then (4.8) holds for each* $\tau \in [l_0, T - l]$.

Let $k \ge 1$ be an integer. Denote by $\mathfrak{A}_k$ the set of all integrands $f \in \mathfrak{A} \cap C^k(R^n)$. For $p = (p_1, \dots p_{2n}) \in \{0, \dots k\}^{2n}$ and $f \in C^k(R^{2n})$ we set

$$|p| = \sum_{i=1}^{2n} p_i, \; D^p f = \partial^{|p|} f / \partial y_1^{p_1} \dots \partial y_{2n}^{p_{2n}}.$$

For the set $\mathfrak{A}_k$ we consider the uniformity which is determined by the following base

$$\begin{aligned} E(N, \epsilon, \lambda) = \{(f, g) \in \mathfrak{A}_k \times \mathfrak{A}_k \colon & |D^p f(x, u) - D^p g(x, u)| \le \epsilon \\ & (u.x \in R^n, \; |x|, |u| \le N, \; p \in \{0, \dots k\}^{2n}, \; |p| \le k), \\ & |f(x, u) - g(x, u)| \le \epsilon \; (u, x \in R^n, \; |x|, |u| \le N), \\ & (|f(x, u)| + 1)(|g(x, u)| + 1)^{-1} \in [\lambda^{-1}, \lambda] \; (x, u \in R^n, \; |x| \le N)\} \end{aligned}$$

where $N > 0, \; \epsilon > 0, \; \lambda > 1$ (see Kelley [11]). It is easy to verify that the uniform space $\mathfrak{A}_k$ is metrizable and complete.

For each integer $k \ge 1$ we define $\mathfrak{M}_k = \mathfrak{M} \cap \mathfrak{A}_k$. Set

$$\mathfrak{A}_0 = \mathfrak{A}, \; \mathfrak{M}_0 = \mathfrak{M}.$$

Let $k \ge 0$ be an integer. Denote by $\bar{\mathfrak{M}}_k$ the closure of $\mathfrak{M}_k$ in $\mathfrak{A}_k$ and consider the topological subspace $\bar{\mathfrak{M}}_k \subset \mathfrak{A}_k$ with the relative topology. In [30] we established the following result.

THEOREM 4.8. *Let* $q \ge 0$ *be an integer. Then there exists a set* $F_q \subset \bar{\mathfrak{M}}_q$ *which is a countable intersection of open everywhere dense subsets of* $\bar{\mathfrak{M}}_q$, *and such that each* $f \in F_q$ *has property B and the following property:*

For each $\epsilon, K > 0$ *there exist a neighborhood* $\mathbf{U}$ *of* f *in* $\mathfrak{A}$ *and numbers* $M > K$, $l_0 > l > 0$, $\delta > 0$ *such that for each* $g \in \mathbf{U}$, *each* $T \ge 2l_0$, *and each a.c. function* $v \colon [0, T] \to R^n$ *which satisfies*

$$|v(0)|, |v(T)| \le K, \; I^g(0, T, v) \le U^g(0, T, v(0), v(T)) + \delta,$$

the relation $|v(t)| \le M$ *holds for all* $t \in [0, T]$ *and*

$$\text{dist}(H(f), \{v(t) \colon t \in [\tau, \tau + l]\}) \le \epsilon \tag{4.9}$$

for each $\tau \in [l_0, T - l_0]$. *Moreover if* $d(v(0), H(f)) \le \delta$, *then (4.9) holds for each* $\tau \in [0, T - l_0]$, *and if* $d(v(T), H(f)) \le \delta$, *then (4.9) holds for each* $\tau \in [l_0, T - l]$.

4.4. Examples. Fix a constant $a > 0$ and set $\psi(t) = t, \; t \in [0, \infty)$. Consider the complete metric space $\mathfrak{A}$ of integrands $f \colon R^n \times R^n \to R^1$ defined in section 4.1.

EXAMPLE 1. Consider an integrand $f(x, u) = |x|^2 + |u|^2, \; x, u \in R^n$. It is easy to see that $f \in \mathfrak{M}_q$ for each integer $q \ge 0$ if the constant a is large enough. We can show that $\Omega(v) = 0$ for every (f)-good function $v \colon [0, \infty) \to R^n$. Therefore the integrand f has property B.

EXAMPLE 2. Fix a number $q > 0$ and consider an integrand $g(x, u) = q|x|^2|x - e|^2 + |u|^2, \; x, u \in R^n$ where $e = (1, 1, \dots 1)$. It is easy to see that $g \in \mathfrak{M}$ if the constant a is large enough. Clearly f does not have the turnpike property.

References

1. B.D.O. Anderson and J.B. Moore, *Linear optimal control*, Prentice-Hall, Englewood Cliffs NJ, 1971.
2. Z. Artstein and A. Leizarowitz, *Tracking periodic signals with overtaking criterion*, IEEE Trans. on Autom. Control AC **30** (1985), 1122-1126.
3. S. Aubry and P.Y. Le Daeron, *The discrete Frenkel-Kontorova model and its extensions*, Physica D **8** (1983), 381-422.
4. V. Bangert, *Geodesic rays, Busemann functions and monotone twist maps*, Calc. Var. **2** (1994), 49-63.
5. W.A. Brock and A. Haurie, *On existence of overtaking optimal trajectories over an infinite horizon*, Math. Op. Res. **1** (1976), 337-346.
6. D.A. Carlson, *The existence of catching-up optimal solutions for a class of infinite horizon optimal control problems with time delay*, SIAM Journal on Control and Optimization **28** (1990), 402-422.
7. D.A. Carlson, A. Jabrane and A. Haurie, *Existence of overtaking solutions to infinite dimensional control problems on unbounded time intervals*, SIAM Journal on Control and Optimizaton **25** (1987), 1517-1541.
8. D.A. Carlson, A. Haurie and A. Leizarowitz, *Infinite horizon optimal control*, Springer-Verlag, Berlin, 1991.
9. B.D. Coleman, M. Marcus and V.J. Mizel, *On the thermodynamics of periodic phases*, Arch. Rational Mech. Anal. **117** (1992), 321-347.
10. D. Gale, *On optimal development in a multisector economy*, Rev. of Econ. Studies **34** (1967), 1-19.
11. J.L. Kelley, *General topology*, Van Nostrand, Princeton NJ, 1955.
12. A. Leizarowitz, *Infinite horizon autonomous systems with unbounded cost*, Appl. Math. and Opt. **13** (1985), 19-43.
13. A. Leizarowitz, *Existence of overtaking optimal trajectories for problems with convex integrands*, Math. Op. Res. **10** (1985), 450-461.
14. A. Leizarowitz, *Optimal trajectories on infinite horizon deterministic control systems*, Appl. Math. and Opt. **19** (1989), 11-32.
15. A. Leizarowitz and V.J. Mizel, *One dimensional infinite horizon variational problems arising in continuum mechanics*, Arch. Rational Mech. Anal. **106** (1989), 161-194.
16. V.L. Makarov and A.M. Rubinov, *Mathematical theory of economic dynamics and equilibria*, Nauka, Moscow, 1973; English trans. Springer-Verlag, New York, 1977.
17. M. Marcus, *Uniform estimates for variational problems with small parameters*, Arch. Rational Mech. Anal. **124** (1993), 67-98.
18. J. Moser, *Recent developments in the theory of Hamiltonian systems*, SIAM Review **28** (1986), 459-485.
19. J. Moser, *Minimal solutions of variational problems on a torus*, Ann. Inst. H. Poincare, Anal. non lineare **3** (1986), 229-272.
20. R.T. Rockafellar, *Saddle points of Hamiltonian systems in convex problems of Lagrange*, Journal of Optimization Theory and Applications **12** (1973), 367-389.
21. A.M. Rubinov, *Economic dynamics*, J. Soviet Math. **26** (1984), 1975-2012.
22. C.C. von Weizsacker, *Existence of optimal programs of accumulation for an infinite horizon*, Rev. Econ. Studies **32** (1965), 85-104.
23. A.J. Zaslavski, *Ground states in Frenkel-Kontorova model*, Math. USSR Izvestiya **29** (1987), 323-354.
24. A.J. Zaslavski, *Optimal programs on infinite horizon 1*, SIAM Journal on Control and Optimization **33** (1995), 1643-1660.
25. A.J. Zaslavski, *Optimal programs on infinite horizon 2*, SIAM Journal on Control and Optimization **33** (1995), 1661-1686.
26. A.J. Zaslavski, *The existence of periodic minimal energy configurations for one dimensional infinite horizon variational problems arising in continuum mechanics*, Journal of Mathematical Analysis and Applications **194** (1995), 459-476.

27. A.J. Zaslavski, *The existence and structure of extremals for a class of second order infinite horizon variational problems*, Journal of Mathematical Analysis and Applications **194** (1995), 660-696.
28. A.J. Zaslavski, *Dynamic properties of optimal solutions of variational problems*, Nonlinear Analysis: Theory, Methods and Applications **27** (1996), 895-932.
29. A.J. Zaslavski, *Structure of extremals for one-dimensional variational problems arising in continuum mechanics*, Journal of Mathematical Analysis and Applications **198** (1996), 893-921.
30. A.J. Zaslavski, *Turnpike property for extremals of variational problems with vector-valued functions*, Preprint (1995).

DEPARTMENT OF MATHEMATICS, TECHNION-ISRAEL INSTITUTE OF TECHNOLOGY, 32000 HAIFA, ISRAEL.

Selected Titles in This Series

(Continued from the front of this publication)

175 **Paul J. Sally, Jr., Moshe Flato, James Lepowsky, Nicolai Reshetikhin, and Gregg J. Zuckerman, Editors,** Mathematical aspects of conformal and topological field theories and quantum groups, 1994

174 **Nancy Childress and John W. Jones, Editors,** Arithmetic geometry, 1994

173 **Robert Brooks, Carolyn Gordon, and Peter Perry, Editors,** Geometry of the spectrum, 1994

172 **Peter E. Kloeden and Kenneth J. Palmer, Editors,** Chaotic numerics, 1994

171 **Rüdiger Göbel, Paul Hill, and Wolfgang Liebert, Editors,** Abelian group theory and related topics, 1994

170 **John K. Beem and Krishan L. Duggal, Editors,** Differential geometry and mathematical physics, 1994

169 **William Abikoff, Joan S. Birman, and Kathryn Kuiken, Editors,** The mathematical legacy of Wilhelm Magnus, 1994

168 **Gary L. Mullen and Peter Jau-Shyong Shiue, Editors,** Finite fields: Theory, applications, and algorithms, 1994

167 **Robert S. Doran, Editor,** C^*-algebras: 1943–1993, 1994

166 **George E. Andrews, David M. Bressoud, and L. Alayne Parson, Editors,** The Rademacher legacy to mathematics, 1994

165 **Barry Mazur and Glenn Stevens, Editors,** p-adic monodromy and the Birch and Swinnerton-Dyer conjecture, 1994

164 **Cameron Gordon, Yoav Moriah, and Bronislaw Wajnryb, Editors,** Geometric topology, 1994

163 **Zhong-Ci Shi and Chung-Chun Yang, Editors,** Computational mathematics in China, 1994

162 **Ciro Ciliberto, E. Laura Livorni, and Andrew J. Sommese, Editors,** Classification of algebraic varieties, 1994

161 **Paul A. Schweitzer, S. J., Steven Hurder, Nathan Moreira dos Santos, and José Luis Arraut, Editors,** Differential topology, foliations, and group actions, 1994

160 **Niky Kamran and Peter J. Olver, Editors,** Lie algebras, cohomology, and new applications to quantum mechanics, 1994

159 **William J. Heinzer, Craig L. Huneke, and Judith D. Sally, Editors,** Commutative algebra: Syzygies, multiplicities, and birational algebra, 1994

158 **Eric M. Friedlander and Mark E. Mahowald, Editors,** Topology and representation theory, 1994

157 **Alfio Quarteroni, Jacques Periaux, Yuri A. Kuznetsov, and Olof B. Widlund, Editors,** Domain decomposition methods in science and engineering, 1994

156 **Steven R. Givant,** The structure of relation algebras generated by relativizations, 1994

155 **William B. Jacob, Tsit-Yuen Lam, and Robert O. Robson, Editors,** Recent advances in real algebraic geometry and quadratic forms, 1994

154 **Michael Eastwood, Joseph Wolf, and Roger Zierau, Editors,** The Penrose transform and analytic cohomology in representation theory, 1993

153 **Richard S. Elman, Murray M. Schacher, and V. S. Varadarajan, Editors,** Linear algebraic groups and their representations, 1993

152 **Christopher K. McCord, Editor,** Nielsen theory and dynamical systems, 1993

151 **Matatyahu Rubin,** The reconstruction of trees from their automorphism groups, 1993

150 **Carl-Friedrich Bödigheimer and Richard M. Hain, Editors,** Mapping class groups and moduli spaces of Riemann surfaces, 1993

149 **Harry Cohn, Editor,** Doeblin and modern probability, 1993

(See the AMS catalog for earlier titles)